The Biology of Krüppel-like Factors

Ryozo Nagai • Scott L. Friedman
Masato Kasuga
Editors

The Biology of Krüppel-like Factors

Editors
Ryozo Nagai, M.D., Ph.D.
Professor
Department of Cardiovascular Medicine
The University of Tokyo Graduate School of Medicine
7-3-1 Hongo, Bunkyo-ku
Tokyo 113-8655
Japan

Scott L. Friedman, M.D.
Fishberg Professor and Chief
Division of Liver Diseases
Mount Sinai School of Medicine
Box 1123, 1425 Madison Avenue
Room 1170C
New York, NY 10029
USA

Masato Kasuga, M.D., Ph.D.
Director-General
Research Institute
International Medical Center of Japan
1-21-1 Toyama, Shinjuku-ku
Tokyo 162-8655
Japan

ISBN 978-4-431-87774-5 e-ISBN 978-4-431-87775-2
DOI 10.1007/978-4-431-87775-2
Springer Tokyo Berlin Heidelberg New York

Library of Congress Control Number: 2009928831

Cover illustration: Solution structure of the DNA-binding domain of Krüppel-like factor. The second (orange) and the third (blue) fingers of the three zinc fingers bind double-stranded DNA (show in stick model). The DNA-binding domain of EGR1/Zif268 (dark brown) is superimposed for structural comparison. (From Chapter 2, courtesy of S. Yokoyama)

Printed on acid-free paper

Springer is part of Springer Science+Business Media (www.springer.com)

Preface

The Krüppel-like factor (KLF) transcription family plays important roles in the development and malfunction of cells and tissues, and this very fact has stimulated increasing interest across the spectrum from basic biology to translational applications in medicine. KLFs have been studied extensively, yet many aspects of these molecules remain to be clarified, e.g., regulation of KLF gene expression at developmental stages or in cell–cell interactions; the interplay with cofactors amidst the transcriptional network; and their tertiary structures. This monograph brings together articles that review our current understanding of members of the KLF family, establishing a landmark as the first comprehensive publication on KLFs to cover basic biology, medical science, and translational research. It has been assembled to document not only the knowledge shared by the most active investigators in the field of KLF from around the world, but also their enthusiasm and energy to propel us into the future. This book project grew out of discussions at the 1st International Symposium on the Biology of Krüppel-like Factors held at the University of Tokyo, May 6–7, 2008. The symposium was part of the 21st Century Center of Excellence Program "Study of Diseases Caused by Environment–Genome Interactions" (Program Leader, R. Nagai) sponsored by the Japanese Ministry of Education, Science and Technology. This meeting, conducted in a collegial atmosphere, was highly informative and consolidated emerging themes that characterize KLFs. Reflecting one of the important goals of the meeting, the gathering also sparked new friendships that we hope will promote cooperation among investigators to further accelerate progress. Such collaborative efforts should yield a more comprehensive understanding of this family of factors by consolidating our collective expertise. The editors thank the authors for their invaluable contributions to this book, which we hope will stimulate current investigators and draw new talent into this important field.

March 2009

Ryozo Nagai
Scott L. Friedman
Masato Kasuga

Contents

Part 1
Overview

Chapter 1
Krüppel-like Factors: Ingenious Three Fingers Directing Biology and Pathobiology

Ryozo Nagai, Ichiro Manabe, and Toru Suzuki

Abstract Krüppel-like transcription factors (KLFs) participate in diverse physiological and pathological processes, such as cell growth, cell differentiation, tumorigenicity, metabolism, inflammation, and tissue remodeling in response to diverse external stress. The importance of KLFs has recently been appreciated as detailed mechanisms of their molecular functions have been rapidly unraveled. However, many questions remain to be addressed: for instance, (1) how is gene expression of KLFs regulated—in a developmental stage-specific manner or in terms of cell–cell interaction; (2) how do KLFs interplay with other cofactors amid the transcriptional network; and (3) need to explore the tertiary structure of KLFs. Given the importance of KLFs in disease biology, extensive investigations on KLFs are expected to lead to identification of therapeutic targets for many diseases.

Introduction

Mammalian Krüppel-like transcription factors (KLF) have recently gained recognition as critical regulators in cell proliferation and differentiation during normal development as well as in many disease states. KLFs constitute the KLF-half of the Sp1/KLF family of transcription factors, characteristically containing three consecutive and conserved cysteine and histidine (C2H2)-type zinc fingers in the DNA-binding domain located at the carboxyl terminal end. The paired cysteine and histidine in the DNA-binding domain are critical in spatially coordinating and anchoring the zinc atom. KLFs consist of at least 17 members that recognize GC- and GT-rich sequences and can either transactivate or repress target genes; and individual factors can act as both an activator and a repressor in a context-dependent manner.

Discovery of KLFs with DNA-binding properties similar to Sp1 have led us to reconsider the oversimplified notion that GC-boxes and CACCC-boxes (GT-boxes)

R. Nagai (✉), I. Manabe, and T. Suzuki
Department of Cardiovascular Medicine, Graduate School of Medicine,
The University of Tokyo, 7-3-1 Hongo, Bunkyo-ku, Tokyo 113-8655, Japan

R. Nagai et al. (eds.), *The Biology of Krüppel-like Factors*,
DOI 10.1007/978-4-431-87775-2_1, © Springer 2009

are unique binding sites for Sp1, one of the first transcription factors to be cloned (Kadonaga et al. 1987). From early studies on Sp1, it was widely held that basal transcription levels of "housekeeping genes" are maintained by Sp1, which regulates GC-boxes or GT-boxes in promoter regions. However, KLF members are involved in a variety of cellular functions and even compete with Sp1 for their DNA-binding sites. Many gene promoters with GC- or GT-boxes are not necessarily constitutively active but are finely tuned by KLFs in cells during development and differentiation or in response to physical or metabolic stress. KLF members participate in regulation of gene transcription within a network of Sp/KLF factors through interaction with intracellular signal molecules and transcriptional cofactors, including nuclear receptors, histone chaperones, oncogenes, and tumor suppressors. Physiological and pathophysiological functions of KLF members have been studied extensively, and many important findings are rapidly accumulating. We present an overview here on the current understanding of the KLF family in biology and disease.

KLFs in Development, Morphogenesis, and Differentiation

The ancestral gene of KLF is Krüppel in *Drosophila*, which is a factor activated in the center of the embryoid body during early development and involved in segmentation of the thorax and abdomen. In mammals, many KLF members participate in embryogenesis and fetal development. Homozygous KLF5 knockout mice ($KLF5^{-/-}$) die at early embryonal stage before E8.5 (Shindo et al. 2002). $KLF2^{-/-}$ embryos die between E12.5 and E14.5, and $KLF4^{-/-}$ mice die at birth (Kuo et al. 1997; Segre et al. 1999).

The importance of KLFs in embryonic stem (ES) cell biology has recently been shown by Yamanaka et al., who found that pluripotent ES-like cells can be generated from murine or human fibroblasts by the introduction of four genes (i.e., KLF4, Oct3/4, Sox2, and c-Myc (Takahashi et al. 2007; Takahashi and Yamanaka 2006)). Yamanaka's group further demonstrated that KLF2 and KLF5 can replace KLF4 in combination with the three other genes to convert fibroblasts into ES-like cells. Jian et al. also recently reported that KLF2, 4, and 5 constitute a core circuitry in self-renewal of ES cells in which simultaneous depletion of KLF2, 4, and 5 results in ES cell differentiation (Jiang et al. 2008). These three KLFs cooperatively activate self-renewal genes and suppress differentiation. Parisi et al., on the other hand, indicated that depletion of KLF5 alone resulted in differentiation of mouse ES cells, which did not occur by depletion of KLF2 or KLF4, suggesting a central role of KLF5 to maintain ES cells in the undifferentiated state (Parisi et al. 2008). KLF5 is reported to play a role in maintenance of pluripotency of ES cells by promoting phosphorylation of Akt1, and early embryonic lethality of $KLF5^{-/-}$ mice is due to an implantation defect (Ema et al. 2008).

KLF members regulate differentiation of various cell types during development and normal growth. KLF1, the founder factor of the KLF family and also known as EKLF, plays a critical role in erythropoiesis (Miller and Bieker 1993). KLF1 is not required for erythroid commitment but promotes maturation of erythrocytes by inducing the β-globin gene or genes essential for integrity of the erythrocyte membrane, cytoskeleton, and heme synthesis (Hodge et al. 2006; Perkins et al. 1995). KLF2 regulates the embryonic globin gene, whereas KLF13 is involved in the maturation of erythrocytes but its target genes remain elusive (Basu et al. 2005; Gordon et al. 2008). KLF6 also regulates hematopoiesis in the yolk sac (Matsumoto et al. 2006).

Adipocyte differentiation is another system intensively investigated in the context of transcriptional regulation by KLFs. KLF2, 3, 4, 5, 6, and 15 reportedly regulate adipogenesis (Birsoy et al. 2008; Li et al. 2005; Mori et al. 2005; Oishi et al. 2005; Sue et al. 2008; Wu et al. 2005); however, the mechanisms as to how these KLF members regulate adipogenesis are diverse. KLF2 and KLF3 are abundantly expressed in 3T3-L1 preadipocytes and are downregulated during differentiation. KLF2 directly inhibits PPARγ, and KLF3 represses C/EBPα together with co-repressor CtBP. KLF5 and KLF15 promote adipogenesis through transactivating the PPARγ gene; however, KLF5 expression is elevated during early differentiation and reduced in mature adipocytes, whereas KLF15 is upregulated later during adipocyte differentiation. KLF4 together with Krox20 is an early regulator of adipogenesis because they cooperatively activate C/EBPβ. Expression of KLF6 is increased during preadipocyte differentiation and in mature adipocytes. KLF6 together with HDAC3 inhibits Dlk1, which is a transmembrane protein containing an epidermal growth factor repeat domain and promotes preadipocyte proliferation. These studies clearly indicate that KLFs are involved in adipocyte differentiation, but how actions of these factors are orchestrated and whether they cooperatively interact is not known.

KLFs have also been implicated in epithelial cell differentiation (McConnell et al. 2007). In the intestinal epithelium, KLF4 and KLF5 show opposing roles in terms of proliferation and differentiation. KLF4 is expressed in terminally differentiated epithelial cells in the villi and upper crypt. KLF5, on the other hand, is localized to the proliferating epithelial cells at the base of the crypt. KLF4 is induced upon activation of the adenomatous polyposis coli (APC) gene and interacts with β-catenin, thus inhibiting its signaling (Dang et al. 2001). KLF4 also activates genes encoding negative regulators of the cell cycle, such as p27(Kip1), and suppresses cyclin D1 and cyclin B, which promote the cell cycle (Wei et al. 2008). KLF4 regulates the Sprr gene clusters and the keratin families contributing to epidermal barrier integrity (Turksen and Troy 2002). KLF5, to the contrary, is a positive regulator of cell proliferation with transformation activities. KLF5 regulates growth factor genes, such as platelet-derived growth factor (PDGF), fibroblast growth factor (FGF), and vascular endothelial growth factor (VEGF), in addition to the transforming growth factor–bone morphogenetic protein (TGF-BMP) signaling pathway (Wan et al. 2008).

Gene targeting of KLFs in mice has demonstrated their critical roles in the differentiation of many other cells: KLF2 in thymocytes, monocytes (Carlson et al. 2006), and vascular smooth muscle cells (SMCs) (Kuo et al. 1997); KLF4 in monocytes (Alder et al. 2008), testicular Sertoli cells (Godmann et al. 2008), corneal epithelial and epidermal cells (Swamynathan et al. 2007); KLF5 in SMCs and respiratory epithelial cells (Shindo et al. 2002; Wan et al. 2008); KLF7 in olfactory sensory neurons (Kajimura et al. 2007); and KLF9 in intestinal epithelial cells (Simmen et al. 2007). KLF13 plays important roles in cardiac morphogenesis, as knockdown of the gene in the *Xenopus* embryo results in atrial septal defect and hypotrabeculation (Lavallee et al. 2006). Physical and functional interaction of KLF13 with GATA4, a transcriptional effector of cardiac development, may in part be responsible for this phenotype. However, the significance of KLF13 in the mammalian heart has not been reported to date.

KLFs in Cell Proliferation, Apoptosis, and Oncogenesis

KLF4, KLF5, and KLF6 have been extensively studied in cell growth. KLF4 and KLF5 exhibit contrasting effects on cell growth in epithelial cells (Ghaleb et al. 2005). KLF5 is markedly induced in proliferating SMCs and fibroblasts, and it promotes proliferation of those cells, whereas KLF4 inhibits proliferation. Induction of KLF5 has been shown to be mediated by Erk-1 and the Wnt-1 signaling pathway (Kawai-Kowase et al. 1999; Ziemer et al. 2001). Introduction of KLF5 into NIH 3T3 and intestinal epithelial cells promotes growth and induces a transformed phenotype as indicated by anchorage-independent growth (Nandan et al. 2004). KLF5 positively regulates not only growth factor genes but also cell cycle promoting genes such as cyclin D1, cyclin B1, and dk1/Cdc2 (Nandan et al. 2005).

In KLF5$^{+/-}$ mice, proliferation of vascular SMCs and cardiac fibroblasts, which occurred in response to mechanical injury and angiotensin II treatment, was markedly diminished (Shindo et al. 2002). Furthermore, infection of KLF5$^{+/-}$ mice with *Citrobacter rodentium*, a gram-negative bacterium, attenuated a hyperproliferative response of epithelium in the colon (McConnell et al. 2008).

Tumorigenicity of KLF5 is still controversial. KLF5 has been implicated in progression of colorectal cancer (Ghaleb and Yang 2008), although there has been a report showing that expression of KLF5 is reduced in intestinal tumors (Bateman et al. 2004). However, at least in colorectal cancer with mutated (activated) KRAS, KLF5 has been implicated in tumor progression because inhibition of KLF5 expression by mitogen-activated protein kinase/extracellular signal-regulated kinase (MEK) inhibitors or KLF5-specific small interfering RNA led to reduced proliferation and transformation (Nandan et al. 2008).

Antiapoptotic activity of KLF5 is mediated in part by interaction with a 24-kDa proteolytic fragment of poly(ADP-ribose) polymerase-1 (PARP-1), a nuclear enzyme important in apoptosis. KLF5, particularly when acetylated, binds a 24-kDa proteolytic fragment of PARP-1 with high affinity and suppresses its apoptotic activities (Suzuki et al. 2007).

KLF4 and KLF6, on the other hand, are generally considered tumor suppressors. KLF4 inhibits the transition between the G_1 and S phases of the cell cycle, which has been shown to occur by coordinated regulation of expression of numerous cell cycle regulatory genes including induction of p21/WAF1 and G1/S checkpoint genes (Zhang et al. 2000). The significance of KLF4 as a tumor suppressor has been extensively investigated in vitro and in vivo. KLF4 is activated by the tumor suppressor APC; and when $KLF4^{+/-}$ mice were crossbred with $APC^{min/+}$ mice, which develop adenomas in the intestine, they developed more intestinal adenomas than the $APC^{min/+}$ mice (Ghaleb et al. 2007). These reports indicate a role for KLF4 as a tumor suppressor in the gut. Conversely, experiments in cutaneous squamous epithelial cells showed KLF4 to be growth promoting because conditional overexpression of KLF4 in the basal layer of mouse skin led to hyperplasia, dysplasia, and squamous cell carcinoma (Foster et al. 2005). KLF4 levels are increased in mammary carcinoma and oropharyngeal squamous cell carcinoma (Foster et al. 2000), and KLF4 can act as a context-dependent oncogene through suppression of p53 (Rowland et al. 2005).

KLF6 is ubiquitously expressed and inducible by various stimuli. KLF6 is also considered a tumor suppressor. KLF6 promotes G_1 cell cycle arrest by inhibiting cell cycle-related gene expression; KLF6 upregulates expression of the p21/WAF1 gene and represses the cyclin D1 gene, thus disrupting cyclin D1-cyclin-dependent kinase (cdk) 4 complexes (Benzeno et al. 2004; Shie et al. 2000). KLF6 has been studied in detail in human prostate cancer in which the KLF6 locus is frequently deleted or mutated (Narla et al. 2001). Loss of heterozygosity (LOH) of the KLF6 gene is found in multiple cancers, including colorectal, hepatocellular, lung, and ovarian carcinomas (Narla et al. 2007; Watanabe et al. 2008). Furthermore, glioblastoma tumorigenicity could be suppressed by expression of KLF6 both in vitro and in vivo (Kimmelman et al. 2004). However, a role for KLF6 as a tumor suppressor is still controversial because silencing KLF6 leads to inhibition of cell proliferation and sensitization to apoptosis (Sirach et al. 2007). One explanation for this contradiction is the generation of splice variants of the KLF6 gene (KLF6-SV1-3) (Yea et al. 2008). KLF6-SV1 and SV2 are localized in the cytoplasm and antagonize the transcriptional activities of KLF6. Targeting KLF6-SV1 has been shown to induce spontaneous apoptosis in prostate cancer cells; and high levels of KLF6-SV1 expression in prostate tumor are associated with a low survival rate (Narla et al. 2008).

KLF2, KLF8, KLF10, and KLF11 have also been implicated in tumor growth. KLF2 inhibits cell growth by upregulating cell cycle checkpoint genes such as p21/WAF1 and WEE1. In ovarian cancer, KLF2 has been demonstrated to repress WEE1, which facilitates tumor cells to undergo apoptosis (Wang et al. 2005). KLF8 is known to be expressed in several types of human cancer. KLF8 upregulates cyclin D1 and is required for v-Src-induced transformation in NIH3T3 cells (Wang and Zhao 2007). KLF10 and KLF11 are TGF-β-responsive genes (Ribeiro et al. 1999). KLF11 has been implicated in inhibition of cell growth in vitro and in vivo and in neoplastic transformation. The oxidative stress scavengers SOD2 and Catalase1 are target genes of KLF11 and are downregulated in transgenic mice for KLF11. A complex of KLF11 and co-repressor mSin3a suppresses cell growth by repressing TGF-β-induced transcription from the Smad7 promoter (Fernandez-Zapico et al. 2003).

Phosphorylation of KLF11 by Erk-MAPK inactivates this pathway in pancreatic cancer cells with oncogenic Ras mutations (Ellenrieder et al. 2004). Despite these effects of KLF11, however, deletion of the KLF11 gene did not show any phenotypes in development and growth (Song et al. 2005).

KLFs in Inflammation

KLF2, KLF4, KLF5, and KLF13 are known to be involved in modulating inflammation and/or differentiation of immune cells. KLF2 is expressed in lymphocytes, monocytes, and endothelial cells, playing a key role in thymocyte and T-cell migration by regulating sphingosine-1-phosophate receptor S1P1 (Carlson et al. 2006). KLF2 in monocytes and endothelial cells inhibits proinflammatory activation of these cells. Inhibition of KLF2 expression in monocytes by short interfering RNA increases inflammatory gene expression, whereas overexpression of KLF2 inhibits lipopolysaccharide (LPS)-mediated induction of proinflammatory cytokines and reduces phagocytosis, which are mediated by inhibiting the transcriptional activity of NF-κB (Das et al. 2006). KLF2 in endothelial cells also inhibits endothelial activation in response to proinflammatory stimuli (SenBanerjee et al. 2004). KLF2 is thought to be atheroprotective, as $KLF2^{+/-}$ /$ApoE^{-/-}$ mice show increased diet-induced atherosclerosis (Atkins et al. 2008).

KLF4 functions as a regulator of monocyte differentiation. Cre-excision of KLF4 in hematopoietic stem cells (HSCs) resulted in a reduced number of monocytes and an increase in granulocyte formation, whereas overexpression of KLF4 in HSCs resulted in differentiation to monocytes (Feinberg et al. 2007). In $KLF4^{-/-}$ chimeras, which were generated by transplanting $KLF4^{-/-}$ fetal liver cells into irradiated wild-type mice, showed a lack of circulating inflammatory ($CD115^{+}Gr1^{+}$) monocytes and reduced numbers of resident ($CD115^{+}Gr1^{-}$) monocytes (Alder et al. 2008). In macrophages, KLF4 expression was induced by proinflammatory cytokines such as interferon-γ (IFN-γ), LPS, and tumor necrosis factor-α (TNF-α), and overexpression of KLF4 activated macrophage as shown by iNOS induction (Feinberg et al. 2005).

KLF5 regulates various genes involved in cell growth and angiogenesis, whereas proinflammatory cytokines, growth factors, angiotensin II, and S1P induce KLF5 expression (Shindo et al. 2002; Usui et al. 2004). $KLF5^{+/-}$ mice showed reduced chronic inflammation when angiotensin II was infused or a foreign body was placed around the artery (Shindo et al. 2002). KLF5 binds p50 subunit of NF-κB and promotes PDGF-A gene activation by KLF5 alone (Aizawa et al. 2004). Furthermore, overexpression of KLF5 in fibroblasts induced S100A, a potent proinflammatory protein (unpublished data). In intestinal epithelial cells, KLF5 mediates a LPS-induced proinflammatory response (Chanchevalap et al. 2006).

KLF13 is involved in the development of both B and T cells at multiple stages. In $KLF13^{-/-}$ mice, peripheral T cell activation was impaired. B-cell maturation in bone marrow was also impaired as revealed by partial arrest of B cells at the transition from $CD43^{+}$ to $CD43^{-}$ pre-B cells (Outram et al. 2008).

KLFs in Phenotypic Modulation of Cell and Tissue Remodeling

Phenotypic modulation of cells underlies various diseases. In the cardiovascular system, for example, external stress such as pressure overload or angiotensin II loading induces changes in gene expression of cells, leading to hyperplasia of vascular SMCs, hypertrophy of cardiac myocytes, and tissue fibrosis. Structural rearrangement in organs is referred to as *tissue remodeling*.

The role of KLFs in tissue remodeling has been most extensively studied in the cardiovascular system. KLF5 is involved in phenotypic modulation of vascular SMCs and fibroblasts. KLF5 regulates genes of cell growth-related genes such as PDGF-A and PDGF-B, TGF-β, and SMemb/NMHC-B, a molecular marker of dedifferentiated SMCs (Fujiu et al. 2005; Shindo et al. 2002; Usui et al. 2004; Wan et al. 2008; Watanabe et al. 1999). Introduction of short interference RNA into SMCs inhibits phenotypic modulation of the cell and increases expression of SMC differentiation markers (Liu et al. 2005). In KLF5$^{+/-}$ mice under various pathological conditions, hypertrophy and fibrosis in the heart, neointimal hyperplasia in the artery, fibrosis in the kidney, and angiogenesis in implanted cancers were mitigated compared to that in wild-type mice (Fujiu et al. 2005; Shindo et al. 2002). In the cardiovascular system, a complex of RARα-RXR in SMCs and fibroblasts binds KLF5 as a coactivator and mediates KLF5-induced phenotypic modulation of cells and tissue remodeling (Shindo et al. 2002).

KLF4 has been thought to be a repressor of SMC differentiation markers in cultured SMCs such as smooth muscle α-actin, smooth muscle myosin heavy chain, and SM22α (Liu et al. 2005). KLF4 promotes phenotypic modulation of SMCs in vitro when stimulated with PDGF-BB. However, KLF4 can be a growth-promoting transcription factor in a context-dependent manner. Interestingly, conditional targeting of the KLF4 gene in vivo suppressed phenotypic modulation of SMCs but accelerated SMC proliferation in response to vascular injury (Yoshida et al. 2008). KLF15 has also been implicated in SMC proliferation following vascular injury because KLF15$^{-/-}$ mice developed exaggerated neointimal formation (Wang et al. 2008a).

Null mutants of KLF2 show apparently normal angiogenesis and vasculogenesis, but null embryos die in utero because of hemorrhaging and abnormal blood vessel structure (Kuo et al. 1997). They show endothelial cell necrosis, thin tunica media, and a reduction in pericytes and differentiated SMCs. KLF2 in endothelial cells is inducible by laminar flow and 3-hydroxy-3-methylglutaryl coenzyme A (HMG-CoA) reductase inhibitors (statins). Statins upregulate heme oxygenase-1 (HO-1), endothelial nitric oxide synthase (eNOS), and thrombomodulin, each of which has an antiinflammatory and an antiatherosclerotic effect on the vasculature (Ali et al. 2007; Haldar et al. 2007; SenBanerjee et al. 2004; van Thienen et al. 2006).

In liver fibrosis, activation and proliferation of satellite cells are thought to play a critical role. KLF6 is induced in activated hepatic satellite cells and has been suggested to be involved in hepatic fibrosis (Friedman 2006; Ratziu et al. 1998). Another mechanism of organ fibrosis is epithelial-mesenchymal transition (EMT), which is mediated by TGF-β. In renal fibrosis due to high glucose levels, elevated

KLF6 in renal proximal tubules has been suggested to induce EMT (Holian et al. 2008). KLF8, a downstream effector of focal adhesion kinase (FAK), has been implicated to play a crucial role in EMT in mammary gland epithelial cell lines. Aberrant expression of KLF8 in invasive human cancer is correlated with a decrease in E-cadherin expression; suppression of KLF8 in cancer cells restored E-cadherin expression and inhibited invasiveness of the cells (Wang et al. 2007).

In the presence of cardiac hypertrophy, gene expression in the myocardium is altered; that is, there is reduced expression of differentiation markers and induction of fetal genes. KLF10 is involved in the TGF-β–Smad pathway, and knockout of the gene results in hypertrophic cardiomyopathy (Subramaniam et al. 2007). KLF15 is also known to play a role in cardiac phenotypes, as knockout of the gene results in cardiac hypertrophy (Fisch et al. 2007).

KLFs in Metabolism

Roles of KLFs in adipocyte differentiation were described earlier in the chapter. In *Caenorhabditis elegans*, suppression of KLF1 gene expression by RNA interference results in increased fat accumulation in the intestine (Hashmi et al. 2008). In $KLF5^{+/-}$ mice did not develop obesity with a high-calorie diet as much as did wild-type mice, although they fed more (Oishi et al. 2005). This is due in part to enhanced lipid oxidation and energy uncoupling in skeletal muscles. SUMOylated KLF5 is associated with co-repressors containing PPARδ; it represses CPT1b, UCP2, and UCP3 in skeletal muscles, thereby suppressing lipid metabolism (Oishi et al. 2008).

$KLF15^{-/-}$ mice developed severe hypoglycemia after an overnight fast, which is caused by defective amino acid catabolism in liver and skeletal muscles. KLF15 regulates gene expression of alanine aminotransferase (ALT), which converts alanine to pyruvate (Gray et al. 2007). KLF15 also regulates insulin-sensitive glucose transporter GLUT4 in both adipocytes and muscle tissues, which seems to be due to synergistic activation of the GLUT4 gene by KLF15 and MEF2A (Gray et al. 2002).

3,5,3′-Triiodothyronine (T_3) is a biologically active thyroid hormone that regulates the basal metabolic rate in vertebrates. T_3 is generated from prohormone thyroxine (T_4) by deiodination enzymes Dio1 and Dio2. Dio1 is known to catalyze both activation and inactivation of T_4. Expression of the Dio1 gene is strongly upregulated by KLF9, which is more enhanced in the presence of HNF4α and GATA4 (Ohguchi et al. 2008).

Transcriptional Regulatory Mechanisms of KLFs and Pharmacological Control

KLFs and Sp1 have common binding sites (i.e., GC-rich rich sites or CACCC-boxes) and activate promoter reporter constructs containing these sites in in vitro promoter assays. However, these binding sites in vivo are selectively discriminated by individual factors. The mechanisms whereby Sp/KLF family members exhibit their specific biological functions through these similar DNA-binding sequences remain to be elucidated. To our current knowledge, formation of a transcription complex that consists of transcription factors with co-activators/co-repressors and chromatin-associated factors is a key regulatory mechanism for gene expression in a context-dependent manner. Protein modification of transcription factors is another mechanism of gene expression that may affect binding activities of transcription factors with cofactors. KLF5 has been thoroughly investigated in several laboratories, including ours, and provides an example of transcriptional regulation in which a member of KLF is incorporated. Therefore, we focus our discussion here on KLF5. In SMCs and fibroblasts, KLF5 forms a complex with RARα-RXR and activates the PDGF-A gene. KLF5 associates with the p50 subunit of NF-κB in phorbol ester-induced pathological conditions in SMCs, which promotes KLF5-induced PDGF-A gene activation and possibly underlies the transition from acute to chronic inflammation (Aizawa et al. 2004). RARα-RXR-induced activation of KLF5 is repressed by the RARα agonist Am80 but further activates RARα antagonist LE135 (Shindo et al. 2002). All-trans retinoic acid (ATRA), Am80, and acyclic retinoid NIK-333 induce SMC differentiation and inhibit dedifferentiation in which KLF5 and possibly KLF4 are involved (Fujiu et al. 2005; Kada et al. 2007, 2008; Wang et al. 2008b).

During adipocyte differentiation KLF5 binds C/EBPα and C/EBPβ and upregulates PPARγ_2, which is essential for adipocyte differentiation (Oishi et al. 2005). In skeletal muscles, a complex of KLF5 and PPARδ represses the UCP2 gene. C/EBPα activates UCP2 promoter in both adipocytes and skeletal muscle cells. C/EBPα-induced activation of UCP2 promoter is further activated by KLF5 in adipocytes but inhibited in skeletal muscle cells (Oishi et al. 2008).

In skeletal muscles, SUMOylated KLF5 forms a transcriptionally repressive complex with C/EBPβ, unliganded PPARδ and corepressors, NCoR and SMRT, resulting in repression of the UCP2, UCP3, and CPT1b gene expression under basal conditions. The presence of a PPARδ ligand, GW501516, changes the composition of this complex by recruiting desumoylation enzyme SENP1 and acetylation enzyme CBP, which then induces chromatin remodeling and activation of UCP2, UCP3, and CPT1b genes (Oishi et al. 2008). SUMOylation of KLF5 is also indicated to enhance nuclear localization and promote anchorage-independent growth of HCT116 colon cancer cells (Du et al. 2008). SUMOylation of KLFs leads to enhancement of their transrepression activities as known for KLF1 (Perdomo et al. 2005; Siatecka et al. 2007) and KLF8 (Wei et al. 2006).

Acetylation of KLFs enhances transcriptional activity. KLF5 is acetylated at a lysine residue proximal to the DNA binding domain by p300/CBP, and HeLa cells transfected with nonacetylatable KLF5 are sensitive to TNFα-induced apoptosis (Miyamoto et al. 2003). Histone deacetylase1 (HDAC1), on the other hand, negatively regulates the transcriptional activity of KLF5 (Matsumura et al. 2005). KLF5 activity is also inhibited by binding to SET/TAF-1β, an oncogenic protein that participates in chromatin assembly (Miyamoto et al. 2003). KLF5 interacts with a histone chaperone, acidic nuclear phosphoprotein 32B (ANP32B), leading to transcriptional repression of a KLF5-downstream gene. Recruitment of ANP32B onto the promoter region requires KLF5 and results in promoter region-specific histone incorporation and inhibition of histone acetylation by ANP32B (Munemasa et al. 2008).

Perspective

In recent years, many investigations on KLFs have been performed in laboratories worldwide. However, how KLFs show regulatory functions through binding to similar GC-rich sites or CACC-boxes remains an enigma. Multiple KLFs expressed in individual cells participate in differentiation of various cell types, some of which exert opposite functions. Furthermore, individual CACC-boxes existing in the same promoter region may respond to different types of stimuli as shown in the p21/Waf1 promoter (Gartel and Tyner 1999; Lu et al. 2000; Wang et al. 2000). It is important to note that DNA-binding characteristics likely differ in the in vivo context of chromatin DNA in vivo in contrast to the naked DNA state often used for biochemical experiments. One important example using transgenic mice showed that EKLF/KLF1 preferentially binds the β-globin locus site in vivo that had been shown to bind both EKLF and Sp1 in biochemical studies (Gillemans et al. 1998). Therefore, the KLF family apparently constitutes a complex network through which biological diversity and stress responses are regulated. However, whether individual KLF genes are independently regulated or gene expression of KLFs is coordinated through cross-talk among family members, as shown for KLF1, KLF3, and KLF8, requires more extensive examination (Eaton et al. 2008).

The mechanisms of action as a transcriptional factor are of primary importance in KLF research. Splice-variants of KLF6 transcripts have been reported to inhibit the native form of KLF6 (Narla et al. 2008). Whether similar splice-variants exist in other KLFs is not known and needs to be clarified. KLFs are subjected to posttranslational modifications as well—such as acetylation, phosphorylation, ubiquitination, and sumoylation—which at least in part are thought to affect their binding with transcriptional cofactors, thus mediating the differential biological effects of KLFs. Furthermore, KLFs interact with chromatin-associated factors including histone chaperones, acetylases/deacetylases, and nucleosome-remodeling enzymes, which allow the KLF transcription factor complex to access specific genes in packaged chromatin. KLFs may also posttranscriptionally modify other transcription factors and cofactors (Oishi et al. 2008). To understand the functions of KLFs, it is necessary to continue comprehensive analyses on each KLF member from these aspects. At the same time, molecular structures of KLFs need to be determined. In this volume, Chapter 2, contributed by Shigeyuki Yokoyama and colleagues, is devoted to the structures of KLFs. They reveal structures of DNA-binding domains and the protein-binding domains of KLFs. However, the structures of the N-terminal regions as a complex with co-activators or co-repressors have not been clarified. In analogy, Sp1 has been subjected to structural analysis of its cofactor complex that acts through the regulatory region, and similar analyses are anxiously awaited for KLFs to understand their similarities and differences (Taatjes et al. 2002). It is important to determine the structures of the KLF–cofactor complex and examine how they recognize specific GC-boxes or CACC-boxes in the DNA sequences.

Many members of the KLF family are ubiquitously expressed, and their expression is closely associated with cell growth or growth inhibition, phenotypic modulation of cells, and tumorigenesis in which interactions between parenchymal cells and stromal cells (including inflammatory/immune cells and/or vascular cells) are implicated as playing critical roles. KLFs are involved in disease processes, including cardiovascular disease, certain types of cancer, and metabolic syndrome, in which the proinflammatory response to tissue stress or energy excess is underlying. It is therefore worthwhile to delete KLF genes in a cell type-specific manner and analyzing responses to external stress. Such experiments will provide new molecular and cellular insight into tumor growth and tissue remodeling.

Identification of low-molecular-weight compounds that can modulate the KLF–cofactor complex is another important research target given that this will lead to the development of new therapies. We discuss on this subject in detail in Chapter 18.

Research on the KLF family is still wide open and comprehensive understanding of family members will certainly help expand into new fields of molecular genetics, developmental cell biology and disease biology.

References

Aizawa K, Suzuki T, Kada N et al (2004) Regulation of platelet-derived growth factor-A chain by Kruppel-like factor 5: new pathway of cooperative activation with nuclear factor-kappaB. J Biol Chem 279:70–76.

Alder J K, Georgantas R W, 3rd, Hildreth R L et al (2008) Kruppel-like factor 4 is essential for inflammatory monocyte differentiation in vivo. J Immunol 180:5645–5652.

Ali F, Hamdulay S S, Kinderlerer A R et al (2007) Statin-mediated cytoprotection of human vascular endothelial cells: a role for Kruppel-like factor 2-dependent induction of heme oxygenase-1. J Thromb Haemost 5:2537–2546.

Atkins G B, Wang Y, Mahabeleshwar G H et al (2008) Hemizygous deficiency of Kruppel-like factor 2 augments experimental atherosclerosis. Circ Res 103:690–693.

Basu P, Morris P E, Haar J L et al (2005) KLF2 is essential for primitive erythropoiesis and regulates the human and murine embryonic beta-like globin genes in vivo. Blood 106:2566–2571.

Bateman N W, Tan D, Pestell R G et al (2004) Intestinal tumor progression is associated with altered function of KLF5. J Biol Chem 279:12093–12101.

Benzeno S, Narla G, Allina J et al (2004) Cyclin-dependent kinase inhibition by the KLF6 tumor suppressor protein through interaction with cyclin D1. Cancer Res 64:3885–3891.

Birsoy K, Chen Z, and Friedman J (2008) Transcriptional regulation of adipogenesis by KLF4. Cell Metab 7:339–347.

Carlson C M, Endrizzi B T, Wu J et al (2006) Kruppel-like factor 2 regulates thymocyte and T-cell migration. Nature 442:299–302.

Chanchevalap S, Nandan M O, McConnell B B et al (2006) Kruppel-like factor 5 is an important mediator for lipopolysaccharide-induced proinflammatory response in intestinal epithelial cells. Nucleic Acids Res 34:1216–1223.

Dang D T, Mahatan C S, Dang L H et al (2001) Expression of the gut-enriched Kruppel-like factor (Kruppel-like factor 4) gene in the human colon cancer cell line RKO is dependent on CDX2. Oncogene 20:4884–4890.
Das H, Kumar A, Lin Z et al (2006) Kruppel-like factor 2 (KLF2) regulates proinflammatory activation of monocytes. Proc Natl Acad Sci U S A 103:6653–6658.
Du J X, Bialkowska A B, McConnell B B et al (2008) SUMOylation Regulates Nuclear Localization of Kruppel-like Factor 5. J Biol Chem 283:31991–32002.
Eaton S A, Funnell A P, Sue N et al (2008) A network of Kruppel-like Factors (Klfs). Klf8 is repressed by Klf3 and activated by Klf1 in vivo. J Biol Chem 283:26937–26947.
Ellenrieder V, Buck A, Harth A et al (2004) KLF11 mediates a critical mechanism in TGF-beta signaling that is inactivated by Erk-MAPK in pancreatic cancer cells. Gastroenterology 127:607–620.
Ema M, Mori D, Niwa H et al (2008) Kruppel-like factor 5 is essential for blastocyst development and the normal self-renewal of mouse ESCs. Cell Stem Cell 3:555–567.
Feinberg M W, Cao Z, Wara A K et al (2005) Kruppel-like factor 4 is a mediator of proinflammatory signaling in macrophages. J Biol Chem 280:38247–38258.
Feinberg M W, Wara A K, Cao Z et al (2007) The Kruppel-like factor KLF4 is a critical regulator of monocyte differentiation. Embo J 26:4138–4148.
Fernandez-Zapico M E, Mladek A, Ellenrieder V et al (2003) An mSin3A interaction domain links the transcriptional activity of KLF11 with its role in growth regulation. Embo J 22:4748–4758.
Fisch S, Gray S, Heymans S et al (2007) Kruppel-like factor 15 is a regulator of cardiomyocyte hypertrophy. Proc Natl Acad Sci U S A 104:7074–7079.
Foster K W, Frost A R, McKie-Bell P et al (2000) Increase of GKLF messenger RNA and protein expression during progression of breast cancer. Cancer Res 60:6488–6495.
Foster K W, Liu Z, Nail C D et al (2005) Induction of KLF4 in basal keratinocytes blocks the proliferation-differentiation switch and initiates squamous epithelial dysplasia. Oncogene 24:1491–1500.
Friedman S L (2006) Transcriptional regulation of stellate cell activation. J Gastroenterol Hepatol 21 Suppl 3:S79–83.
Fujiu K, Manabe I, Ishihara A et al (2005) Synthetic retinoid Am80 suppresses smooth muscle phenotypic modulation and in-stent neointima formation by inhibiting KLF5. Circ Res 97:1132–1141.
Gartel A L, and Tyner A L (1999) Transcriptional regulation of the p21((WAF1/CIP1)) gene. Exp Cell Res 246:280–289.
Ghaleb A M, McConnell B B, Nandan M O et al (2007) Haploinsufficiency of Kruppel-like factor 4 promotes adenomatous polyposis coli dependent intestinal tumorigenesis. Cancer Res 67:7147–7154.
Ghaleb A M, Nandan M O, Chanchevalap S et al (2005) Kruppel-like factors 4 and 5: the yin and yang regulators of cellular proliferation. Cell Res 15:92–96.
Ghaleb A M, and Yang V W (2008) The Pathobiology of Kruppel-like Factors in Colorectal Cancer. Curr Colorectal Cancer Rep 4:59–64.
Gillemans N, Tewari R, Lindeboom F et al (1998) Altered DNA-binding specificity mutants of EKLF and Sp1 show that EKLF is an activator of the beta-globin locus control region in vivo. Genes Dev 12:2863–2873.
Godmann M, Katz J P, Guillou F et al (2008) Kruppel-like factor 4 is involved in functional differentiation of testicular Sertoli cells. Dev Biol 315:552–566.
Gordon A R, Outram S V, Keramatipour M et al (2008) Splenomegaly and modified erythropoiesis in KLF13-/- mice. J Biol Chem 283:11897–11904.
Gray S, Feinberg M W, Hull S et al (2002) The Kruppel-like factor KLF15 regulates the insulin-sensitive glucose transporter GLUT4. J Biol Chem 277:34322–34328.
Gray S, Wang B, Orihuela Y et al (2007) Regulation of gluconeogenesis by Kruppel-like factor 15. Cell Metab 5:305–312.

Haldar S M, Ibrahim O A, and Jain M K (2007) Kruppel-like Factors (KLFs) in muscle biology. J Mol Cell Cardiol 43:1–10.
Hashmi S, Ji Q, Zhang J et al (2008) A Kruppel-like factor in *Caenorhabditis elegans* with essential roles in fat regulation, cell death, and phagocytosis. DNA Cell Biol 27:545–551.
Hodge D, Coghill E, Keys J et al (2006) A global role for EKLF in definitive and primitive erythropoiesis. Blood 107:3359–3370.
Holian J, Qi W, Kelly D J et al (2008) Role of Kruppel-like factor 6 in transforming growth factor-beta1-induced epithelial-mesenchymal transition of proximal tubule cells. Am J Physiol Renal Physiol 295:F1388–1396.
Jiang J, Chan Y S, Loh Y H et al (2008) A core Klf circuitry regulates self-renewal of embryonic stem cells. Nat Cell Biol 10:353–360.
Kada N, Suzuki T, Aizawa K et al (2007) Acyclic retinoid inhibits neointima formation through retinoic acid receptor beta-induced apoptosis. Arterioscler Thromb Vasc Biol 27:1535–1541.
Kada N, Suzuki T, Aizawa K et al (2008) Acyclic retinoid inhibits functional interaction of transcription factors Kruppel-like factor 5 and retinoic acid receptor-alpha. FEBS Lett 582:1755–1760.
Kadonaga J T, Carner K R, Masiarz F R et al (1987) Isolation of cDNA encoding transcription factor Sp1 and functional analysis of the DNA binding domain. Cell 51:1079–1090.
Kajimura D, Dragomir C, Ramirez F et al (2007) Identification of genes regulated by transcription factor KLF7 in differentiating olfactory sensory neurons. Gene 388:34–42.
Kawai-Kowase K, Kurabayashi M, Hoshino Y et al (1999) Transcriptional activation of the zinc finger transcription factor BTEB2 gene by Egr-1 through mitogen-activated protein kinase pathways in vascular smooth muscle cells. Circ Res 85:787–795.
Kimmelman A C, Qiao R F, Narla G et al (2004) Suppression of glioblastoma tumorigenicity by the Kruppel-like transcription factor KLF6. Oncogene 23:5077–5083.
Kuo C T, Veselits M L, Barton K P et al (1997) The LKLF transcription factor is required for normal tunica media formation and blood vessel stabilization during murine embryogenesis. Genes Dev 11:2996–3006.
Lavallee G, Andelfinger G, Nadeau M et al (2006) The Kruppel-like transcription factor KLF13 is a novel regulator of heart development. Embo J 25:5201–5213.
Li D, Yea S, Li S et al (2005) Kruppel-like factor-6 promotes preadipocyte differentiation through histone deacetylase 3-dependent repression of DLK1. J Biol Chem 280:26941–26952.
Liu Y, Sinha S, McDonald O G et al (2005) Kruppel-like factor 4 abrogates myocardin-induced activation of smooth muscle gene expression. J Biol Chem 280:9719–9727.
Lu S, Jenster G, and Epner D E (2000) Androgen induction of cyclin-dependent kinase inhibitor p21 gene: role of androgen receptor and transcription factor Sp1 complex. Mol Endocrinol 14:753–760.
Matsumoto N, Kubo A, Liu H et al (2006) Developmental regulation of yolk sac hematopoiesis by Kruppel-like factor 6. Blood 107:1357–1365.
Matsumura T, Suzuki T, Aizawa K et al (2005) The deacetylase HDAC1 negatively regulates the cardiovascular transcription factor Kruppel-like factor 5 through direct interaction. J Biol Chem 280:12123–12129.
McConnell B B, Ghaleb A M, Nandan M O et al (2007) The diverse functions of Kruppel-like factors 4 and 5 in epithelial biology and pathobiology. Bioessays 29:549–557.
McConnell B B, Klapproth J M, Sasaki M et al (2008) Kruppel-like factor 5 mediates transmissible murine colonic hyperplasia caused by *Citrobacter rodentium* infection. Gastroenterology 134:1007–1016.
Miller I J, and Bieker J J (1993) A novel, erythroid cell-specific murine transcription factor that binds to the CACCC element and is related to the Kruppel family of nuclear proteins. Mol Cell Biol 13:2776–2786.
Miyamoto S, Suzuki T, Muto S et al (2003) Positive and negative regulation of the cardiovascular transcription factor KLF5 by p300 and the oncogenic regulator SET through interaction and acetylation on the DNA-binding domain. Mol Cell Biol 23:8528–8541.

Mori T, Sakaue H, Iguchi H et al (2005) Role of Kruppel-like factor 15 (KLF15) in transcriptional regulation of adipogenesis. J Biol Chem 280:12867–12875.

Munemasa Y, Suzuki T, Aizawa K et al (2008) Promoter region-specific histone incorporation by the novel histone chaperone ANP32B and DNA-binding factor KLF5. Mol Cell Biol 28:1171–1181.

Nandan M O, Chanchevalap S, Dalton W B et al (2005) Kruppel-like factor 5 promotes mitosis by activating the cyclin B1/Cdc2 complex during oncogenic Ras-mediated transformation. FEBS Lett 579:4757–4762.

Nandan M O, McConnell B B, Ghaleb A M et al (2008) Kruppel-like factor 5 mediates cellular transformation during oncogenic KRAS-induced intestinal tumorigenesis. Gastroenterology 134:120–130.

Nandan M O, Yoon H S, Zhao W et al (2004) Kruppel-like factor 5 mediates the transforming activity of oncogenic H-Ras. Oncogene 23:3404–3413.

Narla G, DiFeo A, Fernandez Y et al (2008) KLF6-SV1 overexpression accelerates human and mouse prostate cancer progression and metastasis. J Clin Invest 118:2711–2721.

Narla G, Heath K E, Reeves H L et al (2001) KLF6, a candidate tumor suppressor gene mutated in prostate cancer. Science 294:2563–2566.

Narla G, Kremer-Tal S, Matsumoto N et al (2007) In vivo regulation of p21 by the Kruppel-like factor 6 tumor-suppressor gene in mouse liver and human hepatocellular carcinoma. Oncogene 26:4428–4434.

Ohguchi H, Tanaka T, Uchida A et al (2008) Hepatocyte nuclear factor 4alpha contributes to thyroid hormone homeostasis by cooperatively regulating the type 1 iodothyronine deiodinase gene with GATA4 and Kruppel-like transcription factor 9. Mol Cell Biol 28:3917–3931.

Oishi Y, Manabe I, Tobe K et al (2008) SUMOylation of Kruppel-like transcription factor 5 acts as a molecular switch in transcriptional programs of lipid metabolism involving PPAR-delta. Nat Med 14:656–666.

Oishi Y, Manabe I, Tobe K et al (2005) Kruppel-like transcription factor KLF5 is a key regulator of adipocyte differentiation. Cell Metab 1:27–39.

Outram S V, Gordon A R, Hager-Theodorides A L et al (2008) KLF13 influences multiple stages of both B and T cell development. Cell Cycle 7:2047–2055.

Parisi S, Passaro F, Aloia L et al (2008) Klf5 is involved in self-renewal of mouse embryonic stem cells. J Cell Sci 121:2629–2634.

Perdomo J, Verger A, Turner J et al (2005) Role for SUMO modification in facilitating transcriptional repression by BKLF. Mol Cell Biol 25:1549–1559.

Perkins A C, Sharpe A H, and Orkin S H (1995) Lethal beta-thalassaemia in mice lacking the erythroid CACCC-transcription factor EKLF. Nature 375:318–322.

Ratziu V, Lalazar A, Wong L et al (1998) Zf9, a Kruppel-like transcription factor up-regulated in vivo during early hepatic fibrosis. Proc Natl Acad Sci U S A 95:9500–9505.

Ribeiro A, Bronk S F, Roberts P J et al (1999) The transforming growth factor beta(1)-inducible transcription factor TIEG1, mediates apoptosis through oxidative stress. Hepatology 30:1490–1497.

Rowland B D, Bernards R, and Peeper D S (2005) The KLF4 tumour suppressor is a transcriptional repressor of p53 that acts as a context-dependent oncogene. Nat Cell Biol 7:1074–1082.

Segre J A, Bauer C, and Fuchs E (1999) Klf4 is a transcription factor required for establishing the barrier function of the skin. Nat Genet 22:356–360.

SenBanerjee S, Lin Z, Atkins G B et al (2004) KLF2 Is a novel transcriptional regulator of endothelial proinflammatory activation. J Exp Med 199:1305–1315.

Shie J L, Chen Z Y, Fu M et al (2000) Gut-enriched Kruppel-like factor represses cyclin D1 promoter activity through Sp1 motif. Nucleic Acids Res 28:2969–2976.

Shindo T, Manabe I, Fukushima Y et al (2002) Kruppel-like zinc-finger transcription factor KLF5/BTEB2 is a target for angiotensin II signaling and an essential regulator of cardiovascular remodeling. Nat Med 8:856–863.

Siatecka M, Xue L, and Bieker J J (2007) Sumoylation of EKLF promotes transcriptional repression and is involved in inhibition of megakaryopoiesis. Mol Cell Biol 27:8547–8560.

Simmen F A, Xiao R, Velarde M C et al (2007) Dysregulation of intestinal crypt cell proliferation and villus cell migration in mice lacking Kruppel-like factor 9. Am J Physiol Gastrointest Liver Physiol 292:G1757–1769.

Sirach E, Bureau C, Peron J M et al (2007) KLF6 transcription factor protects hepatocellular carcinoma-derived cells from apoptosis. Cell Death Differ 14:1202–1210.

Song C Z, Gavriilidis G, Asano H et al (2005) Functional study of transcription factor KLF11 by targeted gene inactivation. Blood Cells Mol Dis 34:53–59.

Subramaniam M, Hawse J R, Johnsen S A et al (2007) Role of TIEG1 in biological processes and disease states. J Cell Biochem 102:539–548.

Sue N, Jack B H, Eaton S A et al (2008) Targeted disruption of the basic Kruppel-like factor gene (Klf3) reveals a role in adipogenesis. Mol Cell Biol 28:3967–3978.

Suzuki T, Nishi T, Nagino T et al (2007) Functional interaction between the transcription factor Kruppel-like factor 5 and poly(ADP-ribose) polymerase-1 in cardiovascular apoptosis. J Biol Chem 282:9895–9901.

Swamynathan S K, Katz J P, Kaestner K H et al (2007) Conditional deletion of the mouse Klf4 gene results in corneal epithelial fragility, stromal edema, and loss of conjunctival goblet cells. Mol Cell Biol 27:182–194.

Taatjes D J, Naar A M, Andel F, 3rd et al (2002) Structure, function, and activator-induced conformations of the CRSP coactivator. Science 295:1058–1062.

Takahashi K, Tanabe K, Ohnuki M et al (2007) Induction of pluripotent stem cells from adult human fibroblasts by defined factors. Cell 131:861–872.

Takahashi K, and Yamanaka S (2006) Induction of pluripotent stem cells from mouse embryonic and adult fibroblast cultures by defined factors. Cell 126:663–676.

Turksen K, and Troy T C (2002) Permeability barrier dysfunction in transgenic mice overexpressing claudin 6. Development 129:1775–1784.

Usui S, Sugimoto N, Takuwa N et al (2004) Blood lipid mediator sphingosine 1-phosphate potently stimulates platelet-derived growth factor-A and -B chain expression through S1P1-Gi-Ras-MAPK-dependent induction of Kruppel-like factor 5. J Biol Chem 279:12300–12311.

van Thienen J V, Fledderus J O, Dekker R J et al (2006) Shear stress sustains atheroprotective endothelial KLF2 expression more potently than statins through mRNA stabilization. Cardiovasc Res 72:231–240.

Wan H, Luo F, Wert S E et al (2008) Kruppel-like factor 5 is required for perinatal lung morphogenesis and function. Development 135:2563–2572.

Wang B, Haldar S M, Lu Y et al (2008a) The Kruppel-like factor KLF15 inhibits connective tissue growth factor (CTGF) expression in cardiac fibroblasts. J Mol Cell Cardiol 45:193–197.

Wang C, Han M, Zhao X M et al (2008b) Kruppel-like factor 4 is required for the expression of vascular smooth muscle cell differentiation marker genes induced by all-trans retinoic acid. J Biochem 144:313–321.

Wang C H, Tsao Y P, Chen H J et al (2000) Transcriptional repression of p21((Waf1/Cip1/Sdi1)) gene by c-jun through Sp1 site. Biochem Biophys Res Commun 270:303–310.

Wang F, Zhu Y, Huang Y et al (2005) Transcriptional repression of WEE1 by Kruppel-like factor 2 is involved in DNA damage-induced apoptosis. Oncogene 24:3875–3885.

Wang X, and Zhao J (2007) KLF8 transcription factor participates in oncogenic transformation. Oncogene 26:456–461.

Wang X, Zheng M, Liu G et al (2007) Kruppel-like factor 8 induces epithelial to mesenchymal transition and epithelial cell invasion. Cancer Res 67:7184–7193.

Watanabe K, Ohnishi S, Manabe I et al (2008) KLF6 in nonalcoholic fatty liver disease: role of fibrogenesis and carcinogenesis. Gastroenterology 135:309–312.

Watanabe N, Kurabayashi M, Shimomura Y et al (1999) BTEB2, a Kruppel-like transcription factor, regulates expression of the SMemb/Nonmuscle myosin heavy chain B (SMemb/NMHC-B) gene. Circ Res 85:182–191.

Wei D, Kanai M, Jia Z et al (2008) Kruppel-like factor 4 induces p27Kip1 expression in and suppresses the growth and metastasis of human pancreatic cancer cells. Cancer Res 68:4631–4639.

Wei H, Wang X, Gan B et al (2006) Sumoylation delimits KLF8 transcriptional activity associated with the cell cycle regulation. J Biol Chem 281:16664–16671.

Wu J, Srinivasan S V, Neumann J C et al (2005) The KLF2 transcription factor does not affect the formation of preadipocytes but inhibits their differentiation into adipocytes. Biochemistry 44:11098–11105.

Yea S, Narla G, Zhao X et al (2008) Ras promotes growth by alternative splicing-mediated inactivation of the KLF6 tumor suppressor in hepatocellular carcinoma. Gastroenterology 134:1521–1531.

Yoshida T, Kaestner K H, and Owens G K (2008) Conditional deletion of Kruppel-like factor 4 delays downregulation of smooth muscle cell differentiation markers but accelerates neointimal formation following vascular injury. Circ Res 102:1548–1557.

Zhang W, Geiman D E, Shields J M et al (2000) The gut-enriched Kruppel-like factor (Kruppel-like factor 4) mediates the transactivating effect of p53 on the p21WAF1/Cip1 promoter. J Biol Chem 275:18391–18398.

Ziemer L T, Pennica D, and Levine A J (2001) Identification of a mouse homolog of the human BTEB2 transcription factor as a beta-catenin-independent Wnt-1-responsive gene. Mol Cell Biol 21:562–574.

Part 2
Molecular Control of Krüppel-like Factor Function

Chapter 2
Molecular Structures of Krüppel-like Factors

Toshio Nagashima, Fumiaki Hayashi, Takashi Umehara, and Shigeyuki Yokoyama

Abstract The Krüppel-like factor (KLF) family regulates several biological processes, such as self-renewal, proliferation, differentiation, development, and tissue-selectively restricted events of a cell at the transcriptional level. The KLF family has a highly conserved array of three C2H2-type zinc fingers with similarity to *Drosophila* Krüppel at the C-terminus, comprising a GC-rich DNA-binding domain, to mediate activation and/or repression of transcription. In contrast, the N-terminal regions of KLFs contain several distinct domains that are required for binding to chromatin-associated proteins, such as CtBP or Sin3A. We describe the structure–function aspects of KLFs, with a primary focus on the DNA-binding domains and the protein-binding domains.

Introduction: Overview of KLF Proteins

DNA-Binding Domain

C2H2-type zinc fingers are found in various transcription factors (Wolfe et al. 1999). In most cases, the C2H2 zinc finger is repeated from twice to more than 30 times in the protein (Iuchi 2001). The zinc fingers of KLF are repeated three times, and the architecture and sequences are well conserved among the 17 human KLF proteins (Fig. 1). The first (ZF1), second (ZF2), and third (ZF3) zinc fingers are composed of 25, 25, and 23 amino acids, respectively. The linkers of each zinc finger are conserved in all of the KLF proteins, including five amino acid residues, TGEKP. Like most of the C2H2-type zinc fingers, the KLF zinc fingers bind DNA

T. Nagashima, F. Hayashi, T. Umehara, and S. Yokoyama
RIKEN Systems and Structural Biology Center, 1-7-22 Suehiro-cho, Tsurumi, Yokohama 230-0045, Japan

S. Yokoyama (✉)
Department of Biophysics and Biochemistry, Graduate School of Science, The University of Tokyo, Bunkyo-ku, Tokyo 113-0033, Japan

R. Nagai et al. (eds.), *The Biology of Krüppel-like Factors*,
DOI 10.1007/978-4-431-87775-2_2, © Springer 2009

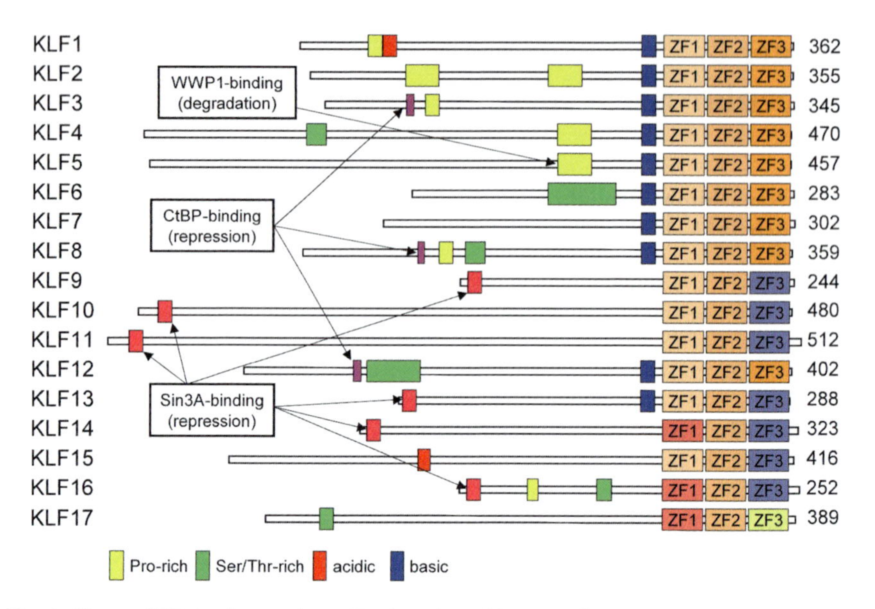

Fig. 1 Human KLF family members. The domain architecture of each KLF is depicted based on the sequence conservation among species. The three C2H2 zinc fingers are shown as *ZF1*, *ZF2*, and *ZF3*. Pro-rich, Ser/Thr-rich, acidic, and basic regions are *yellow*, *green*, *red*, and *blue rectangles*, respectively. CtBP-binding regions and Sin3A-binding regions are indicated by *arrows*. The amino acid length of each KLF protein is shown on the right

and recognize a certain sequence. The amino acids located at positions −1, 2, 3, and 6 from the N-terminal residue of the α-helix directly interact with a DNA base. The DNA recognition modes of these amino acid residues were predicted from a previous structural analysis of non-KLF C2H2 zinc fingers (described below) (Wolfe et al. 1999).

Pro-rich, Ser/Thr-rich, Acidic, and Basic Regions

Transcription factors often have a Pro-rich region that is recognized by the SH3 domain or the WW domain (Macias et al. 2002). The SH3 domain, found in a variety of intracellular or membrane-associated proteins, binds to the sequence motif PXXP, where X is any residue (Mayer 2001). Similarly, the WW domain, found in signal transduction pathways, binds to the Pro-rich sequence motif [A/P]PP[A/P]Y. Such Pro-rich regions are found in the N-terminal regions of KLF1 to KLF5, KLF8, and KLF16 (Fig. 1). The Pro-rich region of KLF5 interacts with the WW domain of the proteolysis-related protein ubiquitin ligase WWP1 (Chen et al. 2005). This interaction modulates the activity of KLF5 in gene regulation by ubiquitination and degradation (Chen et al. 2005).

In addition to the Pro-rich region, there is a Ser/Thr-rich region in KLF4, KLF6, KLF8, KLF12, KLF16, and KLF17 (Fig. 1), The exact function of the Ser/Thr-rich region in the KLFs has not been elucidated.

In several KLFs, an acidic amino acid (Asp/Glu)-rich region (acidic region) exists in the N-terminal region (Fig. 1, shown by a red rectangle). The acidic region is thought to be required for transcriptional activation (Kaczynski et al. 2003; Ptashne 1988).

A basic amino acid (Arg/Lys)-rich region (basic region) can be found just before the repeated zinc fingers in several KLFs (Fig. 1, shown by a blue rectangle). The basic region might function as a nuclear localization signal (NLS) (Kaczynski 2003). However, the basic region of KLF1 is not essential for its nuclear localization, and several basic residues in the repeated zinc fingers are instead critical determinants for its nuclear localization (Pandya and Townes 2002). Furthermore, the basic region is absent in several KLFs, such as KLF9, KLF10, KLF11, KLF14, KLF15, KLF16, and KLF17 (Fig. 1), although the KLFs putatively function as DNA-binding transcription factors in the nucleus. The basic region may be utilized for interactions with other nuclear proteins.

CtBP-Binding Region

KLFs are known to interact with several chromatin-associated factors. Among these factors, C-terminal binding protein (CtBP) plays a key role in transcriptional repression (Criqui-Filipe et al. 1999; Schaeper et al. 1995; Turner and Crossley 1998). KLF3, KLF8, and KLF12 contain a consensus sequence, PXDL(S/T), in their N-terminal regions (Fig. 1). This sequence is thought to bind to the C-terminal binding protein (CtBP), which is known as an E1A-binding transcriptional co-repressor with sequence similarity to 2-hydroxy acid dehydrogenases (Chinnadurai 2002; Turner and Crossley 1998). Several transcription factors recruit the CtBP co-repressor (Shi et al. 2003; Turner and Crossley 2001). These transcription factors, including the KLF family members, directly bind to the co-repressor through the consensus amino acids PXDL(S/T) (Chinnadurai 2002; Turner and Crossley 1998; van Vliet et al. 2000). The KLF family members containing the CtBP-binding motif thus repress transcription in collaboration with CtBP (Kaczynski 2003; Lomberk and Urrutia 2005; Turner and Crossley 2001; van Vliet et al. 2000).

Sin3A-Binding Region

The Sin3 co-repressor functions as a platform for a co-repressor complex comprising multiple proteins (Silverstein and Ekwall 2005). Sin3 recruits transcription factors, such as KLFs and histone deacetylases (HDAC) (Silverstein and Ekwall 2005).

KLF9, KLF10, KLF11, KLF13, KLF14, and KLF16 contain a hydrophobic consensus sequence, ϕϕXXLϕX(M/I)X (where ϕ is a hydrophobic residue), in their N-terminal regions (Fig. 1). The ϕϕXXLϕX(M/I)X motif is known to bind the transcriptional co-repressor mSin3A (Kaczynski et al. 2001; Zhang et al. 2001). The Sin3A co-repressor complex is one of the best characterized HDAC-associated regulator complexes identified from yeasts to humans (Silverstein and Ekwall 2005). The KLF members containing the Sin3A-binding motif play a key role in repressing their target genes through their association with mSin3A (Kaczynski et al. 2003; Lomberk and Urrutia 2005).

DNA-Binding Domains of KLFs

Putative DNA Recognition Sequences

The C2H2-type zinc fingers of the KLF members consist of 25, 25, and 23 amino acids, in order. Each zinc finger holds one zinc ion between a β-hairpin at the N-terminus and an α-helix at the C-terminus. An example of a KLF zinc finger structure is shown in Figure 2 (Nagashima et al. unpublished data). The zinc ion is coordinated by two Cys residues on the β-hairpin and two His residues on the α-helix (Fig. 2A, B). Each zinc finger recognizes three base pairs in double-stranded DNA, and thus the three zinc finger repeats of the KLFs recognize nine base pairs in total. The DNA recognition mode of the KLF zinc fingers can be predicted based on the structural analysis of the Zif268–DNA complex (Wolfe et al. 1999). The four amino acids at positions −1, 2, 3, and 6 from the N-terminal end of the α-helix play the most critical role in direct base recognition (Fig. 2A,B). Most of these four amino acid residues are conserved among the KLFs (Fig. 2C). In ZF1, Ala and Ser, which presumably have little preference for DNA bases, appear at the 6th position (Fig. 2C). Furthermore, the Lys at position −1 also has no preference for DNA bases. Collectively, ZF1 should prefer the sequence 5′-NGN-3′, where N is any base (Fig. 2D). As for ZF2, the four positions are completely conserved among the KLFs, and ZF2 should prefer the sequence 5′-GCG-3′ (Fig. 2D). In the case of ZF3, Leu, Lys, and Gln appear at the 6th position (Fig. 2C). The Leu residue would have no preference for a DNA base, whereas Lys and Gln prefer G or T, and A, respectively. The ZF3 fingers of KLF1 to KLF8, and KLF12 should thus prefer the sequence 5′-NGG-3′. In the same manner, the ZF3 fingers of KLF9, KLF10, KLF11, KLF13, KLF14, KLF15, and KLF16 should prefer 5′-(G/T)GG-3′, and the ZF3 finger of KLF17 may prefer 5′-AGG-3′ (Fig. 2D). These predictions of the DNA recognition, shown in Figure 2D, are essentially consistent with numerous reports on the sequence preferences of KLFs to the CACCC-box (GC-box) in promoter or enhancer regions (Izmailova et al. 1999; Jiang et al. 2008; Nielsen et al. 1998; Philipsen and Suske 1999).

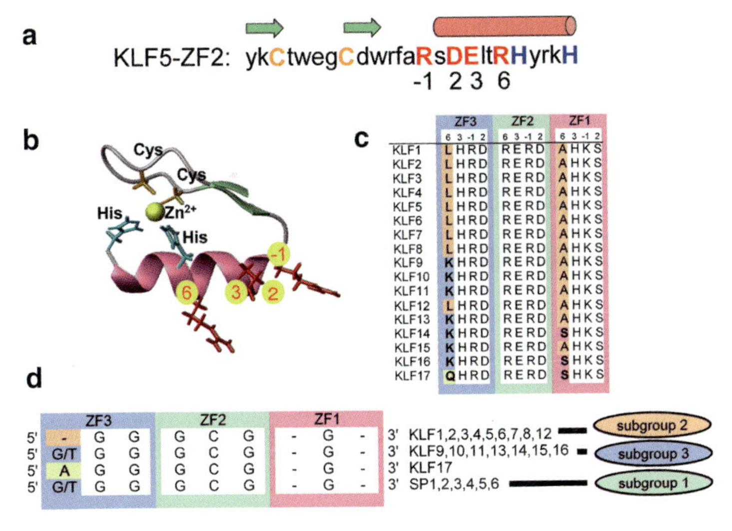

Fig. 2 Prediction of the DNA recognition sequences of the KLF family members. **A** Sequence of a typical KLF finger structure (the second finger of KLF5). The two Cys and two His residues that coordinate a zinc ion are indicated in *yellow* and *blue letters*, respectively. The two β-strands and the α-helix are indicated as *green arrows* and a *red cylinder*, respectively. The putative DNA recognition residues are shown as *red letters* (positions −1, 2, 3, and 6). **B** Typical KLF finger structure (the second finger of KLF5). **C** Alignment of the putative DNA recognition sequences of the KLFs. The residues differing among the KLF family members are highlighted. **D** Putative DNA recognition sequences of KLFs. Note that the 5′-end bases of the DNA sequence have diverged between the KLF subgroups

Structures of DNA-Binding Domains

There is presently no structural information available for DNA-bound KLFs. We have analyzed several solution structures of KLFs in complex with DNA and completed the nuclear magnetic resonance (NMR) structure calculations on one of the KLF zinc fingers in complex with a GC-box DNA (Nagashima et al. unpublished). Without the double-stranded DNA, we found that the two tandem KLF zinc fingers are independently structured, and the two zinc fingers are randomly oriented in solution (Fig. 3, left). The fold of each C2H2 zinc finger is rigid, and the linker between the fingers has a random conformation (Fig. 3). Each zinc finger binds one zinc ion coordinated by two Cys and two His residues in the same manner as the other C2H2-type zinc fingers (Figs. 2B, 3). In contrast, when the KLF protein is incubated with double-stranded GC-box DNA, the zinc fingers dock on the major groove of the predicted DNA sequence with a defined spatial alignment (Fig. 3, right). As a result, the linker between the zinc fingers is fixed on the DNA (Fig. 3). The conformational change that occurs upon binding to the double-stranded DNA appears to lead to specific, high-affinity recognition of the DNA sequence.

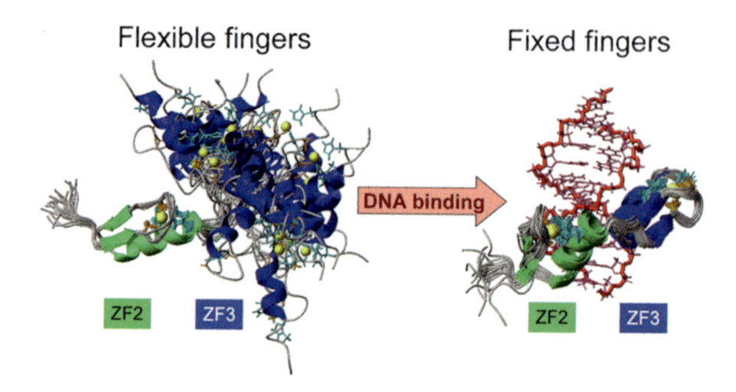

Fig. 3 Solution structures of a typical KLF zinc finger domain. Twenty solution structures of a tandem KLF zinc finger are superposed. *Green* and *blue helices* indicate the second and third zinc fingers, respectively. Without DNA, each KLF zinc finger is structured, but the two fingers connected by a linker sequence are randomly oriented (*left*). The entire domain containing the two zinc fingers becomes structured upon DNA binding (*right*)

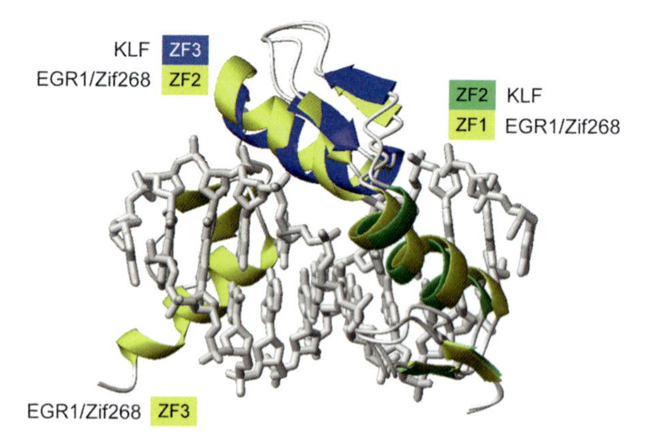

Fig. 4 Superposition of the KLF and EGR1/Zif268 structures

The solution structure of the zinc fingers of the KLF family revealed the structural basis for recognition of the specific GC-box DNA sequence. Concerning prediction of the DNA recognition, the NMR structure confirmed that positions –1, 2, 3, and 6 of the zinc finger recognized the DNA sequence by direct interactions (Nagashima et al. unpublished).

Comparison to the EGR1/Zif268 Structure

The solution structure of the KLF–DNA complex resembles the crystal structure of the EGR1/Zif268 zinc fingers in complex with DNA (Elrod-Erickson et al. 1996). ZF1, ZF2, and ZF3 of EGR1/Zif268 recognize the DNA sequences GCG, TGG, and GCG, respectively (Elrod-Erickson et al. 1998). The sequence alignment and structural comparison between KLF and EGR1/Zif268 indicate that the DNA recognition modes of ZF2 and ZF3 of the KLFs are almost the same as those of ZF1 and ZF2 of EGR1/Zif268, respectively (Fig. 4) (Nagashima et al. unpublished).

This implies that the KLFs and EGR1/Zif268 bind the same DNA sequence, 5′-(T/G)GGGCG-3′, while the recognition of one DNA triplet is different between the KLFs and EGR1/Zif268.

Protein-Binding Domains of KLFs

CtBP-Binding Domain

CtBP interacts with the repression domains of transcription factors (e.g., KLF3, KLF8, KLF12) and with repressor complexes containing histone deacetylases (Chinnadurai 2002; Turner and Crossley 2001). The complex structures of CtBP3 and the PIDLSKK peptide were reported (Nardini et al. 2003). Some KLFs share sequence homology with this peptide (Fig. 5B). In the CtBP-binding motif PXDLS,

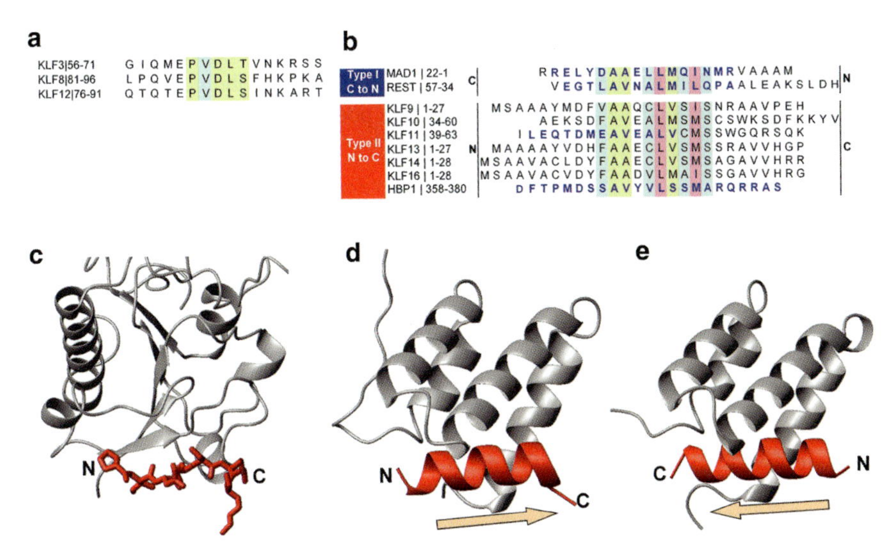

Fig. 5 Typical protein-binding motifs within KLFs. The structures of the Sin3B and CtBP3 motifs are indicated. **A** Sequence alignment of the CtBP-binding motif of KLFs. The CtBP-binding motifs are colored *yellow* and *cyan*. **B** Sequence alignment of the region including the Sin3-binding motif. Amino acid codes colored *blue* indicate that structural information is available. Rows in *cyan*, *magenta*, and *yellow* indicate hydrophilic interaction, hydrophobic interaction, and important hydrophobic interaction with the pocket on the PAH domain, respectively. Note that types I and II have opposite directions of the amino acid sequence, and their binding to the PAH domain of Sin3 occurs in opposite helical orientations. **C** Complex structure of CtBP3 and the designed CtBP3-binding motif peptide. CtBP3 and the peptide are colored *gray* and *red*, respectively (PDB-ID: 1HL3). **D** Complex structure of PAH1 of Sin3B and the transcription factor REST (Nomura et al. 2005) (PDB-ID: 2CYZ). The PAH domain forms a four-helix-bundle (*gray*), and REST forms one helix (*red*) on the pocket of the PAH domain. The *arrow* indicates the orientation of the REST helix (type I). **E** Complex structure of PAH1 of Sin3A and the Sin3-associated peptide (SAP25) (Sahu et al. 2008) (PDB-ID: 2RMS). This figure is shown from the same viewpoint as in **D**. Note that the SAP25 helix is placed in the opposite orientation (type II) on the PAH domain

X is often Leu or Val. Leu or another hydrophobic residue at X docks on a hydrophobic patch of CtBP, and the Asp and Ser residues of the motif form hydrogen bonds. KLF3, KLF8, and KLF12 are known to repress their target genes through the enzymatic activities of CtBP and HDAC (Lomberk and Urrutia 2005).

Sin3-Binding Motif

Sin3 consists of a repeated PAH domain and an HDAC-interacting motif. The PAH domain has a four-helix-bundle fold (Sahu et al. 2008). The complex structures between PAH and the Sin3-binding motif in several transcription factors revealed that the Sin3-binding motif forms an α-helix in a pocket on the PAH domain (Nomura et al. 2005; Sahu et al. 2008; Swanson et al. 2004). Interestingly, the PAH domain can bind to two types of peptide in opposite directions (Swanson et al. 2004). In the group containing MAD1 and the REST transcription factor, the binding motif is ϕxxϕϕxAAxxϕ(E/D), where ϕ is a hydrophobic residue. On the other hand, in the opposite direction of the binding mode, the group containing KLF and the HBP transcription factor has the motif A(A/V)xϕϕxxϕ. The hydrophobic residues in the binding motif are located within a hydrophobic pocket on the PAH domain in the helix bundle.

Other Potential Functional Domains in KLF5

In terms of protein–protein interactions, KLF5 is the most extensively characterized among the KLF family members. In carcinogenesis, KLF5 modulates cell proliferation, differentiation, and cell cycle regulation through direct interactions with several nuclear factors, such as WWP1 (Chen et al. 2005), Poly (ADP-ribose) polymerase-1 (PARP-1) (Suzuki et al. 2007), histone deacetylase HDAC1 (Matsumura et al. 2005), histone acetyltransferase p300 (Miyamoto et al. 2003), the histone chaperones SET (Miyamoto et al. 2003) and ANP32B (Munemasa et al. 2008), and the transcription activator STAT1 (Du et al. 2007). However, structural information about these complexes is still needed.

WWP1 functions as an E3 ubiquitin ligase for KLF5, and it plays a role in the degradation of KLF5 (Chen et al. 2005). WWP1 recognizes KLF5 as a substrate through a direct interaction between the WW domain of WWP1 and the NLTPPPSY motif of KLF5 (Chen et al. 2005). KLF5 may form a complex with WWP1 in the same manner as the complex between the WW domain and the motif peptide (Macias et al. 2002). PARP-1, which plays a key role in cell death and survival, directly interacts with the acetylated basic region of KLF5, and this interaction regulates apoptosis (Suzuki et al. 2007). The p300 histone acetyltransferase and its negative regulator SET also interact with the zinc finger region of KLF5 (Miyamoto et al. 2003). The histone chaperone ANP32B interacts with the KLF5 zinc fingers, which leads to transcriptional repression of a KLF5 target gene (Munemasa et al. 2008). STAT1 interacts with the N-terminal region of KLF5, which increases the transcriptional activity of KLF5 (Du et al. 2007). The interaction between the histone deacetylase HDAC1 and the basic region and ZF1 of KLF5 suppresses the transactivation of KLF5 (Matsumura et al. 2005).

Conclusions

The DNA sequence recognized by the DNA-binding domains of the KLFs can be predicted from several known structures of non-KLF zinc fingers. Basically, KLF1 to KLF8 and KLF12 should prefer 5′-NGGGCGNGN-3′; KLF9 to KLF11, KLF13, KLF14, KLF15, and KLF16 should prefer 5′-(G/T)GGGCGNGN-3′; and KLF17 should prefer 5′-AGGGCGNGN-3′. Our ongoing structural analysis of the KLFs revealed the structural resemblance to known C2H2 zinc fingers, the DNA-binding mode of the KLFs, and the structural basis for the sequence preference. In contrast, the N-terminal regions of the KLFs are variable, with several different characteristic sequences and/or protein-binding sites. However, no structural information on KLFs complexed with interactive proteins has been reported. To understand the molecular mechanisms of the KLF activities in the processes of self-renewal, proliferation, differentiation, and development of a cell, structural analyses of the KLF complexes must be performed.

References

Chen C, Sun X, Guo P, Dong XY, Sethi P, Cheng X, Zhou J, Ling J, Simons JW, Lingrel JB, Dong JT (2005) Human Krüppel-like factor 5 is a target of the E3 ubiquitin ligase WWP1 for proteolysis in epithelial cells. J Biol Chem 280:41553–41561

Chinnadurai G (2002) CtBP, an unconventional transcriptional co-repressor in development and oncogenesis. Mol Cell 9:213–224

Criqui-Filipe P, Ducret C, Maira S-M, Wasylyk B (1999) Net, a negative Ras-switchable TCF, contains a second inhibition domain, the CID, that mediates repression through interactions with CtBP and de-acetylation. EMBO J 18:3392–3403

Du JX, Yun CC, Bialkowska A, Yang VW (2007) Protein inhibitor of activated STAT1 interacts with and up-regulates activities of the pro-proliferative transcription factor Krüppel-like factor 5. J Biol Chem 282:4782–4793

Elrod-Erickson M, Rould MA, Nekludova L, Pabo CO (1996) Zif268 protein-DNA complex refined at 1.6Å: a model system for understanding zinc finger-DNA interactions. Structure 4:1171–1180

Elrod-Erickson M, Benson TE, Pabo CO (1998) High-resolution structures of variant Zif268-DNA complexes: implications for understanding zinc finger-DNA recognition. Structure 6:451–464

Iuchi S (2001) Three Classes of C2H2 zinc finger proteins. Cell Mol Life Sci 58:625–635

Izmailova ES, Wieczorek E, Perkins EB, Zehner ZE (1999) A GC-box is required for expression of the human vimentin gene. Gene 235:69–75

Jiang J, Chan YS, Loh YH, Cai J, Tong GQ, Lim CA, Robson P, Zhong S, Ng HH. (2008) A core Klf circuitry regulates self-renewal of embryonic stem cells. Nat Cell Biol 10:353–360

Kaczynski J, Zhang J-S, Ellenrieder V, Conley A, Duenes T, Kester H, van der Burg B, Urrutia R (2001) The Sp1-like protein BTEB3 inhibits transcription via the basic transcription element box by interacting with mSin3A and HDAC-1 co-repressors and competing with Sp1. J Biol Chem 276:36749–36756

Kaczynski J, Cook T, Urrutia R (2003) Sp1- and Krüppel-like transcription factors. Genome Biol 4:206

Lomberk G, Urrutia R (2005) The family feud: turning off SP1 by Sp1-like KLF proteins. Biochem J 392:1–11

Macias MJ, Wiesner S, Sudol M (2002) WW and SH3 domains, two different scaffolds to recognize proline-rich ligands. FEBS Lett 513:30–37

Matsumura T, Suzuki T, Aizawa K, Munemasa Y, Muto S, Horikoshi M, Nagai R (2005) The deacetylase HDAC1 negatively regulates the cardiovascular transcription factor Krüppel-like factor 5 through direct interaction. J Biol Chem 230:12123–12129

Mayer BJ (2001) SH3 domains: complexity in moderation. J Cell Sci 114:1253–1263

Miyamoto S, Suzuki T, Muto S, Aizawa K, Kimura A, Mizuno Y, Nagino T, Imai Y, Adachi N, Horikoshi M, Nagai R (2003) Positive and negative regulation of the cardiovascular transcription factor KLF5 by p300 and the oncogenic regulator SET through interaction and acetylation on the DNA-binding domain. Mol Cell Biol 23:8528–8541

Munemasa Y, Suzuki T, Aizawa K, Miyamoto S, Imai Y, Matsumura T, Horikoshi M, Nagai R (2008) Promoter region-specific histone incorporation by the novel histone chaperone ANP32B and DNA-binding factor KLF5. Mol Cell Biol 28:1171–1181

Nardini M, Spano S, Cericola C, Pesce A, Massaro A, Corda D, Bolognesi M (2003) CtBP/BARS: a dual-function protein involved in transcription co-repression and Golgi membrane fission. EMBO J 22:3122–3130

Nielsen SJ, Præstegaard M, Jørgensen HF, Clark BFC (1998) Different Sp1 family members differentially affect transcription from the human elongation factor 1 A-1 gene promoter. Biochem J 333:511–517

Nomura M, Uda-Tochio H, Murai K, Mori N, Nishimura Y (2005) The neural repressor NRSF/REST binds the PAH1 domain of the Sin3 co-repressor by using its distinct short hydrophobic helix. J Mol Biol 354:903–915

Pandya K, Townes TM (2002) Basic residues within the Krüppel zinc finger DNA binding domains are the critical nuclear localization determinants of EKLF/KLF-1. J Biol Chem 277:16304–16312

Philipsen S, Suske G (1999) A tale of three fingers: the family of mammalian SP/XKLF transcription factors. Nucleic Acids Res 27:2991–3000

Ptashne M (1988) How eukaryotic transcriptional activators work. Nature 335:683–689

Sahu SC, Swanson KA, Kang RS, Huang K, Brubaker K, Ratcliff K, Radhakrishnan I (2008) Conserved themes in target recognition by the PAH1 and PAH2 domains of the Sin3 transcriptional co-repressor. J Mol Biol 375:144–1456

Schaeper U, Boyd JM, Verma S, Uhlmann E, Subramanian T, Chinnadurai G (1995) Molecular cloning and characterization of a cellular phosphoprotein that interacts with a conserved C-terminal domain of adenovirus E1A involved in negative modulation of oncogenic transformation. Proc Natl Acad Sci USA 92:10467–10471

Shi Y, Sawada J, Sui G, Affar el B, Whetstine JR, Lan F, Ogawa H, Luke MP, Nakatani Y, Shi Y (2003) Coordinated histone modifications mediated by a CtBP co-repressor complex. Nature 422:735–738

Silverstein RA, Ekwall K (2005) Sin3: a flexible regulator of global gene expression and genome stability. Curr Genet 47:1–17

Suzuki T, Nishi T, Nagino T, Sasaki K, Aizawa K, Kada N, Sawaki D, Munemasa Y, Matsumura T, Muto S, Sata M, Miyagawa K, Horikoshi M, Nagai R (2007) Functional interaction between the transcription factor Krüppel-like factor 5 and Poly(ADP-ribose) polymerase-1 in cardiovascular apoptosis. J Biol Chem 282:9895–9901

Swanson KA, Knoepfler PS, Huang K, Kang RS, Cowley SM, Laherty CD, Eisenman RN, Radhakrishnan I (2004) HBP1 and Mad1 repressors bind the Sin3 co-repressor PAH2 domain with opposite helical orientations. Nat Struct Mol Biol 11:738–746

Turner J, Crossley M (1998) Cloning and characterization of mCtBP2, a co-repressor that associates with basic Krüppel-like factor and other mammalian transcriptional regulators. EMBO J 17:5129–5140

Turner J, Crossley M (2001) The CtBP family: enigmatic and enzymatic transcriptional co-repressors. BioEssays 23:683–690

van Vliet J, Turner J, Crossley M (2000) Human Krüppel-like factor 8: a CACCC-box binding protein that associates with CtBP and represses transcription. Nucleic Acids Res 28:1955–1962

Wolfe SA, Nekludova L, Pabo CO (1999) DNA recognition by Cys2His2 zinc finger proteins. Annu Rev Biophys Biomol Struct 29:183–212

Zhang JS, Moncrieffe MC, Kaczynski J, Ellenrieder V, Prendergast FG, Urrutia R (2001) A conserved α-helical motif mediates the interaction of Sp1-like transcriptional repressors with the co-repressor mSin3A. Mol Cell Biol 21:5041–5049

Chapter 3
Krüppel-like Factor Proteins and Chromatin Dynamics

Navtej S. Buttar, Gwen A. Lomberk, Gaurang S. Daftary, and Raul A. Urrutia

Abstract Krüppel-like factors (KLFs) are transcription regulatory proteins. Members of this protein family are characterized by a highly conserved C-terminus that has three zinc finger domains that bind to GC-rich sequences in DNA. The N-terminal domains of these proteins contain regulatory regions that can activate or repress transcription in a context-specific manner. KLFs interact with a wide range of co-activators or co-repressors to accomplish their transcription regulatory function. These interactions provide a complex stage for the chromatin dynamics to unfold and regulate diverse biological functions. This chapter focuses on expanding our understanding of molecular mechanisms of transcription regulation by KLFs and their impact on chromatin dynamics.

Introduction

Krüppel-like factors (KLFs) constitute a family of diverse transcription regulatory proteins characterized by an N-terminal domain that contains transcriptional regulatory motifs (Bieker 2001; Black et al. 2001; Cook et al. 1999; Cook and Urrutia 2000; Turner and Crossley 1998) and a highly conserved C-terminus that has three Cys2His2 zinc finger domains to bind to DNA. The members of this family bind to similar, yet distinct GC-rich target sequences, and they function either as activators or repressors (Bieker 2001; Lomberk and Urrutia 2005; Turner and Crossley 1998). KLF activator proteins, such as KLF1 and KLF4, function by interacting with histone acetylases, requiring interaction with the co-activator CBP/p300 and p300/CBP-associated factor (PCAF) (Geiman et al. 2000; Zhang et al. 2001a). On the other hand, KLF repressor proteins form a repressor complex with CtBP through use of a canonical PVDLS/T motif, a CtBP-interacting domain

N.S. Buttar, G.A. Lomberk, G.S. Daftary, and R.A. Urrutia (✉)
Chromatin Dynamics and Epigenetics Laboratory, Gastrointestinal Research Unit,
Division of Gastroenterology and Hepatology, Mayo Clinic Rochester,
200 First Street, SW, Rochester, MN 55905, USA

R. Nagai et al. (eds.), *The Biology of Krüppel-like Factors*,
DOI 10.1007/978-4-431-87775-2_3, © Springer 2009

(KLF3 and KLF8) (Turner and Crossley 1998; van den Ent et al. 1993), or they utilize the SIN3A-HDAC complex, which binds to a SIN3A-interacting domain, or SID (KLF10, KLF-11, KLF-13, KLF-16). Such complex formation has profound effects on chromatin dynamics, which affect virtually all known biological processes governing normal and abnormal mammalian development, differentiation, survival, and aging. Among others, histones play a central role in the chromatin dynamics as their N-terminal tails are subject to covalent modifications by the opposing actions of histone acetyltransferases (HATs) and histone deacetylases (HDACs), as well as other enzymatic activities. This reversible acetylation along with other histone modifications—collectively known as the Histone Code—alter either focal or global chromatin domains and thereby influence the activation or repression of gene transcription. In this chapter, we discuss the molecular mechanisms that have emerged during last two decades that have shed light on the transcriptional regulation by KLF family members and their effect on chromatin dynamics.

Classification of KLF Proteins and Their Co-repressor/ Co-activator Interactions

The discovery of Sp1 as a transcriptional regulator led to an extensive search, during last two decades, for proteins that are structurally and functionally related to Sp1. During the early 1990s, our laboratory characterized two novel, transforming growth factor-β (TGF-β)-inducible Sp1-like proteins—transcription factors TIEG1 and TIEG2—which are now known as KLF10 and KLF11, respectively (Cook et al. 1998; Tachibana et al. 1997). Since then, database comparisons and library screening in addition to extensive work performed by many dedicated laboratories worldwide have revealed the existence of 24 different proteins that share a remarkable similarity with Sp1 within their zinc finger domains and also bind to GC-rich sequences to regulate gene expression, thus belonging to the KLF/Sp1-like family of proteins (Buttar et al. 2006; Kaczynski et al. 2003; Lomberk and Urrutia 2005). For instance, we and several others have shown that both KLF10 and KLF11 proteins bind to and regulate the function of promoters that contain Sp1-like sequences (Cao et al. 2005; Fernandez-Zapico et al. 2003; Lomberk et al. 2008; Neve et al. 2005; Subramaniam et al. 2007; Wang et al. 2007; Zhang et al. 2007). The existence of a family of Sp1-like proteins posed several important biological questions regarding their function. For example, do Sp1-like proteins: (1) have redundant or distinct roles in mammalian cell physiology; (2) form homodimers and/or heterodimers; (3) work in a cell type-specific manner; (4) participate in a hierarchical cascade of gene expression; and/or (5) antagonize each other's functions to fine-tune specific cellular processes?

Several seminal publications with promising results by our group and several others over last two decades have helped move this field forward in addressing these questions. The ubiquitous expression of Sp1 in murine cells suggests that most, if not all, mammalian cells require Sp1 for proper function. The validity of this hypothesis was supported by finding that the knockout of this gene leads to gross morphological defects in a large number of tissues (Liu et al. 1996; Marin et al. 1997). In contrast, several other members of the Sp1-like family (e.g., KLF1, KLF2, Sp4) are expressed in a tissue-enriched manner (Abdelrahim et al. 2004; Carlson et al. 2006; Lin et al. 2005; Nielsen et al. 1998; Safe and Abdelrahim 2005; Watanabe et al. 1998). This selective pattern of expression raises the possibility that these proteins have a cell-specific function. This is particularly true in the case of KLF1 knockouts, where there are selective defects in erythropoiesis (Drissen et al. 2005; Eaton et al. 2008; Funnell et al. 2007). Because several Sp1-like proteins that recognize identical DNA sequences can be co-expressed in a single mammalian cell, it raised the question of redundancy or distinct transcriptional regulatory activity. By utilizing the biochemical comparison paradigm, we identified that whereas Sp1 acts as a potent transcriptional activator on reporter plasmids carrying GC boxes, KLF10 and KLF11 proteins act as transcriptional repressors (Cao et al. 2005; Fernandez-Zapico et al. 2003; Kaczynski et al. 2002; Neve et al. 2005). Moreover, there is a competition between Sp1-like proteins that function as "on" or "off" switches for similar promoters (Kaczynski et al. 2001; Sogawa et al. 1993; Turner and Crossley 1999). Another interesting finding is that certain members of the Sp/KLF family can activate transcription if the promoter contains multiple GC boxes but behave as repressors on promoters containing a single copy of this sequence (Imataka et al. 1992). This supports the notion that the regulation of gene expression by Sp1-like proteins may depend on promoter context. Overall, the discovery of Sp1-like transcriptional repressors, in addition to those members that activate transcription, has represented a significant step in the transcriptional field, and it challenged the early paradigm of "Sp1 activates all GC-rich sites." As modeled by the current, more accurate paradigm, GC-rich sites are not necessarily the target of Sp1 in isolation; rather, these sites may be activated or repressed depending on the family member by which it is recognized. Collective studies in this field have emphasized the complex nature of the biological effects generated by the existence of various KLF proteins, which in large part are dictated by the co-activators and co-repressors that facilitate the chromatin dynamics occurring on a given promoter.

To first understand the function ascribed to these transcription factors and subsequently their effect on chromatin dynamics, it is important briefly to revisit a few basic structural properties of these proteins. At least three domains are required for any family member of these Sp/KLF transcription factors: the DNA-binding zinc finger domain, a nuclear localization signal (NLS) domain, and a transcriptional regulatory domain. Within the DNA-binding domain, comprised of three Cys_2His_2 zinc finger motifs each of 25–30 amino acid residues, the sequence identity among the family members is higher than 65%, again emphasizing a role in the regulation of similar promoters (Kaczynski et al. 2003). Some zinc finger proteins recognize DNA sequences slightly different from the one that is predicted from its amino acid sequence, which is likely to be due to a mechanism of cooperative binding when the interaction of one finger with DNA modifies the selectivity of another finger. Similarly, DNA recognition by these zinc fingers may reflect a "wobbler effect" similar to the one that operates during peptide synthesis. Regardless of the exact mechanism, currently Sp1-like proteins have been divided into two groups based on the selectivity between two highly similar GC-rich sites, either a CGCCC or CACCC core sequence (Crossley et al. 1996; Hagen et al. 1995; Kingsley and Winoto 1992; Matsumoto et al. 1998; Shields and Yang 1998; Thiesen 1990). Returning to the basic questions of KLF biology that remain to be answered, whether proteins compete against each other for recognition of these two different sites or acquire different binding selectivity by posttranslational modifications or combinations of homo-/hetero dimerization must be determined. Because there are several thousands of these sites genome-wide, this information would be useful to advance a large number of studies that are focused on the mechanism of expression of distinct genes while utilizing caution against adopting an Sp1-centric assumption.

In contrast to the high conservation of the zinc finger domain that defines the members of this family, the structure and function of the transcriptional regulatory domain in the N-terminal portion of the proteins, as well as the location of their NLS, are variable. The location of the NLS can categorize these proteins into two major groups: one containing the signal within the zinc finger domain and the second with the NLS directly upstream of this region (Pandya and Townes 2002; Shields and Yang 1997). Along with the structural variability in the N-terminus, the ability of distinct family members to regulate transcription and subsequently affect cellular processes are divergent as well. For instance, KLF11 behaves as a potent transcriptional repressor, distinguishing it from the powerful transcriptional activation of Sp1 (Cook et al. 1998). The functional distinctions between the members of this family are embedded within the high level of variability in the N-terminal portion of the protein, which contains specific activation and repression domains. These domains, in turn, interact with distinct co-activators and co-repressors, thus regulating the chromatin dynamics and consequently transcription of a promoter in its own unique manner. In summary, although the presence of the similar zinc finger domain classifies these proteins in the KLF family, it is the N-terminal region thatprovides the functional identity to each member.

Structural and Epigenetic Aspects of KLF–Co-activator Interactions

Several members of the Sp/KLF family of transcription factors have been shown to interact with co-activators. Interestingly, the interaction of the Sp/KLF proteins and their co-activators appears to be selective and may contribute to transcriptional specificity. Beyond physical interaction, however, functionally most important is the chromatin remodeling capacity that KLF recruitment of these co-activators to a promoter facilitates. Some Sp/KLF members are able to promote transcription through glutamine-rich regions within their N-terminal domain, such as Sp1 and Sp3, which interact with components of the general transcription factor TAFII130 to recruit the RNA polymerase II complex (Gill et al. 1994). However, because the DNA of promoter regions is not in isolation but, rather, is within the context of the chromatin landscape, generally speaking transcriptionally activating KLF proteins require the assistance of co-activators, which can remodel the chromatin at the target site via complexes containing histone acetyltransferase (HAT) activity, such as p300/CBP and PCAF. CBP (CREB-binding protein) and p300 (EP300, E1A binding protein) are transcriptional co-activators that are structurally and functionally closely related.(Vo and Goodman 2001). Together with an acetylase (p300/CBP associated factor, or PCAF) they have been shown to bind numerous transcription factors. Transcription factor binding by these co-activators generally results in increased target gene expression.

All of these transcriptional co-activators contain intrinsic histone/protein acetyltransferase domains (HAT/PAT). Functionally, histone acetylation at target gene promoters provides a binding site for bromodomain-containing proteins, including the HATs themselves and the SWI/SNF family of chromatin remodelers, allowing structural relaxation of chromatin, thereby facilitating access to transcriptional machinery (Yang 2004). In addition to histone acetylation, p300/CBP and PCAF have been shown to acetylate several transcription factors. The functional significance of transcription factor acetylation is still under investigation, although it has been speculated that it affects their stability, DNA binding, and interaction with other proteins (Chen et al. 1999; Gu et al. 2001).

One KLF family member found to utilize the p300/CBP co-activator complex through direct transcription factor acetylation is KLF1 (erythroid KLF, or EKLF). KLF1 is necessary for the establishment of optimized chromatin structure and high-level expression of the β-globin gene that is characteristic of the erythroid cell lineage (Zhang et al. 2001b). CBP/p300 acetylates KLF1 at residues Lys288 and Lys302. Interestingly, KLF1 acetylation is necessary for transactivation of β-globin expression. Acetylation of KLF1 also facilitates interaction with the SWI/SNF chromatin remodeling complex, which further facilitates target gene expression. In contrast to KLF1, however, the acetyltransferase activity of CBP/p300 is not necessary for KLF13 (FKLF2)-mediated transactivation of human γ-globin expression (Song et al. 2002). KLF13 instead depends on the PCAF acetyltransferase to upregulate γ-globin. Interestingly, although CBP/p300 cooperatively facilitates KLF13 DNA binding and hence γ-globin transactivation along with PCAF, only PCAF (not CBP/p300) acetyltransferase activity is necessary for this response. In addition, CBP and PCAF acetylate KLF13 in its zinc finger domain, causing differential effects (Song et al. 2003). Acetylation by CBP disrupts the DNA binding of KLF13; and the regulation of DNA binding by KLF13 via PCAF and CBP can be synergistic or antagonistic, depending on acetylation status. This is exemplified by the observation that PCAF blocks CBP acetylation and thus prevents CBP disruption of KLF13 binding to DNA, whereas CBP-mediated acetylation of KLF13 prevents PCAF stimulation of KLF13 DNA binding. Selective recruitment of specific co-activator domains in addition to selective utilization of co-activators may be one mechanism conferring target gene specificity of the diverse but structurally related Sp/KLF family members.

A different mechanism underlies regulation of the KLF4-mediated inflammatory response in macrophages (Feinberg et al. 2005). KLF4 acts downstream of pro-inflammatory cytokines—e.g., interferon-γ (IFN-γ) and tumor necrosis factor-α

(TNF-α)—and activates inducible nitric oxide synthase expression (iNOS). In doing so, KLF4 antagonizes the antiinflammatory signal mediated by TGF-β1 and Smad3 via plasminogen activator inhibitor 1 (PAI1). Rather than binding the PAI1 promoter directly, KLF4 competes with Smad3 for CBP/p300 binding. Thus, KLF4 is an indirect trans-repressor of PAI1 by competitive antagonism via co-activator binding.

All of these mechanisms are operative to some extent in the regulation of KLF5 (Matsumura et al. 2005). HDAC1 deacetylase competes with p300 for a common binding site on the first zinc finger of KLF5. HDAC1 has been shown to bind KLF5 directly and diminish its DNA-binding affinity. Conversely, p300 acetylates and activates KLF5-mediated transcription. The HDAC1-mediated reduction in DNA binding causes a decrease in expression of platelet derived growth factor A (PDGF-A), a KLF5 target gene. HDAC1 therefore inhibits KLF5 directly by decreasing DNA binding as well as indirectly by competitively antagonizing binding of acetylase p300. Studies continue to emerge with more KLF family members utilizing CBP/p300 as a co-activator, such as KLF2 (SenBanerjee et al. 2004) and more recently KLF11 (R. Urrutia and R. Stein, unpublished results). These direct interactions with CBP/p300 recruit HAT activity to the site of the promoter and thus presumably facilitate an active chromatin state for transcription (Fig. 1). Further investigations will likely reveal additional KLF members that not only interact with CBP/p300 but with PCAF as well to allow more dynamic control of the chromatin landscape surrounding the Sp/KLF site.

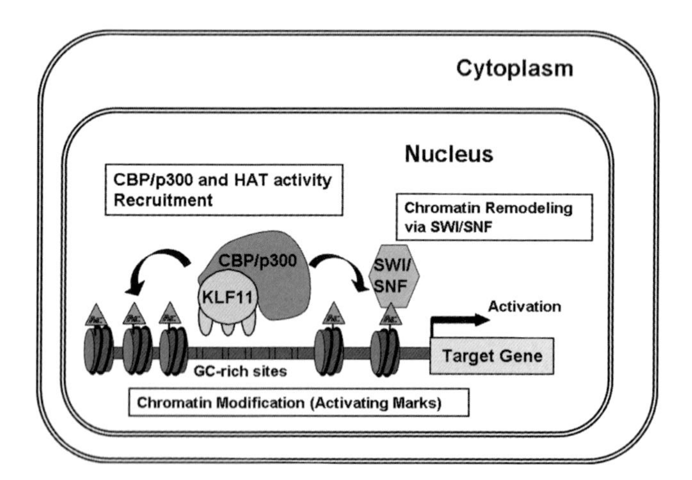

Fig. 1 Chromatin dynamics of KLF-mediated activation. Using KLF11 as an example, this cartoon depicts a model for KLF-mediated activation that involves the recruitment of CBP/p300 to a target gene promoter. The recruitment of CBP/p300 to the promoter also provides histone acetyltransferase (*HAT*) activity, which facilitates modification of surrounding histones to create "active" chromatin with acetylated histones. Addition of acetylated marks to histones signals activation of transcription through recruitment of other bromodomain-containing proteins, such as the SWI/SNF family of chromatin remodelers, allowing structural relaxation of chromatin and, thus, access to transcriptional machinery. A similar mechanism is proposed for the KLF family members that utilize p300/CBP-associated factor

The Sp/KLF family is comprised of distinct but structurally related transcription factors. As our understanding of this family has evolved, it has become clear that no single classification scheme can accurately characterize the function of any member distinctly. Individual KLF transcription factors function as activators or repressors in a context-dependent manner. In addition to co-activator binding, recruitment of selective cofactors in specific complexes further determines the specificity of KLF-directed gene regulation. As mentioned for KLF1 and KLF5, the KLF transcription factors themselves are subject to post-translational modifications, such as acetylation, further augmenting or regulating their specific biological capability. The development of novel inhibitors of specific enzymatic activity of PCAF, CBP, and p300 acetyltransferases will not only the elucidate the role of these cofactor–KLF complexes but also contribute to novel therapeutic possibilities.

Structural and Epigenetic Aspects of KLF-Co-repressor Interactions

In 1998, the primary functional subfamilies of KLF transcriptional repressors were identified and characterized almost concurrently. A subset of KLF family members were found to utilize the C-terminal-binding protein (CtBP) co-repressors (Turner and Crossley 1998), and our laboratory discovered the KLF/TIEG TGF-β-inducible early gene subfamily of transcriptional repressors that function via the Sin3-HDAC system (Cook et al. 1998, 1999; Zhang et al. 2001a). Subsequently, an extended subfamily of Sin3-mediated repressors was described, known as the KLF/BTEBs [BTE (basic transcription element)-binding proteins] (Kaczynski et al. 2001, 2002). Initially, these KLF subfamilies were classified based entirely on structural features; however, because KLF/TIEGs and KLF/BTEBs utilize the same co-repressor system (i.e., Sin3-HDAC), these two groups may actually represent the same functional subfamily. Therefore, to focus on the relation of KLF proteins to chromatin dynamics, we discuss the KLF proteins according to their mechanisms of action—i.e., CtBP- and Sin3-dependent KLF repressors.

Ctbp-Dependent KLF Repressors

The CtBP-dependent KLF repressors include KLF3, KLF8, and KLF12, which have a five-amino-acid motif, PXDLS (Pro-Xaa-Asp-Leu-Ser), that interacts with CtBP (Schuierer et al. 2001; Turner and Crossley 1998; van Vliet et al. 2000). It is noteworthy that outside of this small CtBP-recognition motif, no additional significant similarity occurs in the N-terminal region of these three KLF proteins. Originally characterized as the binding protein of the C-terminal portion of the adenovirus E1A protein, CtBPs are highly evolutionarily conserved and share significant amino acid similarity to NAD-dependent 2-hydroxy acid dehydrogenases

(Boyd et al. 1993; Schaeper et al. 1995). Although the function of CtBPs as transcriptional co-repressors has been well established, their mechanism of action is still emerging (Cook et al. 1998; Nibu et al. 1998a, 1998b; Poortinga et al. 1998; Postigo and Dean 1999). One mechanism for gene silencing via CtBP proteins is through the recruitment of HDACs (Koipally and Georgopoulos 2000; Sundqvist et al. 1998). However, evidence exists for the involvement of additional co-repressors as CtBP transcriptional repression also occurs independently of HDACs (Koipally and Georgopoulos 2000; Meloni et al. 1999; Phippen et al. 2000). Other transcriptional repressor families have been reported to interact with CtBP, including Ikaros and members of polycomb (Koipally and Georgopoulos 2000; Sewalt et al. 1999). Therefore, gene silencing via CtBP appears to occur also through the physical rearrangement of nucleosomes, as several of these interacting proteins are fundamental parts of chromatin-remodeling complexes. During KLF-CtBP-mediated repression, it remains unclear as to which co-repressor CtBP recruits and whether the choice of co-repressor is KLF- or promoter-dependent.

Even though KLF3 was initially assumed to function only as an activator, as shown in studies on a minimal promoter, the achieved activation was still significantly less than other KLF proteins, and this effect required an excess of KLF3 protein (Crossley et al. 1996). Subsequently, KLF3 was found to be a potent repressor that mapped to a domain located within a 74-amino-acid sequence in the N-terminus (Turner and Crossley 1998). Using yeast two-hybrid screening, CtBP2 was identified as the co-repressor for this KLF family member. KLF3 appears to have additional co-repressors as disruption of CtBP interaction does not completely abolish its transcriptional repression. Additional two-hybrid screening revealed an interaction between KLF3 and FHL3, a member of the FHL (four and a half LIM domain) family (Turner et al. 2003). FHL proteins have been implicated in cytoskeletal organization and more recently observed within the nucleus, associated with co-regulation of transcription (Du et al. 2002; McLoughlin et al. 2002; Muller et al. 1991). Thus, similar to the KLF proteins that interact with various co-activators, KLF3 interacts with distinct multiprotein complexes to achieve transcriptional repression of GC-rich promoters.

Identified "in silico" owing to its similarity to KLF3, KLF8 also associates with CtBP through a PVDLS recognition motif (van Vliet et al. 2000). Similar to KLF3, the N-terminus, specifically the CtBP-binding site, of KLF8 is responsible for its transcriptional repression activity but not complete. Again, loss of CtBP binding does not abolish the repression capacity of KLF8, suggesting the existence of additional co-repressors for this KLF protein (van Vliet et al. 2000). Whether a FHL protein also interacts with KLF8, as with KLF3, or other, yet unidentified co-repressors are involved in its full repression activity remains to be determined.

Originally identified from studies of its target gene, KLF12 is a repressor of the activator protein-2α (AP-2α) gene, which also encodes a mammalian transcription factor (Imhof et al. 1999). This repression occurs through a PVDLS sequence located within the N-terminus of KLF12, which facilitates a direct interaction with CtBP1 (Schuierer et al. 2001). Although much of KLF12 repression is associated with its binding to CtBP, the C-terminal portion of KLF12 containing the three zinc fingers is also capable of partial repression of the same promoter, AP-2α, suggesting that the zinc finger domain may provide a site for an additional co-repressor interaction or sterically interfere with activator binding to nearby sequences (Roth et al. 2000). Interestingly, a mechanism of trans-regulation exists between KLF12 and its gene target, as induction of *KLF12* expression leads to subsequent downregulation of AP-2α expression and vice versa with AP-2α acting as a negative regulator of *KLF12* expression (Roth et al. 2000). The biological relevance of this reciprocation remains unknown; however, it is clear that although KLF12 is a CtBP-mediated transcriptional repressor it shows distinct structural and functional differences from KLF3 and KLF8.

Sin3-Dependent KLF Repressors

As mentioned, the KLF/TIEGs and KLF/BTEBs, which include KLF9, KLF10, KLF11, KLF13, and KLF16, utilize the HDAC system to facilitate transcriptional repression through direct interaction with the scaffold co-repressor protein Sin3A (Lomberk and Urrutia 2005). Mammalian Sin3 (mSin3) proteins, part of large, multiprotein complexes capable of local chromatin modification, are orthologues of the Sin3p transcriptional repressor in *Saccharomyces cerevisiae* (Kadosh and Struhl 1998; Kasten et al. 1997; Vidal et al. 1991; Wang et al. 1990). The mSin3-HDAC complexes are composed of many subunits, including mSin3A/mSin3B, HDAC1, HDAC2, RBAP46 [Rb [retinoblastoma protein (Rb)-associated protein 46], RBAP48 (Rb-associated protein 48), SAP18 (Sin3-associated polypeptide 18), and SAP30 (Sin3-associated polypeptide 30) (Hassig et al. 1997; Laherty et al. 1992; Zhang et al. 1997). The Sin3 protein has multiple protein interaction domains, which allows it to function as a central scaffold for assembly of the entire complex. HDAC activity is essential for mediating the repression capacity of the complex, evidenced by significant disruption of repression activity on either mutation of the HDAC binding site in Sin3 or treatment with HDAC inhibitors (Kadosh and Struhl 1998; Sommer et al. 1997). The structure of Sin3 itself is comprised of four evolutionarily conserved imperfect repeats of ~100 residues, each predicted to form a four-helix-bundle fold, known as a paired amphipathic helix (PAH) region (Ayer et al. 1995; Halleck et al. 1995; Wang et al. 1990). To facilitate recruitment to a specific target sequence on a promoter, these PAH domains mediate binding with various transcription factors, such as the KLF proteins.

Initial biochemical characterization of these proteins demonstrated that the N-terminal domains of both KLF/TIEGs (KLF10 and KLF11) contain three distinct transcriptional repressor domains (R1, R2, R3) (Cook et al. 1999). Within KLF11, the repression of reporter gene activity achieved a minimum of 75% in the R1 (amino acids 24–41), R2 (151–162), and R3 (273–351) domains (Cook et al. 1999). Based on low-resolution secondary structure prediction algorithms, the R1 domain had been predicted to adopt an α-helical conformation, which was later confirmed by circular dichroism analysis (Zhang et al. 2001a). Congruent with this idea, proline mutations in the central core of the R1 domain (amino acids 30–39; AVEALVCMSS) disrupted its repression activity (Cook et al. 1999). Subsequent studies using either the wild-type R1 domain or its mutant led to the purification of a 160-KDa R1-binding protein. This high-affinity binding protein was later identified as Sin3a (Zhang et al. 2001a), and the core R1 domain was characterized as the α-helical repression motif SID (Sin3-interacting domain), which is further discussed below. Upon detailed biochemical and functional analyses, the KLF11 SID was found to interact specifically with the PAH2 domain of Sin3a to repress transcription. Subsequent analysis of the N-terminal domains of three other repressor proteins discovered in our laboratory—KLF13, KLF14, KLF16—clearly showed related R1 domains that function as Sin3-interacting-domains (SIDs), analogous to the corresponding domain in KLF10 and KLF11 (Kaczynski et al. 2001, 2002; Zhang et al. 2001a). KLF9, which is the first identified and cloned of this subfamily of KLF proteins that bind to BTE sites (Ratziu et al. 1998), also shares a SID in its N-terminus.(Zhang et al. 2001a). Thus, because the SID is a defining structural and necessary functional feature of these KLF repressors, this subset of KLF members are intricately linked to HDAC-mediated chromatin modification and transcription repression via mSin3A binding (Fig. 2) (Hassig et al. 1997). For instance, we have found that KLF14 represses the TGFβRII promoter via a co-repressor complex containing mSin3A and HDAC2 (Truty et al. 2008). Furthermore, TGF-β pathway

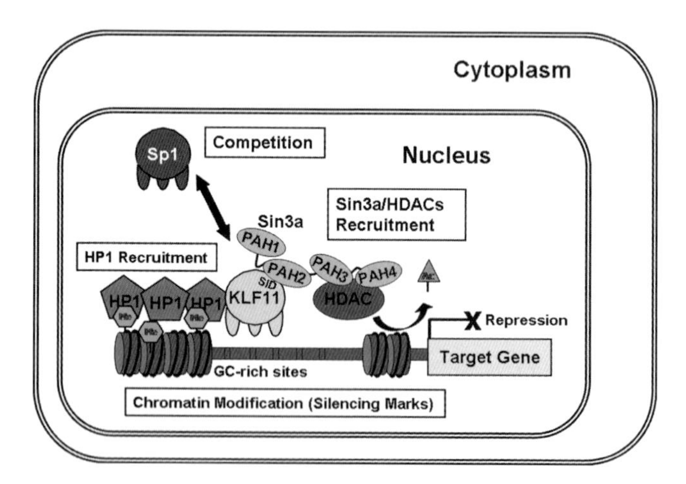

Fig. 2 Chromatin dynamics of KLF-mediated repression. Using KLF11 as an example, this cartoon depicts a model for KLF11-mediated repression that involves the recruitment of histone deacetylases (*Sin3a-HDACs*) to a target gene promoter (through the SID of KLF11 and PAH2 domain of Sin3a) and inhibition of Sp1 binding through competition. The recruitment of Sin3a-HDAC to the promoter facilitates remodeling of surrounding chromatin with silencing marks, namely the deacetylation of histones. (This process would be similar for the CtBP-dependent KLF repressors because at least one mechanism also involves the recruitment of HDACs.) Removal of acetylation signals short-term repression of a target gene and, in addition, primes the histone for receiving additional long-term silencing marks, such as methylation of K9 on histone H3, through the interaction between KLF11 and HP1

activation leads to the recruitment of a KLF14-mSin3A-HDAC2 repressor complex to the TGFβRII promoter, as well as remodeling of chromatin to decrease histone marks that associate with transcriptional activation (e.g., histone acetylation) and increase marks associated with transcriptional silencing (e.g., methylated K20 of histone H4) (Truty et al. 2008). Interestingly, at the time that this domain was discovered in the KLF family, a similar SID was thought to be a unique domain described only for the tumor suppressor and MYC oncogene inhibitors Mad1, Ume6, and Pf1 (Brubaker et al. 2000; Washburn and Esposito 2001; Yochum and Ayer 2001). This discovery marked a high point in the study of tumor-suppressor proteins by demonstrating that the presence of the SID domain is not exclusive to Mad1 but, rather, a more widespread molecular mechanism used by tumor suppressors in epithelial cells.

As mentioned, the SID within KLF proteins has some structural and functional resemblance to the better characterized SID of Mad1, the basic helix–loop–helix protein that dimerizes with Max to antagonize the function of the c-Myc oncoprotein (Zhang et al. 2001a). Additional members of the Mad family, including Mad1, Mad3, Mad4, and Mxi-SR, also have an N-terminal SID, which interacts with Sin3a through its PAH2 domain (Ayer et al. 1995). First found through circular dichroism (CD) and mutational analyses and then confirmed via nuclear magnetic resonance (NMR) structural analysis, the Mad1 SID was found to adopt an amphipathic α-helical conformation (Brubaker et al. 2000; Eilers et al. 1999). These findings also supported the concept hat this α-helical structure binds to the PAH2 domain by docking into a hydrophobic pocket in the base of this four-helix-bundle structure. As the SID interactions of KLF11 and Mad1 are both with the PAH2 domain, we sought to evaluate whether there are structural similarities (Pang et al. 2003; Truty et al. 2008). A comprehensive investigation into SIDs of KLFs and Mad1 repressor proteins suggested that SIDs of both KLFs and Mad1 have the AA/VXXL core consensus and a similar propensity for helix formation, but the two SIDs can be classified into two subtypes on the basis of their sequence—in particular, the residues outside the AA/VXXL core sequence (Zhang et al. 2001a). Even with structural similarities, the affinity of the KLF SID is lower than that of the Mad1 SID for the Sin3a PAH2 domain. This difference in affinities was evaluated with molecular modeling experiments combined with molecular dynamics simulation of the Mad1 SID–PAH2 complex, as compared with the KLF11 SID–PAH2 complex, to substantiate that this is a result of distinct binding mechanisms (Pang et al. 2003). These structural differences between the binding of KLF SID and that of the Mad1 SID offered new insight into transcriptional regulation via KLF repressor proteins.

The discovery of the SID in both the Mad1 and KLF repressor proteins also raised the question as whether these domains function in a constitutive or a regulated manner. Interestingly, we showed that the pro-proliferative EGF-ras-MEK1-ERK1 pathway phosphorylates residues near the SID in KLF11, leading to its dissociation from Sin3a and consequent inactivation of the Sin3-dependent KLF11 repressor function (Ellenrieder et al. 2002). Thus, these data demonstrated that the SID functions in a manner that can be influenced by cell signaling. Because KLF11 is induced by TGF-β to mediate an antiproliferative pathway, its inactivation by

EGF signaling may contribute to its inactivation in cells where this mechanism is hyperactive, such as pancreatic cancer.

Future of KLF Proteins in Chromatin Dynamics

Recently, in *Drosophila melanogaster*, Sp/KLF proteins have been found to bind to a site necessary for the activity of a Polycomb-group response element (PRE) from the *engrailed* gene (Brown et al. 2005). These PREs are DNA elements recognized by the Polycomb-group (PcG) of transcriptional repressors through chromatin modification, suggesting further the complexity of chromatin cofactors involved in KLF-mediated transcriptional activation and/or repression.

In addition, in efforts to determine the interacting partners of the other repressor domains (R2, R3) of KLF11, our laboratory has discovered that KLF11 binds to the Gβ subunit of the heterotrimeric G-protein (Mathison et al. 2008; Zhang et al. 2007). This interaction occurs through the R3 domain of KLF11; and more specifically, mutation of alanine 347 to a serine significantly disrupts the binding between these two proteins. Interestingly, this mutation is naturally occurring in a French family with maturity-onset diabetes of the young (MODY) (Neve et al. 2005). These findings are significant, suggesting that activation of GPCRs can mediate short-term responses via their Gα-subunit and Gβγ complex, and long-term transcriptional effects can be triggered via Gβ translocation into the nucleus and functional cooperation with transcription factors such as KLF11. Concurrently, we also found that KLF11 interacts with the chromatin protein HP1α through the extreme C-terminal tail after the zinc finger domains (Lomberk et al. 2008a, 2008b). HP1 proteins play a role as "gatekeepers" of long-term epigenetic gene silencing that is mediated by histone H3 lysine-9 methylation via recruitment of the G9a or SUV39H1 histone methylases (Lomberk et al. 2006). Therefore, this finding offers a new model for the function of KLF11, which may work not only in transient repression via HDAC but also in long-term repression via histone methylation/HP1 (Fig. 2).

Much remains to be discovered in regard to the role of KLF proteins in chromatin dynamics. Ongoing studies continue to implicate additional nonhistone chromatin proteins in mediating KLF function. We now understand that a gene promoter is not simply a DNA sequence to be recognized by a KLF protein. The promoter is occupied by a complex array of chromatin proteins associated with either an active or a repressed state. If this chromatin state opposes the necessary biological function at a given moment, the chromatin landscape requires modification and remodeling to switch a promoter "on" or "off," which can be accomplished onlyby recruitment of the appropriate co-activators or co-repressors. These transitions between "active" and "inactive" chromatin states are catalyzed by the specific targeting of these megadalton multiprotein co-activators/co-repressors to DNA, which can be directed to a particular sequence only via sequence-specific DNA-binding proteins, such as the KLF proteins.

References

Abdelrahim, M., R. Smith, 3rd, et al. (2004). "Role of Sp proteins in regulation of vascular endothelial growth factor expression and proliferation of pancreatic cancer cells." *Cancer Research* 64(18): 6740–9.

Ayer, D. E., Q. A. Lawrence, et al. (1995). "Mad-Max transcriptional repression is mediated by ternary complex formation with mammalian homologs of yeast repressor Sin3." *Cell* 80(5): 767–76.

Bieker, J. J. (2001). "Kruppel-like factors: three fingers in many pies." *J Biol Chem* 276(37): 34355–8.

Black, A. R., J. D. Black, et al. (2001). "Sp1 and kruppel-like factor family of transcription factors in cell growth regulation and cancer." *Journal of Cellular Physiology* 188(2): 143–60.

Boyd, J. M., T. Subramanian, et al. (1993). "A region in the C-terminus of adenovirus 2/5 E1a protein is required for association with a cellular phosphoprotein and important for the negative modulation of T24-ras mediated transformation, tumorigenesis and metastasis." *Embo J* 12(2): 469–78.

Brown, J. L., D. J. Grau, et al. (2005). "An Sp1/KLF binding site is important for the activity of a Polycomb group response element from the *Drosophila* engrailed gene." *Nucleic Acids Res* 33(16): 5181–9.

Brubaker, K., S. M. Cowley, et al. (2000). "Solution structure of the interacting domains of the Mad-Sin3 complex: implications for recruitment of a chromatin-modifying complex." *Cell* 103(4): 655–65.

Buttar, N. S., M. E. Fernandez-Zapico, et al. (2006). "Key role of Kruppel-like factor proteins in pancreatic cancer and other gastrointestinal neoplasias." *Current Opinion in Gastroenterology* 22(5): 505–11.

Cao, S., M. E. Fernandez-Zapico, et al. (2005). "KLF11-mediated repression antagonizes Sp1/sterol-responsive element-binding protein-induced transcriptional activation of caveolin-1 in response to cholesterol signaling." *J Biol Chem* 280(3): 1901–10.

Carlson, C. M., B. T. Endrizzi, et al. (2006). "Kruppel-like factor 2 regulates thymocyte and T-cell migration." *Nature* 442(7100): 299–302.

Chen, H., R. J. Lin, et al. (1999). "Regulation of hormone-induced histone hyperacetylation and gene activation via acetylation of an acetylase." *Cell* 98(5): 675–86.

Cook, T., B. Gebelein, et al. (1999). "Three conserved transcriptional repressor domains are a defining feature of the TIEG subfamily of Sp1-like zinc finger proteins." *J Biol Chem* 274(41): 29500–4.

Cook, T., B. Gebelein, et al. (1998). "Molecular cloning and characterization of TIEG2 reveals a new subfamily of transforming growth factor-beta-inducible Sp1-like zinc finger-encoding genes involved in the regulation of cell growth." *J Biol Chem* 273(40): 25929–36.

Cook, T. and R. Urrutia (2000). "TIEG proteins join the Smads as TGF-beta-regulated transcription factors that control pancreatic cell growth." *Am J Physiol Gastrointest Liver Physiol* 278(4): G513–21.

Crossley, M., E. Whitelaw, et al. (1996). "Isolation and characterization of the cDNA encoding BKLF/TEF-2, a major CACCC-box-binding protein in erythroid cells and selected other cells." *Mol Cell Biol* 16(4): 1695–705.

Drissen, R., M. von Lindern, et al. (2005). "The erythroid phenotype of EKLF-null mice: defects in hemoglobin metabolism and membrane stability." *Mol Cell Biol* 25(12): 5205–14.

Du, X., P. Hublitz, et al. (2002). "The LIM-only coactivator FHL2 modulates WT1 transcriptional activity during gonadal differentiation." *Biochim Biophys Acta* 1577(1): 93–101.

Eaton, S. A., A. P. Funnell, et al. (2008). "A network of Kruppel-like Factors (Klfs). Klf8 is repressed by Klf3 and activated by Klf1 in vivo." *J Biol Chem* 283(40): 26937–47.

Eilers, A. L., A. N. Billin, et al. (1999). "A 13-amino acid amphipathic alpha-helix is required for the functional interaction between the transcriptional repressor Mad1 and mSin3A." *J Biol Chem* 274(46): 32750–6.

Ellenrieder, V., J. S. Zhang, et al. (2002). "Signaling disrupts mSin3A binding to the Mad1-like Sin3-interacting domain of TIEG2, an Sp1-like repressor." *Embo J* 21(10): 2451–60.
Feinberg, M. W., Z. Cao, et al. (2005). "Kruppel-like factor 4 is a mediator of proinflammatory signaling in macrophages." *Journal of Biological Chemistry* 280(46): 38247–58.
Fernandez-Zapico, M. E., A. Mladek, et al. (2003). "An mSin3A interaction domain links the transcriptional activity of KLF11 with its role in growth regulation." *Embo J* 22(18): 4748–58.
Funnell, A. P., C. A. Maloney, et al. (2007). "Erythroid Kruppel-like factor directly activates the basic Kruppel-like factor gene in erythroid cells." *Mol Cell Biol* 27(7): 2777–90.
Geiman, D. E., H. Ton-That, et al. (2000). "Transactivation and growth suppression by the gut-enriched Kruppel-like factor (Kruppel-like factor 4) are dependent on acidic amino acid residues and protein-protein interaction." *Nucleic Acids Res* 28(5): 1106–13.
Gill, G., E. Pascal, et al. (1994). "A glutamine-rich hydrophobic patch in transcription factor Sp1 contacts the dTAFII110 component of the Drosophila TFIID complex and mediates transcriptional activation." *Proc Natl Acad Sci U S A* 91(1): 192–6.
Gu, J., R. M. Rubin, et al. (2001). "A sequence element of p53 that determines its susceptibility to viral oncoprotein-targeted degradation." *Oncogene* 20(27): 3519–27.
Hagen, G., J. Dennig, et al. (1995). "Functional analyses of the transcription factor Sp4 reveal properties distinct from Sp1 and Sp3." *J Biol Chem* 270(42): 24989–94.
Halleck, M. S., S. Pownall, et al. (1995). "A widely distributed putative mammalian transcriptional regulator containing multiple paired amphipathic helices, with similarity to yeast SIN3." *Genomics* 26(2): 403–6.
Hassig, C. A., T. C. Fleischer, et al. (1997). "Histone deacetylase activity is required for full transcriptional repression by mSin3A." *Cell* 89(3): 341–7.
Imataka, H., K. Sogawa, et al. (1992). "Two regulatory proteins that bind to the basic transcription element (BTE), a GC box sequence in the promoter region of the rat P-4501A1 gene." *Embo J* 11(10): 3663–71.
Imhof, A., M. Schuierer, et al. (1999). "Transcriptional regulation of the AP-2alpha promoter by BTEB-1 and AP-2rep, a novel wt-1/egr-related zinc finger repressor." *Mol Cell Biol* 19(1): 194–204.
Kaczynski, J., T. Cook, et al. (2003). "Sp1- and Kruppel-like transcription factors." *Genome Biol* 4(2): 206.
Kaczynski, J., J. S. Zhang, et al. (2001). "The Sp1-like protein BTEB3 inhibits transcription via the basic transcription element box by interacting with mSin3A and HDAC-1 co-repressors and competing with Sp1." *J Biol Chem* 276(39): 36749–56.
Kaczynski, J. A., A. A. Conley, et al. (2002). "Functional analysis of basic transcription element (BTE)-binding protein (BTEB) 3 and BTEB4, a novel Sp1-like protein, reveals a subfamily of transcriptional repressors for the BTE site of the cytochrome P4501A1 gene promoter." *Biochem J* 366(Pt 3): 873–82.
Kadosh, D. and K. Struhl (1998). "Histone deacetylase activity of Rpd3 is important for transcriptional repression in vivo." *Genes Dev* 12(6): 797–805.
Kasten, M. M., S. Dorland, et al. (1997). "A large protein complex containing the yeast Sin3p and Rpd3p transcriptional regulators." *Mol Cell Biol* 17(8): 4852–8.
Kingsley, C. and A. Winoto (1992). "Cloning of GT box-binding proteins: a novel Sp1 multigene family regulating T-cell receptor gene expression." *Mol Cell Biol* 12(10): 4251–61.
Koipally, J. and K. Georgopoulos (2000). "Ikaros interactions with CtBP reveal a repression mechanism that is independent of histone deacetylase activity." *J Biol Chem* 275(26): 19594–602.
Laherty, C. D., H. M. Hu, et al. (1992). "The Epstein-Barr virus LMP1 gene product induces A20 zinc finger protein expression by activating nuclear factor kappa B." *Journal of Biological Chemistry* 267(34): 24157–60.
Lin, Z., A. Kumar, et al. (2005). "Kruppel-like factor 2 (KLF2) regulates endothelial thrombotic function." *Circulation Research* 96(5): e48–57.
Liu, C., A. Calogero, et al. (1996). "EGR-1, the reluctant suppression factor: EGR-1 is known to function in the regulation of growth, differentiation, and also has significant tumor suppressor

activity and a mechanism involving the induction of TGF-beta1 is postulated to account for this suppressor activity." *Crit Rev Oncog* 7(1–2): 101–25.

Lomberk, G., S. Ilyas, et al. (2008a). "KLF11 Complexes With the Epigenetic Gene Silencer Protein, HP1 to Mediate Tumor Suppressor Activities. ." *Pancreas* 37((4)): 481.

Lomberk, G. and R. Urrutia (2005). "The family feud: turning off Sp1 by Sp1-like KLF proteins." *Biochem J* 392(Pt 1): 1–11.

Lomberk, G., L. Wallrath, et al. (2006). "The Heterochromatin Protein 1 family." *Genome Biol* 7(7): 228.

Lomberk, G., J. Zhang, et al. (2008b). " A New Molecular Model for Regulating the TGFβ Receptor II Promoter in Pancreatic Cells. ." *Pancreas* 36((2)): 223.

Marin, M., A. Karis, et al. (1997). "Transcription factor Sp1 is essential for early embryonic development but dispensable for cell growth and differentiation." *Cell* 89(4): 619–28.

Mathison, A., G. Lomberk, et al. (2008). " The Gβ Subunit of Heterotrimeric G Proteins Links Pancreatic PCR Activation To Long Term Responses.." *Pancreas*. 37: 484.

Matsumoto, N., F. Laub, et al. (1998). "Cloning the cDNA for a new human zinc finger protein defines a group of closely related Kruppel-like transcription factors." *J Biol Chem* 273(43): 28229–37.

Matsumura, T., T. Suzuki, et al. (2005). "The deacetylase HDAC1 negatively regulates the cardiovascular transcription factor Kruppel-like factor 5 through direct interaction." *J Biol Chem* 280(13): 12123–9.

McLoughlin, P., E. Ehler, et al. (2002). "The LIM-only protein DRAL/FHL2 interacts with and is a corepressor for the promyelocytic leukemia zinc finger protein." *J Biol Chem* 277(40): 37045–53.

Meloni, A. R., E. J. Smith, et al. (1999). "A mechanism for Rb/p130-mediated transcription repression involving recruitment of the CtBP corepressor." *Proc Natl Acad Sci U S A* 96(17): 9574–9.

Muller, H. J., C. Skerka, et al. (1991). "Clone pAT 133 identifies a gene that encodes another human member of a class of growth factor-induced genes with almost identical zinc-finger domains." *Proceedings of the National Academy of Sciences of the United States of America* 88(22): 10079–83.

Neve, B., M. E. Fernandez-Zapico, et al. (2005). "Role of transcription factor KLF11 and its diabetes-associated gene variants in pancreatic beta cell function." *Proc Natl Acad Sci U S A* 102(13): 4807–12.

Nibu, Y., H. Zhang, et al. (1998a). "dCtBP mediates transcriptional repression by Knirps, Kruppel and Snail in the Drosophila embryo." *Embo J* 17(23): 7009–20.

Nibu, Y., H. Zhang, et al. (1998b). "Interaction of short-range repressors with Drosophila CtBP in the embryo." *Science* 280(5360): 101–4.

Nielsen, S. J., M. Praestegaard, et al. (1998). "Different Sp1 family members differentially affect transcription from the human elongation factor 1 A-1 gene promoter." *Biochem J* 333 (Pt 3): 511–7.

Pandya, K. and T. M. Townes (2002). "Basic residues within the Kruppel zinc finger DNA binding domains are the critical nuclear localization determinants of EKLF/KLF-1." *J Biol Chem* 277(18): 16304–12.

Pang, Y. P., G. A. Kumar, et al. (2003). "Differential binding of Sin3 interacting repressor domains to the PAH2 domain of Sin3A." *FEBS Lett* 548(1–3): 108–12.

Phippen, T. M., A. L. Sweigart, et al. (2000). "Drosophila C-terminal binding protein functions as a context-dependent transcriptional co-factor and interferes with both mad and groucho transcriptional repression." *J Biol Chem* 275(48): 37628–37.

Poortinga, G., M. Watanabe, et al. (1998). "Drosophila CtBP: a Hairy-interacting protein required for embryonic segmentation and hairy-mediated transcriptional repression." *Embo J* 17(7): 2067–78.

Postigo, A. A. and D. C. Dean (1999). "ZEB represses transcription through interaction with the corepressor CtBP." *Proceedings of the National Academy of Sciences of the United States of America* 96(12): 6683–8.

Ratziu, V., A. Lalazar, et al. (1998). "Zf9, a Kruppel-like transcription factor up-regulated in vivo during early hepatic fibrosis." *Proc Natl Acad Sci U S A* 95(16): 9500–5.
Roth, C., M. Schuierer, et al. (2000). "Genomic structure and DNA binding properties of the human zinc finger transcriptional repressor AP-2rep (KLF12)." *Genomics* 63(3): 384–90.
Safe, S. and M. Abdelrahim (2005). "Sp transcription factor family and its role in cancer." *European Journal of Cancer* 41(16): 2438–48.
Schaeper, U., J. M. Boyd, et al. (1995). "Molecular cloning and characterization of a cellular phosphoprotein that interacts with a conserved C-terminal domain of adenovirus E1A involved in negative modulation of oncogenic transformation." *Proc Natl Acad Sci U S A* 92(23): 10467–71.
Schuierer, M., K. Hilger-Eversheim, et al. (2001). "Induction of AP-2alpha expression by adenoviral infection involves inactivation of the AP-2rep transcriptional corepressor CtBP1." *J Biol Chem* 276(30): 27944–9.
SenBanerjee, S., Z. Lin, et al. (2004). "KLF2 Is a novel transcriptional regulator of endothelial proinflammatory activation." *J Exp Med* 199((10)): 1305–15.
Sewalt, R. G., M. J. Gunster, et al. (1999). "C-Terminal binding protein is a transcriptional repressor that interacts with a specific class of vertebrate Polycomb proteins." *Mol Cell Biol* 19(1): 777–87.
Shields, J. M. and V. W. Yang (1997). "Two potent nuclear localization signals in the gut-enriched Kruppel-like factor define a subfamily of closely related Kruppel proteins." *J Biol Chem* 272(29): 18504–7.
Shields, J. M. and V. W. Yang (1998). "Identification of the DNA sequence that interacts with the gut-enriched Kruppel-like factor." *Nucleic Acids Res* 26(3): 796–802.
Sogawa, K., Y. Kikuchi, et al. (1993). "Comparison of DNA-binding properties between BTEB and Sp1." *J Biochem (Tokyo)* 114(4): 605–9.
Sommer, A., S. Hilfenhaus, et al. (1997). "Cell growth inhibition by the Mad/Max complex through recruitment of histone deacetylase activity." *Curr Biol* 7(6): 357–65.
Song, C. Z., K. Keller, et al. (2003). "Functional interplay between CBP and PCAF in acetylation and regulation of transcription factor KLF13 activity." *J Mol Biol* 329(2): 207–15.
Song, C. Z., K. Keller, et al. (2002). "Functional interaction between coactivators CBP/p300, PCAF, and transcription factor FKLF2." *J Biol Chem* 277(9): 7029–36.
Subramaniam, M., J. R. Hawse, et al. (2007). "Role of TIEG1 in biological processes and disease states." *J Cell Biochem.*
Sundqvist, A., K. Sollerbrant, et al. (1998). "The carboxy-terminal region of adenovirus E1A activates transcription through targeting of a C-terminal binding protein-histone deacetylase complex." *FEBS Lett* 429(2): 183–8.
Tachibana, I., M. Imoto, et al. (1997). "Overexpression of the TGFbeta-regulated zinc finger encoding gene, TIEG, induces apoptosis in pancreatic epithelial cells." *Journal of Clinical Investigation* 99(10): 2365–74.
Thiesen, H. J. (1990). "Multiple genes encoding zinc finger domains are expressed in human T cells." *New Biologist* 2(4): 363–74.
Truty, M., G. Lomberk, et al. (2008). "Silencing of the TGFbeta receptor II by kruppel-like factor 14 underscores the importance of a negative feedback mechanism in TGFbeta signaling. ." *J Biol Chem* Dec 15. [Epub ahead of print].
Turner, J. and M. Crossley (1998). "Cloning and characterization of mCtBP2, a co-repressor that associates with basic Kruppel-like factor and other mammalian transcriptional regulators." *Embo J* 17(17): 5129–40.
Turner, J. and M. Crossley (1999). "Mammalian Kruppel-like transcription factors: more than just a pretty finger." *Trends Biochem Sci* 24(6): 236–40.
Turner, J., H. Nicholas, et al. (2003). "The LIM protein FHL3 binds basic Kruppel-like factor/ Kruppel-like factor 3 and its co-repressor C-terminal-binding protein 2." *J Biol Chem* 278(15): 12786–95.

van den Ent, F. M., A. J. van Wijnen, et al. (1993). "Concerted control of multiple histone promoter factors during cell density inhibition of proliferation in osteosarcoma cells: reciprocal regulation of cell cycle-controlled and bone-related genes." *Cancer Research* 53(10 Suppl): 2399–409.

van Vliet, J., J. Turner, et al. (2000). "Human Krüppel-like factor 8: a CACCC-box binding protein that associates with CtBP and represses transcription." *Nucleic Acids Res* 1;28((9)): 1955–62.

Vidal, M., R. Strich, et al. (1991). "RPD1 (SIN3/UME4) is required for maximal activation and repression of diverse yeast genes." *Mol Cell Biol* 11(12): 6306–16.

Vo, N. and R. H. Goodman (2001). "CREB-binding protein and p300 in transcriptional regulation." *J Biol Chem* 276(17): 13505–8.

Wang, H., I. Clark, et al. (1990). "The Saccharomyces cerevisiae SIN3 gene, a negative regulator of HO, contains four paired amphipathic helix motifs." *Mol Cell Biol* 10(11): 5927–36.

Wang, Z., B. Spittau, et al. (2007). "Human TIEG2/KLF11 induces oligodendroglial cell death by downregulation of Bcl-X(L) expression." *J Neural Transm* 114(7): 867–75.

Washburn, B. K. and R. E. Esposito (2001). "Identification of the Sin3-binding site in Ume6 defines a two-step process for conversion of Ume6 from a transcriptional repressor to an activator in yeast." *Mol Cell Biol* 21(6): 2057–69.

Watanabe, G., C. Albanese, et al. (1998). "Inhibition of cyclin D1 kinase activity is associated with E2F-mediated inhibition of cyclin D1 promoter activity through E2F and Sp1." *Molecular & Cellular Biology* 18(6): 3212–22.

Yang, X. J. (2004). "Lysine acetylation and the bromodomain: a new partnership for signaling." *Bioessays* 26(10): 1076–87.

Yochum, G. S. and D. E. Ayer (2001). "Pf1, a novel PHD zinc finger protein that links the TLE corepressor to the mSin3A-histone deacetylase complex." *Mol Cell Biol* 21(13): 4110–8.

Zhang, J., G. Lomberk, et al. (2007). "The Gb Subunit of Heterotrimeric G Proteins Links GPCR Activation With Transcriptional Regulation by KLF11, a Pancreatic Tumor Suppressor and a Diabetes Gene." *Gastroenterology* 132(4 Suppl2).

Zhang, J. S., M. C. Moncrieffe, et al. (2001a). "A conserved alpha-helical motif mediates the interaction of Sp1-like transcriptional repressors with the corepressor mSin3A." *Mol Cell Biol* 21(15): 5041–9.

Zhang, W., S. Kadam, et al. (2001b). "Site-specific acetylation by p300 or CREB binding protein regulates erythroid Kruppel-like factor transcriptional activity via its interaction with the SWI-SNF complex." *Mol Cell Biol* 21(7): 2413–22.

Zhang, Y., R. Iratni, et al. (1997). "Histone deacetylases and SAP18, a novel polypeptide, are components of a human Sin3 complex." *Cell* 89(3): 357–64.

Chapter 4
Co-regulator Interactions in Krüppel-like Factor Transcriptional Programs

Richard C.M. Pearson, Briony H.A. Jack, Stella H.Y. Lee, Alister P.W. Funnell, and Merlin Crossley

Abstract Krüppel-like transcription factors (KLFs) comprise a family of gene regulatory proteins with diverse roles in cellular proliferation, survival, and differentiation. KLFs contain three characteristic, highly conserved C-terminal zinc fingers that coordinate sequence-specific DNA binding. Despite having highly homologous DNA binding domains, family members are able to regulate the expression of diverse target genes, resulting in both temporal and tissue-specific control of differentiation. To do this, KLFs have evolved distinct N-terminal regulatory domains that allow interaction with various co-regulators. This chapter describes the cofactors that interact with KLFs and outlines how these interactions potentiate or inhibit KLF transcriptional activity, how they help define target gene specificity, and how they dictate whether a gene is activated or repressed.

Introduction

The 17 members of the Krüppel-like factor (KLF) family of gene regulatory proteins share considerable homology in their C-terminal zinc finger DNA-binding domains. This homology means that family members have the potential to bind and regulate an overlapping set of target genes, implying both functional redundancy and cross regulation. However, KLFs also show a high degree of functional specificity that has been achieved partly through the evolution of distinct N-terminal regulatory domains that mediate the binding of different interacting protein

R.C.M. Pearson, B.H.A. Jack, A.P.W. Funnell, and M. Crossley (✉)
School of Molecular and Microbial Biosciences, University of Sydney,
Sydney, NSW 2006, Australia
e-mail: merlinc@usyd.edu.au

S.H.Y. Lee
Victor Chang Cardiac Research Institute, 405 Liverpool St,
Darlinghurst, NSW 2010, Australia

R. Nagai et al. (eds.), *The Biology of Krüppel-like Factors*,
DOI 10.1007/978-4-431-87775-2_4, © Springer 2009

partners. These partners include co-activators, co-repressors, histone-modifying enzymes, and other transcription factors. Interaction with these partners mediates KLF activity by facilitating the assembly of multiprotein complexes at the promoters of target genes, leading to remodeling of chromatin and subsequent activation or repression. This chapter describes the main categories of co-regulators and explains how these interactions contribute to specificity in gene regulation.

Co-regulators That Bind KLF Regulatory Domains

Phylogenetic analysis of the KLF family (van Vliet et al. 2006) reveals that the members can be divided into three groups based on structural homology (Fig. 1). In addition, the KLFs of each subgroup show some functional similarities that can be explained in part by homology in their N-terminal protein interaction domains. This allows each group to recruit similar partners to either activate or repress transcription by similar mechanisms. Group 1 consists of KLFs 1, 2, 4, 5, 6, and

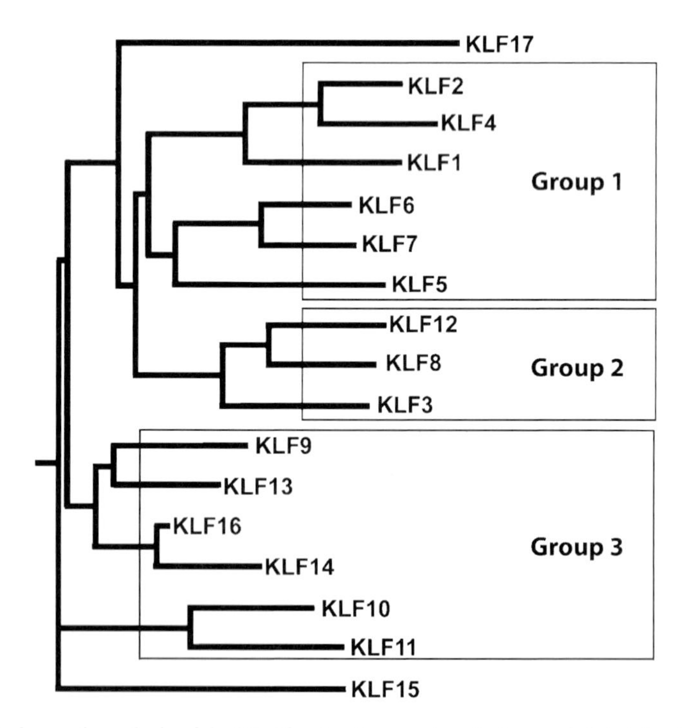

Fig. 1 Phylogenetic analysis of the KLF family. Structural analysis revealed that the KLF family can be divided into three distinct subgroups. The analysis was performed using the bioinformatics tool Phylogeny Analysis, which was provided by the Australian National Genome Information Service (ANGIS). (Adapted from van Vliet et al. 2006)

7. Proteins in this group generally function as activators, although not exclusively. Group 2 contains KLFs 3, 8, and 12, all of which act as transcriptional repressors via recruitment of the co-repressor C-terminal binding protein (CtBP). Group 3 is made up of KLFs 9, 10, 11, 13, 14, and 16. These KLFs are capable of binding the co-repressor Sin3; accordingly, members of this group are generally thought of as inhibitors of gene expression. The following sections describe the co-regulators that interact with each of these subgroups.

Co-regulators That Bind Group 1 KLFs

This group of KLFs contains KLFs 1, 2, 4, 5, 6, and 7. Much of the information about group 1 co-regulators has been deduced from studying the molecular mechanisms of gene regulation by KLF1/EKLF, which has most often been analyzed as a transcriptional activator that drives expression of a number of genes important in erythropoietic differentiation (Drissen et al. 2005; Hodge et al. 2006; Keys et al. 2007; Pilon et al. 2006). The most notable of these is the adult *β-globin* gene (*β-major globin* in mice), the expression of which is highly dependent on KLF1 in definitive erythroid cells. The regulatory DNA elements that govern *β-globin* expression, including the proximal promoter and upstream locus control region (LCR), have been extensively characterized. The *β-globin* locus has thus commonly been used as a paradigm to explore the mode by which KLF1 activates transcription of its target genes. To date, several cofactors of KLF1 have been identified, including various histone-modifying enzymes and chromatin-remodeling complexes.

KLF1 has been shown to interact with several histone acetyltransferases (HATs) in vivo. These include p300, CREB-binding protein (CBP), and p300/CBP-associated factor (PCAF). CBP and p300 acetylate KLF1 in vitro and potentiate KLF1's activation of the β-*globin* promoter in reporter assays (Zhang and Bieker 1998; Zhang et al. 2001b). Furthermore, this potentiation is dependent on the acetyltransferase activities of p300 and CBP. In contrast, PCAF does not detectably acetylate KLF1 in vitro and, in fact, inhibits KLF1's transcriptional activity at the β-*globin* promoter (Zhang and Bieker 1998).

Two p300/CBP acetylation sites have been identified in KLF1: K288 and K302 (Zhang et al. 2001b). Directed mutagenesis has shown that the stimulatory effects of p300 and CBP on KLF1's transcriptional activity are dependent on K288 but not K302. Thus it appears that KLF1's activity is modulated by its acetylation status, whereby acetylation of K288 by p300/CBP increases KLF1's propensity to drive transcription from the β-*globin* promoter. K288 and K302 lie proximal to and within the zinc finger region of Klf1. Their acetylation might therefore be anticipated to have an influence on the DNA-binding activity of KLF1, however, this has not proven to be the case (Zhang et al. 2001b). How then does acetylation of KLF1 augment its transcriptional activity?

One hypothesis stems from the observation that acetylation of K288 and/or K302 increases the affinity with which KLF1 interacts with the cofactor

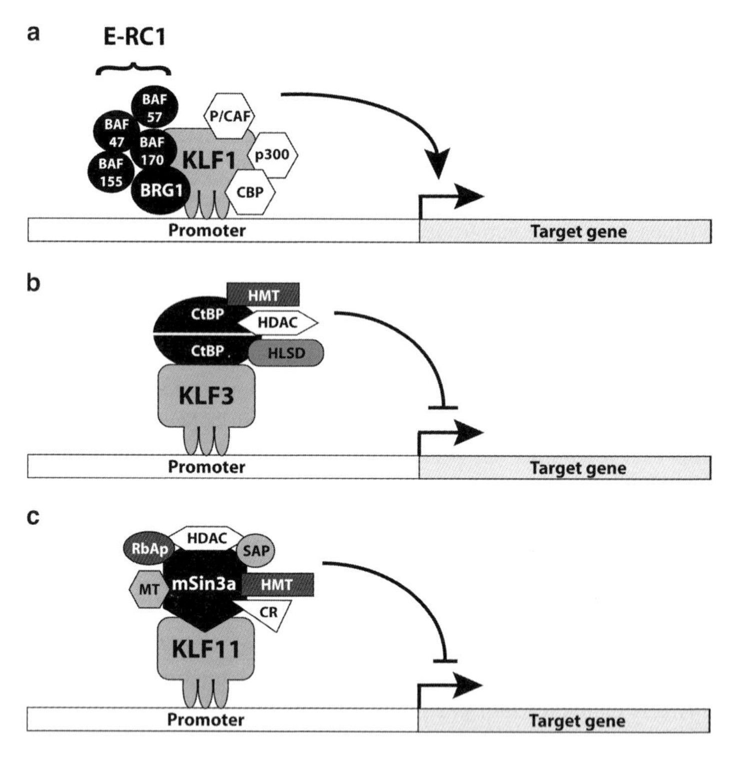

Fig. 2 Co-regulator complexes of group 1, 2, and 3 KLFs. **a** KLF group 1 co-regulator complexes. KLF1 recruits the histone acetyl transferases p300, CBP, and P/CAF and the EKLF co-activator remodeling complex 1 (*E-RC1*) to activate gene expression. Other members of this group may interact with similar complexes. **b** KLF group 2 co-regulator complexes. KLFs 3, 8, and 12 recruit CtBP to mediate gene repression. *CtBP* = C-terminal binding protein; *HMT* = histone methyltransferase; *HDAC* = histone deacetylase; *HLSD* = histone lysine-specific demethylase. **c** KLF group 3 co-regulator complexes. KLFs 9, 10, 11, 13, 14, and 16 interact with Sin3 to mediate gene repression. *SAP* = Sin3-associated protein; *RbAp* = retinoblastoma-associated protein; *MT* = monosaccharide transferase; *CR* = chromatin-remodeling enzyme; *HDAC* = histone deacetylase; *HMT* = histone methyltransferase

Brg1 (brahma-related gene-1), a component of a SWI-SNF-related nucleosome-remodeling complex (Zhang et al. 2001b). This complex is known as the EKLF co-activator remodeling complex 1 (E-RC1) and comprises the subunits Baf170 (Brg1-associated factor, 170kDa), Baf155, Baf47, and Baf57 (Armstrong et al. 1998) (Fig. 2A). To activate transcription from the β-*globin* proximal promoter, KLF1 recruits E-RC1 through its zinc finger domain (Brown et al. 2002; Kadam et al. 2000). This is accompanied by ATP-dependent chromatin remodeling that renders the *β-globin* promoter transcriptionally accessible, as evidenced by the formation of a local DNase I hypersensitive site.

Truncation studies have revealed that whereas the zinc finger domain of KLF1 is sufficient to recruit E-RC1 (Armstrong et al. 1998) a region extending an additional 92 amino acids N-terminal to this domain (to position 164 of hKLF1) is required for complete transcriptional activation of *β-globin* in vivo (Brown et al. 2002). This suggests that residues 164–362 of hKLF1 constitute an activation domain, and in support of this a separate study has shown that a region containing amino acids (aa) 140–358 of hKLF1 is also sufficient for maximal activation from the β-*globin* promoter (Pandya et al. 2001). Another potent activation domain has been defined at the N-terminus of KLF1 (aa 1–139 in hKLF1 and aa 20–124 in mKLF1) that enables a comparable degree of transcriptional activation when fused to the zinc finger DNA-binding domain (Chen and Bieker 1996; Pandya et al. 2001). Although the co-activators that interact with this N-terminal domain remain elusive, it has been shown that casein kinase II (CKII) phosphorylates mKLF1 at T41 and enhances its transactivation capability (Ouyang et al. 1998).

More recently, it has emerged that in some instances KLF1 may function as a transcriptional repressor. KLF1 interacts via its zinc finger domain with the co-repressors Sin3A and histone deacetylase 1 (Chen and Bieker 2001, 2004). In addition, when KLF1 is sumoylated at K74, it is able to recruit the Mi-2β subunit of the NuRD co-repressor complex (Siatecka et al. 2007). KLF1 is able to repress promoter constructs in reporter assays, although only when it is tethered to a target promoter and its zinc finger domain is not engaged in DNA binding (Chen and Bieker 2001, 2004; Siatecka et al. 2007).

Interestingly, K302 has been identified as a residue that is critical for the interaction between KLF1 and Sin3A, and the repressive activity of KLF1 is hampered by mutation of this site (Chen and Bieker 2004). This has led to the postulation that p300/CBP-mediated acetylation of K302 may act as a molecular switch that enables KLF1 to oscillate between being a transcriptional activator and a repressor. Similarly, the dual functionality of KLF1 may be modulated by its sumoylation status. In evidence of this, desumoylation of K74 abrogates the repressive function of KLF1 and causes KLF1 to function as a transcriptional activator (Siatecka et al. 2007). It thus appears likely that KLF1 exerts its varied effects on transcription by exchanging its cofactors as dictated by dynamic posttranslational modifications such as acetylation, phosphorylation, and sumoylation.

In addition to KLF1, it is probable that other members of group 1 also interact with activator complexes containing acetyltransferases. The N-terminal regions of KLF2 and KLF4 have structural similarities to those of KLF1; they contain both activation and repression domains (Anderson et al. 1995; Conkright et al. 2001; Geiman et al. 2000; Yet et al. 1998) and, accordingly, can either stimulate or inhibit transcription of their target genes (Bai et al. 2007; Banerjee et al. 2003; Ghaleb et al. 2005; Rowland et al. 2005; Wang et al. 2005; Wu and Lingrel 2004). Both KLF2 and KLF4 can interact with p300/CBP (Evans et al. 2007; SenBanerjee et al. 2004), and the K288 acetylation site of KLF1 is conserved in KLF2 and KLF4. Furthermore, the function of KLF4 as a transcriptional activator has recently been shown to be enhanced in vivo by p300/CBP-mediated acetylation (Evans et al. 2007). KLFs 5, 6, and 7 also have N-terminal activation domains (Conkright et al. 1999; Koritschoner et al. 1997; Matsumoto et al. 1998). A CBP interaction domain has been mapped for KLF5, although direct acetylation was not observed in vitro (Zhang and Teng 2003). Furthermore, KLF5 is acetylated by p300 in its DNA-binding domain, and co-transfection assays have shown that p300 potentiates KLF5 activation (Miyamoto et al. 2003). Finally, KLF6 can bind CBP and PCAF and is acetylated at several sites (Li et al. 2005).

To date, other than for KLF1, detailed co-regulator complexes have not been defined for the group 1 KLFs. However, it does appear that this group of KLFs is dependent on interaction with acetyltransferases and it is probable that further investigation will reveal additional partners similar to the E-RC1 and Sin3 complexes used by KLF1.

Co-regulators That Bind Group 2 KLFs

Group 2 contains KLFs 3, 8, and 12. Although they do not show extensive sequence similarity outside their C-terminal DNA-binding domains, these KLFs are considered related transcriptional repressors. This is due, at least in part, to their interaction with the co-repressor C-terminal-binding protein (CtBP) (Schuierer et al. 2001; Turner and Crossley 1998; van Vliet et al. 2000). The founding member of the CtBP family, human CtBP1, was first identified as a binding partner of the C-terminal region of adenoviral protein E1A, and this interaction is believed to negatively regulate the oncogenic capacity of the virus (Schaeper et al. 1995). In vertebrates such as zebrafish, *Xenopus*, and mammals, two CtBPs (CtBP1, CtBP2) have been identified (Hildebrand and Soriano 2002). The biological importance of CtBP proteins has been highlighted by gene ablation studies, which have revealed that CtBP2 knockout mice die early in utero owing to severe developmental defects. Compound heterozygotes show intermediate effects, and it is believed that CtBP1 and CtBP2 have both overlapping and unique roles (Hildebrand and Soriano 2002).

The CtBP binding site in E1A has been finely mapped to a small motif consisting of the amino acids PLDLS; and mutational analysis has shown that a DL to AS mutation is sufficient to abrogate the interaction (Schaeper et al. 1998). Similar CtBP binding motifs are present in all of group 2 KLFs; PVDLT is found in both human and mouse KLF3 (Turner and Crossley 1998); PVDLS occurs in human and mouse KLF8 and KLF12 (Schuierer et al. 2001; van Vliet et al. 2000). In KLF3, mutation of the CtBP-binding motif from PVDLT to PVAST (ΔDL) abolishes its interaction with CtBP. However, the lack of CtBP interaction only partially relieves KLF3's repression activity, suggesting that other repression mechanisms may exist (Turner and Crossley, 1998). Indeed, KLF3 has been found to interact with the E2-conjugating enzyme UBC9 and is sumoylated at two lysine residues, K10 and K197. Combined mutation of these sumoylation sites, as well as the PVDLT motif, turned KLF3 from a transcriptional repressor into an activator (Perdomo et al. 2005). Therefore, it is believed that KLF3 represses gene expression through both sumoylation and CtBP interaction. Similar to KLF3, KLF8 retains some repression activity when CtBP binding is abolished (van Vliet et al. 2000). Furthermore, KLF8 can be sumoylated, and it is reported that this sumoylation affects the transcriptional activity of KLF8 (Wei et al. 2006). Sumoylation of KLF12 has not been reported, and it is believed that the interaction with CtBP is sufficient for KLF12's repression activity (Schuierer et al. 2001).

It is proposed that the recruitment of CtBP by group 2 KLFs allows assembly of larger repression complexes, as CtBP is known to bind other co-regulators, such as histone deacetylases (HDACs) 1, 2, 4, 5, and 7 (reviewed by Chinnadurai 2002) (Fig. 2B). It is thought that HDACs facilitate transcriptional repression by deacetylating lysine residues on histones H3 and H4, encouraging chromatin compaction and leading to reduced access by the basal transcriptional machinery (reviewed by Gaston and Jayaraman 2003; Sengupta and Seto 2004). However, the observation that HDAC inhibitors, such as trichostatin A, are unable to relieve CtBP-mediated repression implies that additional repression mechanisms must exist (Koipally and Georgopoulos 2000; Ryu and Arnosti 2003).

Other proteins involved in chromatin modifications, such as the histone methyltransferases EuHMT and G9a and the lysine demethylase LSD-1 have also been found in CtBP repression complexes (Shi et al. 2003) (Fig. 2B). Histone methylation is implicated in both euchromatin and heterochromatin formation (Stewart et al. 2005). Specifically, G9a methylates H3 at lysine 9, and this is associated with gene silencing (Schotta et al. 2002). In contrast, LSD-1 removes methylation on H3 lysine 4 to facilitate gene repression (Shi et al. 2004). It is possible that group 2 KLFs and CtBP repress transcription through the combined actions of the recruited HDACs, histone methyltransferases and histone demethylases. For instance, the HDAC may first remove acetyl groups at residues such as H3 lysine 9, which can then be methylated by the histone methyltransferase G9a to provide a repressive mark. Similarly, the histone demethylase may remove the activating mark, H3 lysine 4, again leading to repression.

Co-regulators That Bind Group 3 KLFs

Group 3 consists of KLFs 9, 10, 11, 13, 14, and 16, all of which interact with Sin3, a co-regulator protein that modulates the expression of genes involved in a wide range of cellular processes. Sin3 was originally isolated and characterized in yeast, where it was implicated in the control of gene expression in several unrelated pathways. In total, Sin3 was discovered seven times and was varyingly described as a transcriptional repressor, a transcriptional activator, and an enhancer of gene silencing (Silverstein and Ekwall 2005). Mammals have two isoforms of Sin3, denoted Sin3a and Sin3b, both of which are highly conserved between species and possess the same basic structure. The Sin3 proteins do not appear to have DNA-binding activity or any intrinsic enzymatic function (Silverstein and Ekwall 2005). Their role is to act as a core around which large protein complexes are assembled by the formation of multiple protein–protein interactions.

Sin3 proteins are large and consist of a number of domains. There are four paired amphipathic α-helix (PAH1-4) domains arranged in tandem, a histone interaction domain located between PAH3 and PAH4, and a highly conserved region at the C-terminus (Ayer et al. 1995; Halleck et al. 1995; Wang and Stillman 1993). These domains provide a scaffold for the assembly of large multiprotein complexes (Fig. 2C). The core complex is made up of Sin3 and the histone deacetylases HDAC1 and HDAC2, which bind to the histone interaction domain of Sin3 (Laherty et al. 1997). The retinoblastoma-associated proteins RbAp46 and RbAp48 function to stabilize the interaction between Sin3 (Hassig et al. 1997), the HDACs, and the nucleosome. Finally, two Sin3-associated proteins, SAP18 and SAP30, provide additional support to the Sin3–HDAC complex (Zhang et al. 1997). In addition to this core complex, other proteins with a range of enzymatic functions can associate with Sin3. Examples include histone methyltransferases (Yang et al. 2003), chromatin remodeling enzymes (Sif et al. 2001), and monosaccharide transferases (Yang et al. 2002). In this way, Sin3 provides a bridge between DNA-binding transcription factors and the enzymes that alter chromatin structure. In different situations Sin3 can recruit different combinations of enzymes to either activate or repress transcription.

For Sin3 proteins to be targeted to gene promoters, they must form interactions with DNA-binding proteins. Numerous transcription factors have been identified as partners of Sin3. The interaction between Sin3 and its partner transcription factors can be mediated via PAH domains 1 and 2, with PAH2 being the most common site for interactions. KLFs 9, 10, 11, 13, and 16 all share a conserved α-helical motif (AA/VXXL), which is located at or near the amino-terminus of the protein and forms the site for Sin3 interaction (Zhang et al. 2001a). This motif was first identified in KLF11 and is highly similar to the Sin3 interaction domain (SID) of the transcription factor Mad1 and to a previously proposed PAH2 interaction consensus sequence (Brubaker et al. 2000). The single helix of the SID motif fits into a cleft within the four-helix bundle of the PAH2 domain of Sin3 (Brubaker et al. 2000). Mutations in this binding motif reduce the ability of these KLFs to interact with Sin3 and repress gene expression (Fernandez-Zapico et al. 2003; Kaczynski et al. 2002; Zhang et al. 2001a).

In all cases described to date, the interactions between KLFs and Sin3 result in the repression of target gene expression. The KLF whose functional interactions with Sin3 are best characterized is KLF11, which interacts with the Sin3 isoform Sin3a. KLF11 regulates the expression of a number of genes involved in cell growth and cancer cell development. An intact SID is required for KLF11-mediated suppression of cellular proliferation and neoplastic transformation. CHO cells transfected with KLF11 display reduced proliferation. Deletion of the SID restores the proliferative capacity of these cells to that of untransfected controls (Fernandez-Zapico et al. 2003). The KLF11–Sin3a interaction is also necessary for the ability of KLF11 to regulate expression of the oxidative stress pathway proteins SOD2 and Catalase1 (Fernandez-Zapico et al. 2003). KLF11 is able to repress the activity of the promoters of both genes in reporter assays, and deletion of the SID significantly reduces this repression. The recruitment of Sin3a by KLF11 plays a crucial role in the development of neural cells. Cell culture studies have shown that a functional SID is required for the promotion of apoptosis and enhancement of transforming growth factor-β (TGF-β) signaling by KLF11 in neural cells (Gohla et al. 2008). The ability of KLF11 to interact with Sin3a can be controlled by posttranslational modifications to KLF11. Phosphorylation of KLF11 at a region adjacent to the SID results in disruption of the interaction between KLF11 and Sin3a and ablation of target gene repression (Ellenrieder et al. 2002).

Although KLFs 9, 10, 13, and 16 all share a conserved SID motif with KLF11, relatively little is known about the interactions of Sin3 and these other KLFs.

KLF9 forms a complex at the growth hormone receptor (GHR) promoter with the transcription factors NF-Y and HMG-Y/I. This protein complex recruits Sin3b to repress activation of the GHR gene (Gowri et al. 2003). KLF13 interacts with both Sin3a and HDAC1 through its SID. This interaction is required for repression of reporter gene expression by KLF13 in transient transfections (Kaczynski et al. 2001). The N-terminal SID of KLF16 interacts with Sin3a and is necessary for the repression of gene expression (Kaczynski et al. 2002).

The Sin3 proteins are key partner proteins of the KLF transcription factor family, with group 3 KLFs recruiting Sin3 to effect gene repression in a number of biological contexts. Sin3 proteins achieve this by providing a framework for the assembly of large repression complexes, which alter chromatin structure, leading to gene silencing.

Cofactors That Posttranslationally Modify KLFs

In addition to the recruitment of co-regulator complexes to target promoters, the transcriptional activity of the KLF family is regulated by interaction with several enzymes that direct posttranslational modifications, such as acetylation, phosphorylation, ubiquitination, and sumoylation.

KLFs 1, 4, 5, 6, and 13 have all been shown to be acetylated, resulting in enhanced trans-activation activity (Evans et al. 2007; Li et al. 2005; Miyamoto et al. 2003; Song et al. 2002; Zhang and Bieker 1998). For example, KLF1 is acetylated on K288 by p300/CBP, allowing interaction with components of a larger chromatin remodeling complex and resulting in activation of the β-globin promoter (Zhang and Bieker 1998; Zhang et al. 2001b). The phosphorylation of KLFs can both enhance and reduce their transcriptional activity. Phosphorylation of KLF1 by casein kinase II at T41 has been shown to enhance its activation potential (Ouyang et al. 1998). Likewise, phosphorylation of KLF5 by protein kinase C at S153 has been shown to play a role in gene activation, possibly by facilitating interaction with CBP (Zhang and Teng 2003). However, the phosphorylation of KLF11 by extracellular signal-regulated kinase/mitogen-activated protein kinase (ERK/MAPK) in pancreatic cancer cells prevents it from interacting with its Sin3 co-repressor, thereby having a negative effect on repression activity (Ellenrieder et al. 2002).

Interaction with ubiquitin ligases can also affect KLF activity by directing degradation via the 26S proteasome. Addition of a single ubiquitin alters protein activity and subsequent polyubiquitination and then targets the protein to the proteasome (Wilkinson 2000). KLFs 1, 2, 5, and 10 are all targets of ubiquitination, leading to proteasomal degradation of the KLF (Chen et al. 2005; Johnsen et al. 2002; Quadrini and Bieker 2006; Zhang et al. 2004). In the case of KLF1, targeted degradation is used to help maintain normal cellular levels of the protein (Quadrini and Bieker 2006). This is in contrast to KLF2, where degradation by the proteasome (Zhang et al. 2004) may allow changes in gene regulation, promoting cellular differentiation in white adipose tissue.

The transcriptional activity of KLFs 1, 3, and 8 is also regulated by sumoylation, a process that involves the addition of the small ubiquitin-like modifier (SUMO) to lysine residues. This is achieved by interaction with sumoylating enzymes in a process similar to that of ubiquitination (Verger et al. 2003). KLF3 recruits the E2 ligase Ubc9 and is sumoylated at K10 and K197, altering repression activity (Perdomo et al. 2005). KLF1 has a single sumoylation site at K74 that appears to play a role in KLF1-mediated repression and the suppression of megakaryocyte differentiation (Siatecka et al. 2007). KLF8 also has a single sumoylation site at K67, close to the CtBP interaction motif. Sumoylation of KLF8 reduces its ability to regulate transcription, and mutation of the sumoylation site can return expression to that of the unsumoylated protein (Wei et al. 2006). These examples demonstrate that sumoylation can both promote and inhibit KLF function.

Transcription Factors That Interact with KLFs

KLF proteins are regulators of cellular differentiation and often interact with other transcription factors involved in the differentiation program. For example, several KLFs play a role in adipogenesis and do so by both regulating key adipogenic genes and interacting with other adipogenic transcription factors, such as C/EBP family members (Mori et al. 2005; Oishi et al. 2005; Sue et al. 2008). KLF5 and KLF15 have both been shown to activate expression of the PPARγ gene, itself a key regulator of adipogenesis. Recruitment of KLF5 and KLF15 to the PPARγ promoter occurs by cooperation with C/EBP proteins, with KLF5 interacting with C/EBPβ/δ and KLF15 with C/EBPα.

Conclusions

Despite having highly homologous DNA-binding domains, KLFs have evolved several mechanisms to ensure specificity in their target gene regulation. This is achieved to some extent by temporal and tissue-specific expression patterns and the existence of KLF regulatory networks. However, it is the interaction with a wide variety of co-regulators that may primarily define how different KLFs operate independently to control gene expression to achieve their biological roles. Their cofactors help determine the cellular concentration of a KLF, its activity, its choice of target gene, and whether it will trans-activate or repress. The co-regulator complexes have been well defined for some KLFs, such as the E-RC1 complex of KLF1, the CtBP complex of KLF3, and the Sin3 complex of KLF11. It is likely that future research will reveal similar or related complexes for other KLFs and also uncover new cofactors that help define KLF function.

References

Anderson KP, Kern CB, Crable SC et al (1995) Isolation of a gene encoding a functional zinc finger protein homologous to erythroid Kruppel-like factor: identification of a new multigene family. Mol Cell Biol 15:5957-5965

Armstrong JA, Bieker JJ, Emerson BM (1998) A SWI/SNF-related chromatin remodeling complex, E-RC1, is required for tissue-specific transcriptional regulation by EKLF in vitro. Cell 95:93-104

Ayer DE, Lawrence QA, Eisenman RN (1995) Mad-Max transcriptional repression is mediated by ternary complex formation with mammalian homologs of yeast repressor Sin3. Cell 80:767-776

Bai A, Hu H, Yeung M et al (2007) Kruppel-like factor 2 controls T cell trafficking by activating L-selectin (CD62L) and sphingosine-1-phosphate receptor 1 transcription. J Immunol 178:7632-7639

Banerjee SS, Feinberg MW, Watanabe M et al (2003) The Kruppel-like factor KLF2 inhibits peroxisome proliferator-activated receptor-gamma expression and adipogenesis. J Biol Chem 278:2581-2584

Brown RC, Pattison S, van Ree J et al (2002) Distinct domains of erythroid Kruppel-like factor modulate chromatin remodeling and transactivation at the endogenous beta-globin gene promoter. Mol Cell Biol 22:161-170

Brubaker K, Cowley SM, Huang K et al (2000) Solution structure of the interacting domains of the Mad-Sin3 complex: implications for recruitment of a chromatin-modifying complex. Cell 103:655-665

Chen C, Sun X, Ran Q et al (2005) Ubiquitin-proteasome degradation of KLF5 transcription factor in cancer and untransformed epithelial cells. Oncogene 24:3319-3327

Chen X, Bieker JJ (1996) Erythroid Kruppel-like factor (EKLF) contains a multifunctional transcriptional activation domain important for inter- and intramolecular interactions. The EMBO journal 15:5888-5896

Chen X, Bieker JJ (2001) Unanticipated repression function linked to erythroid Kruppel-like factor. Mol Cell Biol 21:3118-3125

Chen X, Bieker JJ (2004) Stage-specific repression by the EKLF transcriptional activator. Mol Cell Biol 24:10416-10424

Chinnadurai G (2002) CtBP, an unconventional transcriptional co-repressor in development and oncogenesis. Mol Cell 9:213-224

Conkright MD, Wani MA, Anderson KP et al (1999) A gene encoding an intestinal-enriched member of the Kruppel-like factor family expressed in intestinal epithelial cells. Nucleic Acids Res 27:1263-1270

Conkright MD, Wani MA, Lingrel JB (2001) Lung Kruppel-like factor contains an autoinhibitory domain that regulates its transcriptional activation by binding WWP1, an E3 ubiquitin ligase. J Biol Chem 276:29299-29306

Drissen R, von Lindern M, Kolbus A et al (2005) The erythroid phenotype of EKLF-null mice: defects in hemoglobin metabolism and membrane stability. Mol Cell Biol 25:5205-5214

Ellenrieder V, Zhang JS, Kaczynski J et al (2002) Signaling disrupts mSin3A binding to the Mad1-like Sin3-interacting domain of TIEG2, an Sp1-like repressor. Embo J 21:2451-2460

Evans PM, Zhang W, Chen X et al (2007) Kruppel-like factor 4 is acetylated by p300 and regulates gene transcription via modulation of histone acetylation. J Biol Chem 282:33994-34002

Fernandez-Zapico ME, Mladek A, Ellenrieder V et al (2003) An mSin3A interaction domain links the transcriptional activity of KLF11 with its role in growth regulation. Embo J 22:4748-4758

Gaston K, Jayaraman PS (2003) Transcriptional repression in eukaryotes: repressors and repression mechanisms. Cell Mol Life Sci 60:721-741

Geiman DE, Ton-That H, Johnson JM et al (2000) Transactivation and growth suppression by the gut-enriched Kruppel-like factor (Kruppel-like factor 4) are dependent on acidic amino acid residues and protein-protein interaction. Nucleic Acids Res 28:1106-1113

Ghaleb AM, Nandan MO, Chanchevalap S et al (2005) Kruppel-like factors 4 and 5: the yin and yang regulators of cellular proliferation. Cell Res 15:92-96

Gohla G, Krieglstein K, Spittau B (2008) Tieg3/Klf11 induces apoptosis in OLI-neu cells and enhances the TGF-beta signaling pathway by transcriptional repression of Smad7. J Cell Biochem 104:850-861

Gowri PM, Yu JH, Shaufl A et al (2003) Recruitment of a repressosome complex at the growth hormone receptor promoter and its potential role in diabetic nephropathy. Mol Cell Biol 23:815-825

Halleck MS, Pownall S, Harder KW et al (1995) A widely distributed putative mammalian transcriptional regulator containing multiple paired amphipathic helices, with similarity to yeast SIN3. Genomics 26:403-406

Hassig CA, Fleischer TC, Billin AN et al (1997) Histone deacetylase activity is required for full transcriptional repression by mSin3A. Cell 89:341-347

Hildebrand JD, Soriano P (2002) Overlapping and unique roles for C-terminal binding protein 1 (CtBP1) and CtBP2 during mouse development. Mol Cell Biol 22:5296-5307

Hodge D, Coghill E, Keys J et al (2006) A global role for EKLF in definitive and primitive erythropoiesis. Blood 107:3359-3370

Johnsen SA, Subramaniam M, Monroe DG et al (2002) Modulation of transforming growth factor beta (TGFbeta)/Smad transcriptional responses through targeted degradation of TGFbeta-inducible early gene-1 by human seven in absentia homologue. J Biol Chem 277:30754-30759

Kaczynski J, Zhang JS, Ellenrieder V et al (2001) The Sp1-like protein BTEB3 inhibits transcription via the basic transcription element box by interacting with mSin3A and HDAC-1 co-repressors and competing with Sp1. J Biol Chem 276:36749-36756

Kaczynski JA, Conley AA, Fernandez Zapico M et al (2002) Functional analysis of basic transcription element (BTE)-binding protein (BTEB) 3 and BTEB4, a novel Sp1-like protein, reveals a subfamily of transcriptional repressors for the BTE site of the cytochrome P4501A1 gene promoter. Biochem J 366:873-882

Kadam S, McAlpine GS, Phelan ML et al (2000) Functional selectivity of recombinant mammalian SWI/SNF subunits. Genes & development 14:2441-2451

Keys JR, Tallack MR, Hodge DJ et al (2007) Genomic organisation and regulation of murine alpha haemoglobin stabilising protein by erythroid Kruppel-like factor. British journal of haematology 136:150-157

Koipally J, Georgopoulos K (2000) Ikaros interactions with CtBP reveal a repression mechanism that is independent of histone deacetylase activity. J Biol Chem 275:19594-19602

Koritschoner NP, Bocco JL, Panzetta-Dutari GM et al (1997) A novel human zinc finger protein that interacts with the core promoter element of a TATA box-less gene. J Biol Chem 272:9573-9580

Laherty CD, Yang WM, Sun JM et al (1997) Histone deacetylases associated with the mSin3 co-repressor mediate mad transcriptional repression. Cell 89:349-356

Li D, Yea S, Dolios G et al (2005) Regulation of Kruppel-like factor 6 tumor suppressor activity by acetylation. Cancer Res 65:9216-9225

Matsumoto N, Laub F, Aldabe R et al (1998) Cloning the cDNA for a new human zinc finger protein defines a group of closely related Kruppel-like transcription factors. J Biol Chem 273:28229-28237

Miyamoto S, Suzuki T, Muto S et al (2003) Positive and negative regulation of the cardiovascular transcription factor KLF5 by p300 and the oncogenic regulator SET through interaction and acetylation on the DNA-binding domain. Mol Cell Biol 23:8528-8541

Mori T, Sakaue H, Iguchi H et al (2005) Role of Kruppel-like factor 15 (KLF15) in transcriptional regulation of adipogenesis. J Biol Chem 280:12867-12875

Oishi Y, Manabe I, Tobe K et al (2005) Kruppel-like transcription factor KLF5 is a key regulator of adipocyte differentiation. Cell Metab 1:27-39

Ouyang L, Chen X, Bieker JJ (1998) Regulation of erythroid Kruppel-like factor (EKLF) transcriptional activity by phosphorylation of a protein kinase casein kinase II site within its interaction domain. J Biol Chem 273:23019-23025

Pandya K, Donze D, Townes TM (2001) Novel transactivation domain in erythroid Kruppel-like factor (EKLF). J Biol Chem 276:8239-8243

Perdomo J, Verger A, Turner J et al (2005) Role for SUMO modification in facilitating transcriptional repression by BKLF. Mol Cell Biol 25:1549-1559

Pilon AM, Nilson DG, Zhou D et al (2006) Alterations in expression and chromatin configuration of the alpha hemoglobin-stabilizing protein gene in erythroid Kruppel-like factor-deficient mice. Mol Cell Biol 26:4368-4377

Quadrini KJ, Bieker JJ (2006) EKLF/KLF1 is ubiquitinated in vivo and its stability is regulated by activation domain sequences through the 26S proteasome. FEBS Lett 580:2285-2293

Rowland BD, Bernards R, Peeper DS (2005) The KLF4 tumour suppressor is a transcriptional repressor of p53 that acts as a context-dependent oncogene. Nat Cell Biol 7:1074-1082

Ryu JR, Arnosti DN (2003) Functional similarity of Knirps CtBP-dependent and CtBP-independent transcriptional repressor activities. Nucleic Acids Res 31:4654-4662

Schaeper U, Boyd JM, Verma S et al (1995) Molecular cloning and characterization of a cellular phosphoprotein that interacts with a conserved C-terminal domain of adenovirus E1A involved in negative modulation of oncogenic transformation. Proc Natl Acad Sci U S A 92:10467-10471

Schaeper U, Subramanian T, Lim L et al (1998) Interaction between a cellular protein that binds to the C-terminal region of adenovirus E1A (CtBP) and a novel cellular protein is disrupted by E1A through a conserved PLDLS motif. J Biol Chem 273:8549-8552

Schotta G, Ebert A, Krauss V et al (2002) Central role of Drosophila SU(VAR)3-9 in histone H3-K9 methylation and heterochromatic gene silencing. EMBO J 21:1121-1131

Schuierer M, Hilger-Eversheim K, Dobner T et al (2001) Induction of AP-2alpha expression by adenoviral infection involves inactivation of the AP-2rep transcriptional co-repressor CtBP1. J Biol Chem 276:27944-27949

SenBanerjee S, Lin Z, Atkins GB et al (2004) KLF2 Is a novel transcriptional regulator of endothelial proinflammatory activation. J Exp Med 199:1305-1315

Sengupta N, Seto E (2004) Regulation of histone deacetylase activities. J Cell Biochem 93:57-67

Shi Y, Lan F, Matson C et al (2004) Histone demethylation mediated by the nuclear amine oxidase homolog LSD1. Cell 119:941-953

Shi Y, Sawada J, Sui G et al (2003) Coordinated histone modifications mediated by a CtBP co-repressor complex. Nature 422:735-738

Siatecka M, Xue L, Bieker JJ (2007) Sumoylation of EKLF promotes transcriptional repression and is involved in inhibition of megakaryopoiesis. Mol Cell Biol 27:8547-8560

Sif S, Saurin AJ, Imbalzano AN et al (2001) Purification and characterization of mSin3A-containing Brg1 and hBrm chromatin remodeling complexes. Genes Dev 15:603-618

Silverstein RA, Ekwall K (2005) Sin3: a flexible regulator of global gene expression and genome stability. Curr Genet 47:1-17

Song CZ, Keller K, Murata K et al (2002) Functional interaction between co-activators CBP/p300, PCAF, and transcription factor FKLF2. J Biol Chem 277:7029-7036

Stewart MD, Li J, Wong J (2005) Relationship between histone H3 lysine 9 methylation, transcription repression, and heterochromatin protein 1 recruitment. Mol Cell Biol 25:2525-2538

Sue N, Jack BH, Eaton SA et al (2008) Targeted disruption of the Basic Kruppel-like Factor (Klf3) gene reveals a role in adipogenesis. Mol Cell Biol

Turner J, Crossley M (1998) Cloning and characterization of mCtBP2, a co-repressor that associates with basic Krüppel-like factor and other mammalian transcriptional regulators. EMBO J 17:5129-5140

van Vliet J, Crofts LA, Quinlan KG et al (2006) Human KLF17 is a new member of the Sp/KLF family of transcription factors. Genomics 87:474-482

van Vliet J, Turner J, Crossley M (2000) Human Krüppel-like factor 8: a CACCC-box binding protein that associates with CtBP and represses transcription. Nucleic Acids Res 28:1955-1962

Verger A, Perdomo J, Crossley M (2003) Modification with SUMO. A role in transcriptional regulation. EMBO Rep 4:137-142

Wang F, Zhu Y, Huang Y et al (2005) Transcriptional repression of WEE1 by Kruppel-like factor 2 is involved in DNA damage-induced apoptosis. Oncogene 24:3875-3885

Wang H, Stillman DJ (1993) Transcriptional repression in Saccharomyces cerevisiae by a SIN3-LexA fusion protein. Mol Cell Biol 13:1805-1814

Wei H, Wang X, Gan B et al (2006) Sumoylation delimits KLF8 transcriptional activity associated with the cell cycle regulation. J Biol Chem 281:16664-16671

Wilkinson KD (2000) Ubiquitination and deubiquitination: targeting of proteins for degradation by the proteasome. Semin Cell Dev Biol 11:141-148

Wu J, Lingrel JB (2004) KLF2 inhibits Jurkat T leukemia cell growth via upregulation of cyclin-dependent kinase inhibitor p21WAF1/CIP1. Oncogene 23:8088-8096

Yang L, Mei Q, Zielinska-Kwiatkowska A et al (2003) An ERG (ets-related gene)-associated histone methyltransferase interacts with histone deacetylases 1/2 and transcription co-repressors mSin3A/B. Biochem J 369:651-657

Yang X, Zhang F, Kudlow JE (2002) Recruitment of O-GlcNAc transferase to promoters by co-repressor mSin3A: coupling protein O-GlcNAcylation to transcriptional repression. Cell 110:69-80

Yet SF, McA'Nulty MM, Folta SC et al (1998) Human EZF, a Kruppel-like zinc finger protein, is expressed in vascular endothelial cells and contains transcriptional activation and repression domains. J Biol Chem 273:1026-1031

Zhang JS, Moncrieffe MC, Kaczynski J et al (2001a) A conserved alpha-helical motif mediates the interaction of Sp1-like transcriptional repressors with the co-repressor mSin3A. Mol Cell Biol 21:5041-5049

Zhang W, Bieker JJ (1998) Acetylation and modulation of erythroid Kruppel-like factor (EKLF) activity by interaction with histone acetyltransferases. Proc Natl Acad Sci U S A 95:9855-9860

Zhang W, Kadam S, Emerson BM et al (2001b) Site-specific acetylation by p300 or CREB binding protein regulates erythroid Kruppel-like factor transcriptional activity via its interaction with the SWI-SNF complex. Mol Cell Biol 21:2413-2422

Zhang X, Srinivasan SV, Lingrel JB (2004) WWP1-dependent ubiquitination and degradation of the lung Kruppel-like factor, KLF2. Biochem Biophys Res Commun 316:139-148

Zhang Y, Iratni R, Erdjument-Bromage H et al (1997) Histone deacetylases and SAP18, a novel polypeptide, are components of a human Sin3 complex. Cell 89:357-364

Zhang Z, Teng CT (2003) Phosphorylation of Kruppel-like factor 5 (KLF5/IKLF) at the CBP interaction region enhances its transactivation function. Nucleic Acids Res 31:2196-2208

Part 3
Krüppel-like Factors in Development and Differentiation

Chapter 5
Developmental Expression of Krüppel-like Factors

Yizeng Yang and Jonathan P. Katz

Abstract Krüppel-like factors (KLFs) are members of an emerging family of DNA-binding transcriptional regulators with critical roles in development, differentiation, and a number of other key cellular processes. The KLF family contains at least 17 members, many with overlapping patterns of expression and function, and all linked by a similar DNA-binding element. During development, KLFs may function as transcriptional activators or repressors depending on the cell or tissue context or even the stage of development. Here, we provide a brief introduction to the expression patterns and established roles of the KLFs in development. By examining these patterns and functions, we uncover a number of themes that are explored in detail in ensuing chapters.

Introduction

Members of the KLF family of transcription factors play an essential role during embryonic development and cell-specific lineage differentiation. In many cases, the expression of these factors during embryogenesis is spatiotemporally restricted and regulated. By binding to "CACCC" elements in the regulatory regions of specific target genes, KLFs control multiple intracellular signaling pathways important for cellular proliferation and differentiation, organogenesis, and stem cell commitment. Recently, the role of the KLFs in development has been expanded by the identification of a function for these factors in somatic cell reprogramming to induced pluripotent stem (iPS) cells (Jiang et al. 2008; Okita et al. 2007). Many of the KLFs can be either transcriptional activators or repressors depending on the cell or tissue context or even the developmental stage. In this review, we provide a brief introduction to the expression and the established roles of the KLFs during development (Table 1).

Y. Yang and J.P. Katz (✉)
Division of Gastroenterology, Department of Medicine, University of Pennsylvania School of Medicine, Philadelphia, PA 19104, USA

R. Nagai et al. (eds.), *The Biology of Krüppel-like Factors*,
DOI 10.1007/978-4-431-87775-2_5, © Springer 2009

Table 1 Developmental expression patterns of KLFs

Name (alternative)	Developmental expression pattern	Major gene knockout phenotypes	References
KLF1 (EKLF)	Erythroid cells, yolk sac, fetal liver	Defective hematopoiesis and lethal β-thalassemia around E14	Nuez et al. 1995; Perkins et al. 1995
KLF2 (LKLF)	Blood vessels, lungs, T lymphocytes, erythroid cells, white adipose tissue	Death from severe hemorrhage from E11.5 to 14.5; delayed lung development; defective T-lymphocyte and adipocyte differentiation	Kuo et al. 1997a, 1997b; Wani et al. 1999; Wu et al. 2005
KLF3 (BKLF)	Embryonic hematopoietic tissue, brain, limb buds	Decreased white adipose tissue; report of myeloproliferative disorders and defective hematopoesis	Sue et al. 2008; Turner and Crossley 1999
KLF4 (GKLF, EZF)	Gut epithelia, thymus, skin, testes	Lethal perinatal dehydration due to defective skin barrier; abnormal differentiation of colonic goblet cells	Katz et al. 2002; Segre et al. 1999
KLF5 (IKLF, BTEB2)	Gut epithelia, skin, vascular smooth muscles, heart, skeleton, white adipose tissue	Homozygous mice die at E8.5; heterozygous mice show defects of smooth muscle, arteries, cardiac hypertrophy, and fibrosis in response to stress; deficiencies in white adipose tissue development; skeletal growth retardation	Oishi et al. 2005; Shindo et al. 2002, 2008
KLF6 (Zf9, CPEP)	Extraembryonic tissues, nervous system, cornea, lung buds, ureteric bud, heart, liver, intestinal mucosa; mesenchyme surrounding neural tube and developing brain	Lethal failure of hematopoiesis by E12.5 and poorly organized yolk sac vacularity	Fischer et al. 2001; Laub et al. 2001b; Matsumoto et al. 2006
KLF7 (UKLF)	Predominantly in central and peripheral nervous systems; lower levels throughout embryo at later stages	Neonatal lethality by P3 with defects in selected regions of the nervous system	Laub et al. 2001a, 2005
KLF8	Placenta; otherwise, not described	–	van Vliet et al. 2000
KLF9 (BTEB)	Broadly expressed; high levels in developing brain, thymus, epithelia, and smooth muscle of gut and bladder, vertebrae, and cartilage primordia	Normal lifespan but uterine hypoplasia and defects in parturition in females; impairment of specific behavioral activities; shorter small intestinal villi	Martin et al. 2001; Morita et al. 2003; Simmen et al. 2004, 2007

KLF10 (TIEG1, mGIF)	Broadly expressed	Normal lifespan but osteopenia in females, cardiac hypertrophy in males, defects in the mechanical properties and healing of tendons	Bensamoun et al. 2006; Hawse et al. 2008; Rajamannan et al. 2007; Subramaniam et al. 1995, 2005; Tsubone et al. 2006; Yajima et al. 1997;
KLF11 (FKLF, TIEG2)	Ubiquitous in adult; expression in erythroid cells in fetal liver; other developmental expression not described	Normal development, fertility, and lifespan	Asano et al. 1999; Cook et al. 1998; Song et al. 2005
KLF12 (AP-2rep)	Some expression in developing brain and kidneys; very low levels in adult liver and lung	—	Imhof et al. 1999; Suda et al. 2006
KLF13 (RFLAT-1, BTEB3)	Broadly expressed with high levels temporally in heart, brain, bladder, gut, thymus, epidermis	Some decreased viability by 3 weeks of age; enlarged thymus and spleen; decreased numbers of circulating erythrocytes; increased survival and decreased apoptosis of thymocytes	Gordon et al. 2008; Martin et al. 2001; Scohy et al. 2000; Zhou et al. 2007
KLF14	Ubiquitous expression	–	Parker-Katiraee et al. 2007; Scohy et al. 2000
KLF15 (KKLF)	Minimal cardiac expression during development; other developmental expression not described	Fertile, normal viability; cardiac fibrosis and hypertrophy in response to stress; fasting hypoglycemia	Fisch et al. 2007; Gray et al. 2007; Uchida et al. 2000; Wang et al. 2008
KLF16 (DRRF, BTEB4)	Highly expressed in certain regions of developing brain; lesser expression in thymus, duodenum, kidney, liver, heart, bladder, and lung	—	D'Souza et al. 2002; Hwang et al. 2001
KLF17 (Zfp393)	Spermatids and oocytes; other developmental expression not described	—	van Vliet et al. 2006; Yan et al. 2002

KLFs in Development

KLF1

Klf1 was first identified as an erythroid lineage-specific zinc finger transcription factor, named erythroid Krüppel-like factor (EKLF) (Miller and Bieker 1993). Klf1 plays a essential role in the γ-hemoglobin to β-hemoglobin switch during fetal development, and inactivation of *Klf1* in mice results in defective hematopoiesis in fetal liver and death from anemia, with a deficit in β-globin expression by E16, providing a model for β-thalassemia (Miller and Bieker 1993; Nuez et al. 1995; Perkins et al. 1995). *Klf1* is expressed in the yolk sac as early as E7.5. Following the development of the hematopoietic system, *Klf1* appears in fetal liver and the mesoderm near the hindgut toward the dorsal aorta at E10.5, and by E14.5 *Klf1* is restricted to the fetal liver. Klf1 is not required for yolk sac hematopoiesis or expansion of erythroid progenitors (Perkins et al. 1995) but is required for the last steps of erythroid differentiation (Drissen et al. 2005). Interestingly, *Klf1* is down-regulated in megakaryocytes and inhibits the formation of megakaryocytes while stimulating erythroid differentiation (Frontelo et al. 2007).

KLF2

Klf2 is expressed temporally during early embryonic development and plays an important role in the development of the lungs, blood vessels, T lymphocytes, and adipocytes (Kuo et al. 1997a, 1997b; Wani et al. 1999; Wu et al. 2005). In the adult, *Klf2* is highly expressed in the lung and thus was initially identified as lung Krüppel-like factor (LKLF) (Anderson et al. 1995). Expression of *Klf2* is first seen in vascular endothelial cells throughout the developing mouse embryo at E9.5 (Anderson et al. 1995; Kuo et al. 1997a; Wani et al. 1998). *Klf2* is expressed at high levels between E9.5 and E12.5, especially in the umbilical arteries and veins, a critical time for both angiogenesis and blood vessel wall stabilization. At E14.5, *Klf2* continues to be expressed in the vasculature and appears in the lung buds, vertebral column, and the bony structures of the head and rib cage. By E18.5, *Klf2* is expressed abundantly in the lungs and in blood vessels. Mice with homozygous deletion of *Klf2* exhibit growth retardation and craniofacial abnormalities, and they die between E11.5 and 14.5 from severe intraamniotic and intraembryonic hemorrhage (Kuo et al. 1997a). Notably, blood vessels in these mice have an abnormally thin tunica media. *Klf2* null mice also have defects in their lung development (Wani et al. 1999). In addition, *Klf2* is developmentally induced during single-positive T-lymphocyte maturation, and *Klf2*-deficient T cells are spontaneously activated (Kuo et al. 1997b). *Klf2* also inhibits adipogenesis by maintaining a preadipocyte state (Wu et al. 2005).

KLF3

KLF3, or basic Krüppel-like factor (BKLF), is a highly basic KLF first identified in murine yolk sac and fetal liver erythroid cells (Crossley et al. 1996). *Klf3* is highly expressed in embryonic hematopoietic tissues, brain, and several other tissues. It appears in the midbrain and anterior hindbrain at E8.5 and in the ventral anterior half of the embryo, midbrain–hindbrain junction, ventral midbrain, diencephalon, and forebrain at E9. At E10.5, expression becomes more widespread, with some staining of the developing limb buds (Crossley et al. 1996). *Klf3*-deficient mice are smaller than their littermates, and adipocyte differentiation is altered in murine embryonic fibroblasts from *Klf3* knockout mice (Sue et al. 2008). Myeloproliferative disorders and abnormalities in hematopoiesis have also been reported (Turner and Crossley 1999).

KLF4

KLF4 is highly expressed in postproliferative epithelial cells of the gut and the epidermis and was thus named gut-enriched Krüppel-like factor (GKLF) and epithelial zinc finger (EZF) when it was initially characterized by two independent groups (Garrett-Sinha et al. 1996; Shields et al. 1996). *Klf4* mRNA is found in the epidermal layer of the skin and in epithelial cells in the tongue, palate, esophagus, stomach, and colon of newborn and adult mice; and it is enriched in epithelial cells of the middle to upper colonic crypts, a region of cellular differentiation (Ton-That et al. 1997). *Klf4* transcript is initially low in the whole embryo but begins to rise around E13, peaking on E17, the period in which the intestinal epithelium undergoes major transition from a pseudostratified to a columnar epithelium, before decreasing prior to birth. Klf4 is seen in mesenchymal cells of the nasal prominence and first branchial arch, mesenchymal cells surrounding the cartilaginous primordia of the skeleton, and the metanephric kidney at E11.5, and it is upregulated in thymus epithelium at E18 (Garrett-Sinha et al. 1996; Panigada et al. 1999). Strong Klf4 expression is also seen in postmeiotic germ cells undergoing final differentiation into sperm cells in postnatal mouse testis (Behr and Kaestner 2002). Mice homozygous for a null allele of *Klf4* die shortly after birth due to apparent failure to establish proper skin barrier function and have a 90% decrease in the number of goblet cells in the colon (Katz et al. 2002; Segre et al. 1999). Klf4 has been identified as a critical regulator of pluripotency in embryonic stem cells and recently has been utilized, along with several other factors, to reprogram mouse and human somatic cells directly into pluripotent cells (induced pluripotent stem cells, or iPS cells) (Li et al. 2005b; Okita et al. 2007; Takahashi et al. 2007). This function in stem cells appears to be distinct from the role of KLF4 in a number of adult tissues and cell types (McConnell et al. 2007).

KLF5

KLF5 (BTEB2) was initially cloned from a human placenta cDNA library using rat BTEB cDNA as a probe (Sogawa et al. 1993). Later, the murine homologue was identified and named intestinal-enriched Krüppel-like factor (IKLF) for its high level of expression in intestinal epithelia (Conkright et al. 1999). Temporal changes of *Klf5* expression during embryogenesis indicate that this gene is developmentally regulated (Ohnishi et al. 2000). *Klf5* transcript is abundant in the embryo at E7 (Conkright et al. 1999) and is seen in the developing gastrointestinal tract by E10.5 (Conkright et al. 1999; Ohnishi et al. 2000). Klf5 mRNA is also detected in the E15.5 meninges, E16.5 epithelia of the trachea and bronchi, and the outer layer of the tongue, as well as in the developing epidermis. Progressively, Klf5 expression in the skin and gastrointestinal tract becomes localized to the proliferative compartments, such as the basal layer of the epidermis and the small intestinal crypts (Ohnishi et al. 2000). KLF5 is also abundantly expressed in embryonic vascular smooth muscles and is downregulated in adult vessels (Ogata et al. 2000). Homozygous null mice for *Klf5* die before E8.5, indicating an essential but unclear role for Klf5 in early embryonic development (Shindo et al. 2002). Heterozygotes appear grossly normal, but the arteries exhibit diminished levels of arterial wall thickening, angiogenesis, cardiac hypertrophy, and interstitial fibrosis in response to external stress. In addition, neonatal mice heterozygous for *Klf5* deletion exhibit a marked deficiency in white adipose tissue (Oishi et al. 2005) and show skeletal growth retardation with impaired cartilage matrix degradation (Shinoda et al. 2008). Klf5 also appears to be involved in the maintenance of self-renewal in embryonic stem cells and is expressed in mouse embryonic stem cells, blastocysts, and primordial germ cells (Parisi et al. 2008).

KLF6

KLF6, also known as ZF9 or CPBP, was originally isolated from a cDNA library of human placenta (Koritschoner et al. 1997). Human KLF6 is expressed ubiquitously with a high level in the placenta and adult liver, lung, intestine, and prostate (Blanchon et al. 2001; Narla et al. 2001; Ratziu et al. 1998). KLF6 is also seen in the developing cornea of the 7-week-old fetus, mostly in the cytoplasm, becoming more nuclear after birth (Nakamura et al. 2007). In the mouse, *Klf6* is expressed in extraembryonic tissues at E10.5 and in undifferentiated mesenchyme surrounding the neural tube and brain vesicles by E11.5, with strong expression in the nervous system by E12.5 and low levels in the heart, ureteric bud, and lung buds (Fischer et al. 2001; Laub et al. 2001b). By E14.5, *Klf6* is nearly undetectable except in the ventral horn at the level of the forelimbs. Subsequently, very strong *Klf6* expression is seen between E16.5 and E18.5 in the intestinal mucosa and in the fetal liver between E14 and E20 (Laub et al. 2001b; Ratziu et al. 1998). Like some of the other KLFs, expression of KLF6 plays a role in preadipocyte differentiation, in this case promoting differentiation by inhibiting delta-like 1 (Li et al. 2005a).

Homozygous null mice for *Klf6* die by E12.5; they are small and pale with no obvious liver, have thin, poorly organized yolk sacs, and show significant defects in hematopoiesis (Matsumoto et al. 2006).

KLF7

KLF7 was initially cloned from human vascular endothelial cells by the polymerase chain reaction (PCR) using degenerate oligonucleotides corresponding to the DNA-binding domain of KLF1 (Matsumoto et al. 1998). Given its broad, low-level expression in adult tissues, KLF7 was termed ubiquitous Krüppel-like factor (UKLF). However, the predominant developmental expression of mouse *Klf7* is in postmitotic neuroblasts of the developing central and peripheral nervous systems (Laub et al. 2001a). *Klf5* mRNA is first seen at E9.5 and is maximum around E11.5, with intense expression in the forebrain, midbrain, and hindbrain; the eye; and the trigeminal, geniculate, vestibulocochlear, petrosal, superior, jugular, nodose, accessory, and dorsal root ganglia. *Klf7* expression is maintained in the dorsal root ganglia from E11.5 to E18.5, whereas expression declines in the neural tube and low levels of expression are seen diffusely throughout the embryo. *Klf7* is also expressed in the olfactory epithelium at E16.5 and the neural retina at E17.5. Postnatally, *Klf7* expression is observed in a few regions of the brain but is later confined to the adult cerebellum, olfactory system, and dorsal root ganglia (Laub et al. 2001a). Loss of *Klf7* in mice leads to neonatal lethality, with 98.5% of pups dying within 3 days of birth (Laub et al. 2005). *Klf7* null mice have hypoplastic olfactory bulbs, with defects of axonal projections in the olfactory and visual systems, cerebral cortex, and hippocampus, as well as abnormalities of dendritic organization.

KLF8

Human KLF8 (ZNF741) was first cloned by PCR from K562 cells, a human hematopoietic cell line (van Vliet et al. 2000). KLF8 is broadly expressed in human tissues, with greatest expression in kidney, heart, and placenta. Multiple KLF8 transcripts have been identified, and the relative levels of transcript expression appear to be similar in the various tissues. Little is known to date about *Klf8* expression during development, and a *Klf8* knockout has not been described.

KLF9

Klf9 (BTEB) was initially isolated from a rat liver cDNA library (Imataka et al. 1992). *Klf9* is widely expressed throughout mouse embryonic development, at least as early as E8 (Martin et al. 2001). By E11, *Klf9* is highly expressed in the cephalic mesenchyme of the developing brain, the epithelia and smooth muscle of the gut and bladder, and the skin epidermis. At E16, high levels of *Klf9* are also observed in the thymus and vertebrae cartilage primordia. In the developing cerebral hemispheres, *Klf9* is undetectable from E16 until birth and then rises dramatically into adulthood, suggesting a possible role in neurite outgrowth (Denver et al. 1999). In addition, in mice, *Klf9* expression increases dramatically in Purkinje cells of the cerebellum and in the pyramidal cells of the hippocampus at postnatal day 7, a time when synapses in the brain begin to form (Morita et al. 2003). *Klf9* null mice show a normal lifespan, are fertile, and exhibit no overt pathological defects; their general behavioral activities are unaffected (Morita et al. 2003). However, *Klf9* null mice do show impairments in specific behavioral testing, such as rotorod tests and contextual fear conditioning tests. Ablation of *Klf9* in female mice results in uterine hypoplasia, reduced litter size, and increased incidence of neonatal deaths in offspring, with parturition defects involving the progesterone receptor (Simmen et al. 2004; Zeng et al. 2008). In addition, *Klf9* loss results in an intestinal phenotype, with short small intestinal villi, reduced crypt cell proliferation, decreased migration along the villi, and altered cell lineage allocation (Simmen et al. 2007).

KLF10

KLF10 was identified by differential display PCR from normal human fetal osteoblasts following transforming growth factor-β (TGF-β) treatment and thus is also called TGF-β-inducible early gene 1 (TIEG1) (Subramaniam et al. 1995). Human KLF10 is expressed in keratinocytes; epithelial cells of the placenta, breast, and uterus; osteoblasts and other cells of the bone marrow and cerebellum; skeletal muscle; and pancreas with some cells showing cytoplasmic staining and others a nuclear localization (Subramaniam et al. 1995, 1998). Mouse *Klf10*, also called mGIF, is widely distributed in the adult with high levels in kidney, lung, brain, liver, heart, and testis (Yajima et al. 1997). During development, murine *Klf10* is broadly expressed, including in the cerebral cortex, cerebellar primordium, kidney, intestine, liver, lung, bones, and the differentiating mesenchyme surrounding the nasal cavity and some of the skull (Yajima et al. 1997). *Klf10* null mice initially appeared to be phenotypically normal, with no evidence of alterations in bone formation despite an increase in the number of osteoblasts (Subramaniam et al. 2005). These *Klf10* null osteoblasts display reduced expression of key differentiation markers and a decreased ability to support osteoclast differentiation in vitro. Subsequent analyses revealed that loss of *Klf10* results in severe osteopenia in female animals only, with reduced cancellous and cortical bone and reduced bone strength (Hawse et al. 2008). Conversely, male but not female *Klf10* null mice develop cardiac hypertrophy at 16 months of age (Rajamannan et al. 2007). *Klf10* null mice also show defects in the mechanical properties and healing potential of tendons (Bensamoun et al. 2006; Tsubone et al. 2006).

KLF11

Human KLF11 was first described as TIEG2, another TGF-β-inducible gene that inhibits growth in cultured cells (Cook et al. 1998). Another homologue, FKLF, was cloned from fetal globin-expressing human fetal erythroid cells (Asano et al. 1999). KLF11 is ubiquitously expressed in adult human tissues, with the highest levels in the pancreas and skeletal muscle (Cook et al. 1998). KLF11 is also enriched in erythroid cells, with much higher expression in fetal liver than adult bone marrow (Asano et al. 1999). *Klf11* null mice appear normal at all stages of development, are fertile, and show no abnormalities of hematopoiesis (Song et al. 2005).

KLF12

Klf12, formerly named AP-2rep, is a transcriptional repressor of the AP-2α gene identified by screening a mouse brain cDNA library (Imhof et al. 1999). Overall, *Klf12* is seen in the adult kidney and at very low levels in the adult liver and lung but not in most other adult and embryonic tissues. Some *Klf12* transcripts are seen in brain and kidney at E15.5 and E19.5 (Imhof et al. 1999), and *Klf12* expression in the developing kidney rises at postnatal day 15 (Suda et al. 2006). A knockout mouse for *Klf12* has not been reported.

KLF13

KLF13 (RFLAT-1) was identified as an activator of RANTES (regulated upon activation normal T-cell expressed and secreted), a chemokine for T-cell activation (Song et al. 1999). KLF13 is ubiquitously expressed in human tissues, with two distinct transcripts, and the greatest abundance is seen in peripheral blood lymphocytes and thymus. In the mouse, *Klf13* is widely distributed in adults (Scohy et al. 2000) and embryos (Martin et al. 2001), beginning by at least E8. *Klf13* is expressed in primitive heart at E8 and in atria and ventricles of the developing heart at E11; it is seen less prominently at E13 and E16. *Klf13* is also expressed at high levels in the cephalic mesenchyme of the developing brain, the thymus, vertebrae cartilage primordia, gut, bladder, and epidermis. Whereas *Klf13* in the gut and bladder are expressed throughout the muscle and epithelia at E11 and E13, by E16 the expression is localized to the epithelia. Inactivation of *Klf13* in mice results in decreased viability by 3 weeks after birth, with reduced numbers of circulating erythrocytes, an increase in less mature erythroblasts, prolonged survival of thymocytes due to decreased apoptosis, splenomegaly, and an enlarged thymus (Gordon et al. 2008; Zhou et al. 2007). *Klf13* null mice also show a trend toward reduced numbers of granulocytes and monocytes, suggesting abnormalities in pathways affecting differentiation or survival of hematopoietic cells (Gordon et al. 2008). Although one report indicates an increase in the number of thymocytes, another reports a 30% decrease in thymocyte numbers (Gordon et al. 2008; Zhou et al. 2007). Knockdown of *Klf13* in *Xenopus* embryos leads to atrial septal defects and hypotrabeculation (Lavallee et al., 2006).

KLF14

Klf14 was identified together with mouse *Klf13* using the sequence of the Sp1 zinc finger DNA-binding domain as a probe to screen a mouse EST database (Scohy et al. 2000). Klf14 is ubiquitously expressed in adult tissues. Human *KLF14* has been described as an imprinted gene with monoallelic maternal expression in all embryonic and extraembryonic tissues studied in humans and the mouse (Parker-Katiraee et al. 2007). Expression is seen in placenta and fetal heart, liver, lung, and colon, as well as adult skeletal muscle, colon, stomach, and brain but not the liver or lymphoblasts. A gene knockout of KLF14 has not been described.

KLF15

KLF15 was cloned from a human kidney cDNA library as kidney Krüppel-like factor (KKLF) (Uchida et al. 2000). In human and rat tissues, KLF15 is expressed most abundantly in liver, with moderate levels in kidneys, heart, and skeletal muscle and no expression in bone marrow or lymphoid tissues. *Klf15* is highly expressed in adipocytes and myocytes in vivo and is induced when preadipocytes differentiate into adipocytes (Gray et al. 2002). *Klf15* shows minimal cardiac expression during embryonic development and is barely detectable in the rat heart at postnatal day 3 but reaches adult levels by postnatal day 30 (Fisch et al. 2007). *Klf15* null mice are viable, fertile, and born in expected Mendelian ratios. In response to pressure overload, *Klf15* null mice develop cardiac fibrosis and an eccentric form of cardiac hypertrophy (Fisch et al. 2007; Wang et al. 2008). *Klf15* null mice also develop severe fasting hypoglycemia (Gray et al. 2007).

KLF16

KLF16, also known as dopamine receptor regulating factor (DRRF), was initially cloned from a mouse neuroblastoma cell line (Hwang et al. 2001). In mice, *Klf16* is expressed in multiple adult tissues, including brain, heart, spleen, lung, liver, kidney, and testis. The highest levels of *Klf16* are seen in multiple regions of the brain, including the olfactory tubercle, olfactory bulb, nucleus accumbens, striatum, hippocampal CA1 region, cerebral cortex, dentate gyrus, and amygdala (Hwang et al. 2001). During embryogenesis, the pattern of *Klf16* in brain overlaps that found in the adult and with the expression profile of dopamine receptors (Hwang et al. 2001). *Klf16* is expressed at E12 in regions of the brain and skull and in muscles of the tongue and tail, with moderate expression in the heart and liver (D'Souza et al. 2002). At E14, KLF16 is highly expressed in the olfactory lobe and other regions of the brain, moderately expressed in the liver, and minimally expressed in the lung. At E16, KLF16 expression is seen in the brain, thymus, duodenum, and kidney, with lesser expression in the liver, heart, bladder, and lung. A KLF16 knockout has not yet been described.

KLF17

Human KLF17 is a recently described member of the KLF family, which appears to be the human orthologue of the mouse gene Zfp393 (van Vliet et al. 2006; Yan et al., 2002). Human KLF17 and murine Zfp393 have 54.8% identity at the protein level but have significantly higher similarity in their zinc finger regions (81.5% similarity). Zfp393 is expressed exclusively in the testis and ovary, with specific expression in steps 3–8 spermatids and growing oocytes (Yan et al. 2002). The expression of human KLF17 has not been extensively studied, but based on the sources of human *KLF17* ESTs it appears to be present in testis, brain, and bone, although likely at low levels (van Vliet et al. 2006). A knockout of *Klf17* has not been reported.

Conclusion

The expression patterns of the individual KLFs vary during development and adulthood. A careful examination of the overlapping patterns of expression and function suggest a number of themes for the members of the KLF family. These themes, involving processes such as adipogenesis, cardiac hypertrophy, hematopoiesis, and the pluripotency of stem cells, are explored thoroughly in the ensuing chapters.

References

Anderson, K. P., Kern, C. B., Crable, S. C., and Lingrel, J. B. (1995). Isolation of a gene encoding a functional zinc finger protein homologous to erythroid Kruppel-like factor: identification of a new multigene family. Mol Cell Biol *15*, 5957-5965.

Asano, H., Li, X. S., and Stamatoyannopoulos, G. (1999). FKLF, a novel Kruppel-like factor that activates human embryonic and fetal beta-like globin genes. Mol Cell Biol *19*, 3571-3579.

Behr, R., and Kaestner, K. H. (2002). Developmental and cell type-specific expression of the zinc finger transcription factor Kruppel-like factor 4 (KLF4) in postnatal mouse testis. Mech Dev *115*, 167-169.

Bensamoun, S. F., Tsubone, T., Subramaniam, M., Hawse, J. R., Boumediene, E., Spelsberg, T. C., An, K. N., and Amadio, P. C. (2006). Age-dependent changes in the mechanical properties of tail tendons in TGF-beta inducible early gene-1 knockout mice. J Appl Physiol *101*, 1419-1424.

Blanchon, L., Bocco, J. L., Gallot, D., Gachon, A. M., Lemery, D., Dechelotte, P., Dastugue, B., and Sapin, V. (2001). Co-localization of KLF6 and KLF4 with pregnancy-specific glycoproteins during human placenta development. Mech Dev *105*, 185-189.

Conkright, M. D., Wani, M. A., Anderson, K. P., and Lingrel, J. B. (1999). A gene encoding an intestinal-enriched member of the Kruppel-like factor family expressed in intestinal epithelial cells. Nucleic Acids Res *27*, 1263-1270.

Cook, T., Gebelein, B., Mesa, K., Mladek, A., and Urrutia, R. (1998). Molecular cloning and characterization of TIEG2 reveals a new subfamily of transforming growth factor-beta-inducible Sp1-like zinc finger-encoding genes involved in the regulation of cell growth. J Biol Chem *273*, 25929-25936.

Crossley, M., Whitelaw, E., Perkins, A., Williams, G., Fujiwara, Y., and Orkin, S. H. (1996). Isolation and characterization of the cDNA encoding BKLF/TEF-2, a major CACCC-box-binding protein in erythroid cells and selected other cells. Mol Cell Biol *16*, 1695-1705.

D'Souza, U. M., Lammers, C. H., Hwang, C. K., Yajima, S., and Mouradian, M. M. (2002). Developmental expression of the zinc finger transcription factor DRRF (dopamine receptor regulating factor). Mech Dev *110*, 197-201.

Denver, R. J., Ouellet, L., Furling, D., Kobayashi, A., Fujii-Kuriyama, Y., and Puymirat, J. (1999). Basic transcription element-binding protein (BTEB) is a thyroid hormone-regulated gene in the developing central nervous system. Evidence for a role in neurite outgrowth. J Biol Chem *274*, 23128-23134.

Drissen, R., von Lindern, M., Kolbus, A., Driegen, S., Steinlein, P., Beug, H., Grosveld, F., and Philipsen, S. (2005). The erythroid phenotype of EKLF-null mice: defects in hemoglobin metabolism and membrane stability. Mol Cell Biol *25*, 5205-5214.

Fisch, S., Gray, S., Heymans, S., Haldar, S. M., Wang, B., Pfister, O., Cui, L., Kumar, A., Lin, Z., Sen-Banerjee, S., et al. (2007). Kruppel-like factor 15 is a regulator of cardiomyocyte hypertrophy. Proc Natl Acad Sci U S A *104*, 7074-7079.

Fischer, E. A., Verpont, M. C., Garrett-Sinha, L. A., Ronco, P. M., and Rossert, J. A. (2001). KLF6 is a zinc finger protein expressed in a cell-specific manner during kidney development. J Am Soc Nephrol *12*, 726-735.

Frontelo, P., Manwani, D., Galdass, M., Karsunky, H., Lohmann, F., Gallagher, P. G., and Bieker, J. J. (2007). Novel role for EKLF in megakaryocyte lineage commitment. Blood *110*, 3871-3880.

Garrett-Sinha, L. A., Eberspaecher, H., Seldin, M. F., and de Crombrugghe, B. (1996). A gene for a novel zinc-finger protein expressed in differentiated epithelial cells and transiently in certain mesenchymal cells. J Biol Chem *271*, 31384-31390.

Gordon, A. R., Outram, S. V., Keramatipour, M., Goddard, C. A., Colledge, W. H., Metcalfe, J. C., Hager-Theodorides, A. L., Crompton, T., and Kemp, P. R. (2008). Splenomegaly and modified erythropoiesis in KLF13-/- mice. J Biol Chem *283*, 11897-11904.

Gray, S., Feinberg, M. W., Hull, S., Kuo, C. T., Watanabe, M., Sen-Banerjee, S., DePina, A., Haspel, R., and Jain, M. K. (2002). The Kruppel-like factor KLF15 regulates the insulin-sensitive glucose transporter GLUT4. J Biol Chem *277*, 34322-34328.

Gray, S., Wang, B., Orihuela, Y., Hong, E. G., Fisch, S., Haldar, S., Cline, G. W., Kim, J. K., Peroni, O. D., Kahn, B. B., and Jain, M. K. (2007). Regulation of gluconeogenesis by Kruppel-like factor 15. Cell Metab *5*, 305-312.

Hawse, J. R., Iwaniec, U. T., Bensamoun, S. F., Monroe, D. G., Peters, K. D., Ilharreborde, B., Rajamannan, N. M., Oursler, M. J., Turner, R. T., Spelsberg, T. C., and Subramaniam, M. (2008). TIEG-null mice display an osteopenic gender-specific phenotype. Bone *42*, 1025-1031.

Hwang, C. K., D'Souza, U. M., Eisch, A. J., Yajima, S., Lammers, C. H., Yang, Y., Lee, S. H., Kim, Y. M., Nestler, E. J., and Mouradian, M. M. (2001). Dopamine receptor regulating factor, DRRF: a zinc finger transcription factor. Proc Natl Acad Sci U S A *98*, 7558-7563.

Imataka, H., Sogawa, K., Yasumoto, K., Kikuchi, Y., Sasano, K., Kobayashi, A., Hayami, M., and Fujii-Kuriyama, Y. (1992). Two regulatory proteins that bind to the basic transcription element (BTE), a GC box sequence in the promoter region of the rat P-4501A1 gene. EMBO J *11*, 3663-3671.

Imhof, A., Schuierer, M., Werner, O., Moser, M., Roth, C., Bauer, R., and Buettner, R. (1999). Transcriptional regulation of the AP-2alpha promoter by BTEB-1 and AP-2rep, a novel wt-1/egr-related zinc finger repressor. Mol Cell Biol *19*, 194-204.

Jiang, J., Chan, Y. S., Loh, Y. H., Cai, J., Tong, G. Q., Lim, C. A., Robson, P., Zhong, S., and Ng, H. H. (2008). A core KLF circuitry regulates self-renewal of embryonic stem cells. Nat Cell Biol *10*, 353-360.

Katz, J. P., Perreault, N., Goldstein, B. G., Lee, C. S., Labosky, P. A., Yang, V. W., and Kaestner, K. H. (2002). The zinc-finger transcription factor KLF4 is required for terminal differentiation of goblet cells in the colon. Development *129*, 2619-2628.

Koritschoner, N. P., Bocco, J. L., Panzetta-Dutari, G. M., Dumur, C. I., Flury, A., and Patrito, L. C. (1997). A novel human zinc finger protein that interacts with the core promoter element of a TATA box-less gene. J Biol Chem *272*, 9573-9580.

Kuo, C. T., Veselits, M. L., Barton, K. P., Lu, M. M., Clendenin, C., and Leiden, J. M. (1997a). The LKLF transcription factor is required for normal tunica media formation and blood vessel stabilization during murine embryogenesis. Genes Dev *11*, 2996-3006.

Kuo, C. T., Veselits, M. L., and Leiden, J. M. (1997b). LKLF: A transcriptional regulator of single-positive T cell quiescence and survival. Science *277*, 1986-1990.

Laub, F., Aldabe, R., Friedrich, V., Jr., Ohnishi, S., Yoshida, T., and Ramirez, F. (2001a). Developmental expression of mouse Kruppel-like transcription factor KLF7 suggests a potential role in neurogenesis. Dev Biol *233*, 305-318.

Laub, F., Aldabe, R., Ramirez, F., and Friedman, S. (2001b). Embryonic expression of Kruppel-like factor 6 in neural and non-neural tissues. Mech Dev *106*, 167-170.

Laub, F., Lei, L., Sumiyoshi, H., Kajimura, D., Dragomir, C., Smaldone, S., Puche, A. C., Petros, T. J., Mason, C., Parada, L. F., and Ramirez, F. (2005). Transcription factor KLF7 is important for neuronal morphogenesis in selected regions of the nervous system. Mol Cell Biol *25*, 5699-5711.

Lavallee, G., Andelfinger, G., Nadeau, M., Lefebvre, C., Nemer, G., Horb, M. E., and Nemer, M. (2006). The Kruppel-like transcription factor KLF13 is a novel regulator of heart development. EMBO J *25*, 5201-5213.

Li, D., Yea, S., Li, S., Chen, Z., Narla, G., Banck, M., Laborda, J., Tan, S., Friedman, J. M., Friedman, S. L., and Walsh, M. J. (2005a). Kruppel-like factor-6 promotes preadipocyte differentiation through histone deacetylase 3-dependent repression of DLK1. J Biol Chem *280*, 26941-26952.

Li, Y., McClintick, J., Zhong, L., Edenberg, H. J., Yoder, M. C., and Chan, R. J. (2005b). Murine embryonic stem cell differentiation is promoted by SOCS-3 and inhibited by the zinc finger transcription factor KLF4. Blood *105*, 635-637.

Martin, K. M., Metcalfe, J. C., and Kemp, P. R. (2001). Expression of KLF9 and KLF13 in mouse development. Mech Dev *103*, 149-151.

Matsumoto, N., Kubo, A., Liu, H., Akita, K., Laub, F., Ramirez, F., Keller, G., and Friedman, S. L. (2006). Developmental regulation of yolk sac hematopoiesis by Kruppel-like factor 6. Blood *107*, 1357-1365.

Matsumoto, N., Laub, F., Aldabe, R., Zhang, W., Ramirez, F., Yoshida, T., and Terada, M. (1998). Cloning the cDNA for a new human zinc finger protein defines a group of closely related Kruppel-like transcription factors. J Biol Chem *273*, 28229-28237.

McConnell, B. B., Ghaleb, A. M., Nandan, M. O., and Yang, V. W. (2007). The diverse functions of Kruppel-like factors 4 and 5 in epithelial biology and pathobiology. Bioessays *29*, 549-557.

Miller, I. J., and Bieker, J. J. (1993). A novel, erythroid cell-specific murine transcription factor that binds to the CACCC element and is related to the Kruppel family of nuclear proteins. Mol Cell Biol *13*, 2776-2786.

Morita, M., Kobayashi, A., Yamashita, T., Shimanuki, T., Nakajima, O., Takahashi, S., Ikegami, S., Inokuchi, K., Yamashita, K., Yamamoto, M., and Fujii-Kuriyama, Y. (2003). Functional

analysis of basic transcription element binding protein by gene targeting technology. Mol Cell Biol *23*, 2489-2500.
Nakamura, H., Edward, D. P., Sugar, J., and Yue, B. Y. (2007). Expression of Sp1 and KLF6 in the developing human cornea. Mol Vis *13*, 1451-1457.
Narla, G., Heath, K. E., Reeves, H. L., Li, D., Giono, L. E., Kimmelman, A. C., Glucksman, M. J., Narla, J., Eng, F. J., Chan, A. M., et al. (2001). KLF6, a candidate tumor suppressor gene mutated in prostate cancer. Science *294*, 2563-2566.
Nuez, B., Michalovich, D., Bygrave, A., Ploemacher, R., and Grosveld, F. (1995). Defective haematopoiesis in fetal liver resulting from inactivation of the EKLF gene. Nature *375*, 316-318.
Ogata, T., Kurabayashi, M., Hoshino, Y. I., Sekiguchi, K. I., Kawai-Kowase, K., Ishikawa, S., Morishita, Y., and Nagai, R. (2000). Inducible expression of basic transcription factor-binding protein 2 (BTEB2), a member of zinc finger family of transcription factors, in cardiac allograft vascular disease. Transplantation *70*, 1653-1656.
Ohnishi, S., Laub, F., Matsumoto, N., Asaka, M., Ramirez, F., Yoshida, T., and Terada, M. (2000). Developmental expression of the mouse gene coding for the Kruppel-like transcription factor KLF5. Dev Dyn *217*, 421-429.
Oishi, Y., Manabe, I., Tobe, K., Tsushima, K., Shindo, T., Fujiu, K., Nishimura, G., Maemura, K., Yamauchi, T., Kubota, N., et al. (2005). Kruppel-like transcription factor KLF5 is a key regulator of adipocyte differentiation. Cell Metab *1*, 27-39.
Okita, K., Ichisaka, T., and Yamanaka, S. (2007). Generation of germline-competent induced pluripotent stem cells. Nature *448*, 313-317.
Panigada, M., Porcellini, S., Sutti, F., Doneda, L., Pozzoli, O., Consalez, G. G., Guttinger, M., and Grassi, F. (1999). GKLF in thymus epithelium as a developmentally regulated element of thymocyte-stroma cross-talk. Mech Dev *81*, 103-113.
Parisi, S., Passaro, F., Aloia, L., Manabe, I., Nagai, R., Pastore, L., and Russo, T. (2008). KLF5 is involved in self-renewal of mouse embryonic stem cells. J Cell Sci.
Parker-Katiraee, L., Carson, A. R., Yamada, T., Arnaud, P., Feil, R., Abu-Amero, S. N., Moore, G. E., Kaneda, M., Perry, G. H., Stone, A. C., et al. (2007). Identification of the imprinted KLF14 transcription factor undergoing human-specific accelerated evolution. PLoS Genet *3*, e65.
Perkins, A. C., Sharpe, A. H., and Orkin, S. H. (1995). Lethal beta-thalassaemia in mice lacking the erythroid CACCC-transcription factor EKLF. Nature *375*, 318-322.
Rajamannan, N. M., Subramaniam, M., Abraham, T. P., Vasile, V. C., Ackerman, M. J., Monroe, D. G., Chew, T. L., and Spelsberg, T. C. (2007). TGFbeta inducible early gene-1 (TIEG1) and cardiac hypertrophy: Discovery and characterization of a novel signaling pathway. J Cell Biochem *100*, 315-325.
Ratziu, V., Lalazar, A., Wong, L., Dang, Q., Collins, C., Shaulian, E., Jensen, S., and Friedman, S. L. (1998). Zf9, a Kruppel-like transcription factor up-regulated in vivo during early hepatic fibrosis. Proc Natl Acad Sci U S A *95*, 9500-9505.
Scohy, S., Gabant, P., Van Reeth, T., Hertveldt, V., Dreze, P. L., Van Vooren, P., Riviere, M., Szpirer, J., and Szpirer, C. (2000). Identification of KLF13 and KLF14 (SP6), novel members of the SP/XKLF transcription factor family. Genomics *70*, 93-101.
Segre, J. A., Bauer, C., and Fuchs, E. (1999). KLF4 is a transcription factor required for establishing the barrier function of the skin. Nat Genet *22*, 356-360.
Shields, J. M., Christy, R. J., and Yang, V. W. (1996). Identification and characterization of a gene encoding a gut-enriched Kruppel-like factor expressed during growth arrest. J Biol Chem *271*, 20009-20017.
Shindo, T., Manabe, I., Fukushima, Y., Tobe, K., Aizawa, K., Miyamoto, S., Kawai-Kowase, K., Moriyama, N., Imai, Y., Kawakami, H., et al. (2002). Kruppel-like zinc-finger transcription factor KLF5/BTEB2 is a target for angiotensin II signaling and an essential regulator of cardiovascular remodeling. Nat Med *8*, 856-863.
Shinoda, Y., Ogata, N., Higashikawa, A., Manabe, I., Shindo, T., Yamada, T., Kugimiya, F., Ikeda, T., Kawamura, N., Kawasaki, Y., et al. (2008). Kruppel-like factor 5 causes cartilage degradation through transactivation of matrix metalloproteinase 9. J Biol Chem.

Simmen, F. A., Xiao, R., Velarde, M. C., Nicholson, R. D., Bowman, M. T., Fujii-Kuriyama, Y., Oh, S. P., and Simmen, R. C. (2007). Dysregulation of intestinal crypt cell proliferation and villus cell migration in mice lacking Kruppel-like factor 9. Am J Physiol Gastrointest Liver Physiol *292*, G1757-1769.

Simmen, R. C., Eason, R. R., McQuown, J. R., Linz, A. L., Kang, T. J., Chatman, L., Jr., Till, S. R., Fujii-Kuriyama, Y., Simmen, F. A., and Oh, S. P. (2004). Subfertility, uterine hypoplasia, and partial progesterone resistance in mice lacking the Kruppel-like factor 9/basic transcription element-binding protein-1 (Bteb1) gene. J Biol Chem *279*, 29286-29294.

Sogawa, K., Imataka, H., Yamasaki, Y., Kusume, H., Abe, H., and Fujii-Kuriyama, Y. (1993). cDNA cloning and transcriptional properties of a novel GC box-binding protein, BTEB2. Nucleic Acids Res *21*, 1527-1532.

Song, A., Chen, Y. F., Thamatrakoln, K., Storm, T. A., and Krensky, A. M. (1999). RFLAT-1: a new zinc finger transcription factor that activates RANTES gene expression in T lymphocytes. Immunity *10*, 93-103.

Song, C. Z., Gavriilidis, G., Asano, H., and Stamatoyannopoulos, G. (2005). Functional study of transcription factor KLF11 by targeted gene inactivation. Blood Cells Mol Dis *34*, 53-59.

Subramaniam, M., Gorny, G., Johnsen, S. A., Monroe, D. G., Evans, G. L., Fraser, D. G., Rickard, D. J., Rasmussen, K., van Deursen, J. M., Turner, R. T., et al. (2005). TIEG1 null mouse-derived osteoblasts are defective in mineralization and in support of osteoclast differentiation in vitro. Mol Cell Biol *25*, 1191-1199.

Subramaniam, M., Harris, S. A., Oursler, M. J., Rasmussen, K., Riggs, B. L., and Spelsberg, T. C. (1995). Identification of a novel TGF-beta-regulated gene encoding a putative zinc finger protein in human osteoblasts. Nucleic Acids Res *23*, 4907-4912.

Subramaniam, M., Hefferan, T. E., Tau, K., Peus, D., Pittelkow, M., Jalal, S., Riggs, B. L., Roche, P., and Spelsberg, T. C. (1998). Tissue, cell type, and breast cancer stage-specific expression of a TGF-beta inducible early transcription factor gene. J Cell Biochem *68*, 226-236.

Suda, S., Rai, T., Sohara, E., Sasaki, S., and Uchida, S. (2006). Postnatal expression of KLF12 in the inner medullary collecting ducts of kidney and its trans-activation of UT-A1 urea transporter promoter. Biochem Biophys Res Commun *344*, 246-252.

Sue, N., Jack, B. H., Eaton, S. A., Pearson, R. C., Funnell, A. P., Turner, J., Czolij, R., Denyer, G., Bao, S., Molero-Navajas, J. C., et al. (2008). Targeted disruption of the basic Kruppel-like factor gene (KLF3) reveals a role in adipogenesis. Mol Cell Biol *28*, 3967-3978.

Takahashi, K., Tanabe, K., Ohnuki, M., Narita, M., Ichisaka, T., Tomoda, K., and Yamanaka, S. (2007). Induction of pluripotent stem cells from adult human fibroblasts by defined factors. Cell *131*, 861-872.

Ton-That, H., Kaestner, K. H., Shields, J. M., Mahatanankoon, C. S., and Yang, V. W. (1997). Expression of the gut-enriched Kruppel-like factor gene during development and intestinal tumorigenesis. FEBS Lett *419*, 239-243.

Tsubone, T., Moran, S. L., Subramaniam, M., Amadio, P. C., Spelsberg, T. C., and An, K. N. (2006). Effect of TGF-beta inducible early gene deficiency on flexor tendon healing. J Orthop Res *24*, 569-575.

Turner, J., and Crossley, M. (1999). Basic Kruppel-like factor functions within a network of interacting haematopoietic transcription factors. Int J Biochem Cell Biol *31*, 1169-1174.

Uchida, S., Tanaka, Y., Ito, H., Saitoh-Ohara, F., Inazawa, J., Yokoyama, K. K., Sasaki, S., and Marumo, F. (2000). Transcriptional regulation of the CLC-K1 promoter by myc-associated zinc finger protein and kidney-enriched Kruppel-like factor, a novel zinc finger repressor. Mol Cell Biol *20*, 7319-7331.

van Vliet, J., Crofts, L. A., Quinlan, K. G., Czolij, R., Perkins, A. C., and Crossley, M. (2006). Human KLF17 is a new member of the Sp/KLF family of transcription factors. Genomics *87*, 474-482.

van Vliet, J., Turner, J., and Crossley, M. (2000). Human Kruppel-like factor 8: a CACCC-box binding protein that associates with CtBP and represses transcription. Nucleic Acids Res *28*, 1955-1962.

Wang, B., Haldar, S. M., Lu, Y., Ibrahim, O. A., Fisch, S., Gray, S., Leask, A., and Jain, M. K. (2008). The Kruppel-like factor KLF15 inhibits connective tissue growth factor (CTGF) expression in cardiac fibroblasts. J Mol Cell Cardiol.

Wani, M. A., Means, R. T., Jr., and Lingrel, J. B. (1998). Loss of LKLF function results in embryonic lethality in mice. Transgenic Res *7*, 229-238.

Wani, M. A., Wert, S. E., and Lingrel, J. B. (1999). Lung Kruppel-like factor, a zinc finger transcription factor, is essential for normal lung development. J Biol Chem *274*, 21180-21185.

Wu, J., Srinivasan, S. V., Neumann, J. C., and Lingrel, J. B. (2005). The KLF2 transcription factor does not affect the formation of preadipocytes but inhibits their differentiation into adipocytes. Biochemistry *44*, 11098-11105.

Yajima, S., Lammers, C. H., Lee, S. H., Hara, Y., Mizuno, K., and Mouradian, M. M. (1997). Cloning and characterization of murine glial cell-derived neurotrophic factor inducible transcription factor (MGIF). J Neurosci *17*, 8657-8666.

Yan, W., Burns, K. H., Ma, L., and Matzuk, M. M. (2002). Identification of Zfp393, a germ cell-specific gene encoding a novel zinc finger protein. Mech Dev *118*, 233-239.

Zeng, Z., Velarde, M. C., Simmen, F. A., and Simmen, R. C. (2008). Delayed parturition and altered myometrial progesterone receptor isoform a expression in mice null for Kruppel-like factor 9. Biol Reprod *78*, 1029-1037.

Zhou, M., McPherson, L., Feng, D., Song, A., Dong, C., Lyu, S. C., Zhou, L., Shi, X., Ahn, Y. T., Wang, D., et al. (2007). Kruppel-like transcription factor 13 regulates T lymphocyte survival in vivo. J Immunol *178*, 5496-5504.

Chapter 6
Expanded Role for EKLF/KLF1 Within the Hematopoietic Lineage

James J. Bieker

Abstract Erythroid Krüppel-like factor (EKLF) is a transcriptional activator that was originally identified in 1993 and has since provided a significant window into the carefully orchestrated process of erythroid gene expression, particularly as it plays a critical role in ß-like hemoglobin switching. However, later observations have suggested other roles for this protein. This presentation discusses the lines of evidence that together open up a new hematopoietic horizon for EKLF function.

Introduction

For a number of years my laboratory has been interested in the control of lineage decisions during hematopoiesis, with particular focus on the study of erythroid cell differentiation. We initially took a subtractive cloning approach to this issue, whereby we isolated genes selectively expressed in the murine erythroleukemia (MEL) cell line but not in a monocyte-macrophage (J774) cell line. This led us to the identification of a gene product that contained three TFIIIA-like C2H2 zinc fingers at its carboxy terminus and that we named erythroid Krüppel-like factor (EKLF) because of their high homology to the *Drosophila* gap gene (Miller and Bieker 1993).

Our excitement about this gene product was heightened when we found that its expression is highly restricted to adult bone marrow and spleen and when we realized that the homology of its zinc fingers to a specific subset of C2H2 zinc finger proteins enabled us to predict, from structural arguments, a potential DNA-binding target. This directed us to a conserved element already known to be important for ß-globin gene expression, the CAC box or CACCC element, located ~90 bp upstream of its transcriptional start site. Subsequent molecular and genetic studies have verified that EKLF interacts with this site in vivo, and that EKLF

J.J. Bieker (✉)
Department of Developmental and Regenerative Biology, Mount Sinai School of Medicine, Box 1020, One Gustave L Levy Place, New York, NY 10029, USA

R. Nagai et al. (eds.), *The Biology of Krüppel-like Factors*,
DOI 10.1007/978-4-431-87775-2_6, © Springer 2009

plays a critical and necessary role in the switch to adult ß-globin expression during erythroid ontogeny (Donze et al. 1995; Feng et al. 1994; Im et al. 2005; Nuez et al. 1995; Perkins et al. 1995). It does so by directing the onset of ß-globin expression and establishing the proper three-dimensional chromatin structure across the ß-like globin domain, a large region that includes the far upstream locus control region (LCR) (Armstrong et al. 1998; Drissen et al. 2004).

EKLF has been renamed KLF1 and is now the founding member of a large (17-member) family of transcription factors that play diverse roles in differentiation and development (Bieker 2001).

Although most of these and subsequent studies have focused on EKLF's ability to activate transcription, we had been surprised by finding that EKLF also interacts with co-repressor proteins (Chen and Bieker 2001, 2004), suggesting that other roles lay hidden and remained to be discovered. Together, the activation and repression studies strongly suggested that posttranslational modifications of EKLF play an important role in establishing specific protein–protein interactions, and that these could lead to altered functional effects at both the molecular and cellular levels. For these reasons, we continued to probe EKLF function by concurrently undertaking a series of nominally unrelated studies that, in the end, unexpectedly converged and unraveled a deeper function for EKLF during hematopoiesis.

Four Experimental Avenues That Support an Expanded Role for EKLF/KLF1 During Hematopoiesis

EKLF Promoter Analyses

Examination of the onset and expression pattern of EKLF during development revealed that its mRNA first appears at the neural plate stage (~E7.5), where it is strictly localized to the earliest morphologically identified erythroid cells in the blood islands of the yolk sac, followed by a switch to the fetal lever by E10.5 (Southwood et al. 1996). This led to a series of analyses using varied approaches to determine the control elements responsible for this highly restricted pattern of expression. Initially (Chen et al. 1998), transfection of cloned DNA was used to determine that a 950 base pair (bp) region, located just upstream of the transcription initiation site, was sufficient to generate erythroid-specific expression in transient assays (Fig. 1). This region harbors erythroid-restricted DNAse hypersensitive sites, one of which behaves as a strong enhancer. This core region, in conjunction with the proximal promoter—with its important GATA and CCAAT sequences (Crossley et al. 1994)—accounted for EKLF's tissue-specific expression. The importance of this short region was verified in vivo by the use of transgenic mice, where the 950 piece was sufficient to drive reporter (lacZ) expression specifically to the yolk sac and fetal liver (Xue et al. 2004). Importantly, visualization of thin sections from the yolk sac showed reporter expression to be strictly localized to the hematopoietic,

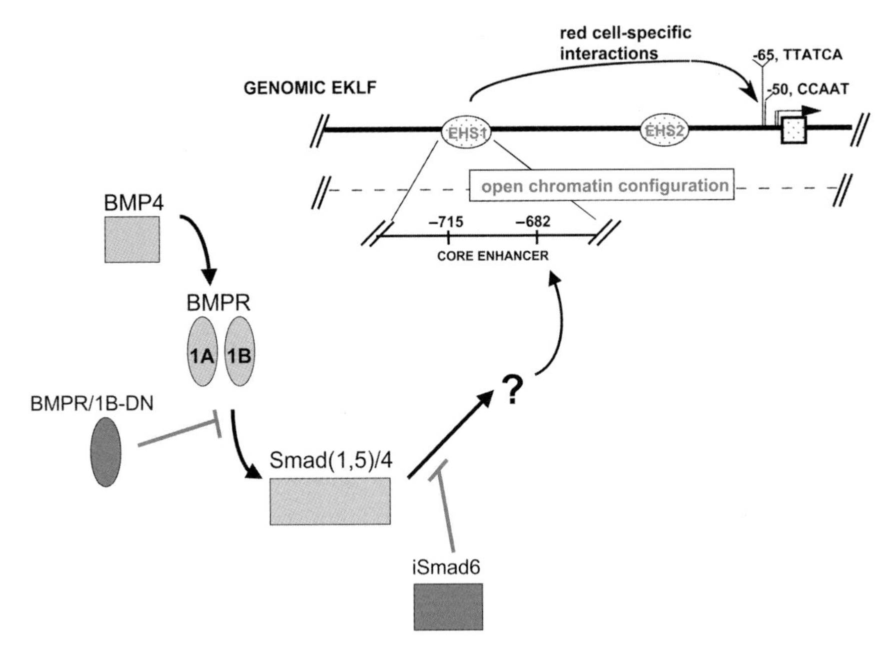

Fig. 1 Regulation of the EKLF promoter, circa 2003. *Top right* The 950-bp region just upstream of the EKLF transcription start site, showing the chromatin structure and location of erythroid hypersensitive sites (*EHS*), conserved promoter motifs, regions responsible for reconstitution of tissue specific expression in transient assays, and the minimal core enhancer element (Chen et al. 1998; Crossley et al. 1994). *Bottom left* Summary of the need for the BMP4/BMP receptor/Smad pathway for EKLF induction, indicating the effects of dominant negative receptor or inhibitory Smad6 on the process (Adelman et al. 2002). The *question mark* in the middle suggests that these experiments had not determined whether the BMP4 pathway exerted its effects on the EKLF promoterin in a direct or indirect manner

not the endothelial or visceral endoderm, compartment. Finally, embryonic stem (ES) cells differentiating in serum-free medium were used to show that BMP4 is necessary and sufficient to induce EKLF expression as embryoid bodies (EBs) are being formed (Adelman et al. 2002). The involvement for the Smad pathway in this process was directly implicated by showing that interference by constitutive expression of a dominant negative BMP receptor or of the inhibitory Smad6 obviated EKLF expression even in the presence of serum. Together, these studies suggested that EKLF promoter regulation is controlled by a proximal 950 bp region that responds to BMP4 signals mediated by (likely) Smad1 or Smad5 (Fig. 1). However, the kinetics of its induction did not allow us to determine if the Smad1/5 effect was direct or was mediated indirectly via another protein (or a combination of the two).

This set the stage for a more detailed analysis of the EKLF promoter (Lohmann and Bieker 2008). A seven species alignment of a 30 kB genomic region encompassing the mammalian KLF1 gene indicated that the most significant homology, in addition to the protein coding exons, resided in the same 1 kB proximal promoter region that had been functionally mapped. A short peak of homology also resided within the first intron. However, a more detailed view of these regions not only revealed transcription factor-binding motifs but also that their architectural layout was exquisitely conserved across species. Of particular interest were Gata and Smad binding motifs, located precisely within the upstream enhancer and proximal promoter sequences previously mapped by functional assays, and within the intronic region. Their functional importance were tested using a novel application of the Kyba/Daley ES/EB differentiating system (Fig. 2A), whereby an EKLF promoter/GFP transgene was developed that faithfully recapitulated the onset of endogenous EKLF expressing during EB differentiation. Mutation of conserved Gata and Smad sites verified that this assay was sensitive enough to distinguish their importance for directing the onset, maintenance, or optimal levels of transcription.

The proteins that interact with these regions were identified in two ways. First, chromatin immunoprecipitation (ChIP) of GATA proteins revealed a switch in their occupancy when comparing early versus late times of EB differentiation, corresponding to GATA2 occupancy in the progenitor stage, followed by GATA1 occupancy after lineage commitment. Second, the Kyba/Daley system was again modified and used to generate a doxycycline-inducible shRNA line (Fig. 2B) directed against Smad5 expression, which verified its critical importance for EKLF expression. Together with the promoter analyses, these data enable us to propose a two-tiered mechanism for transcriptional regulation of EKLF, providing low levels

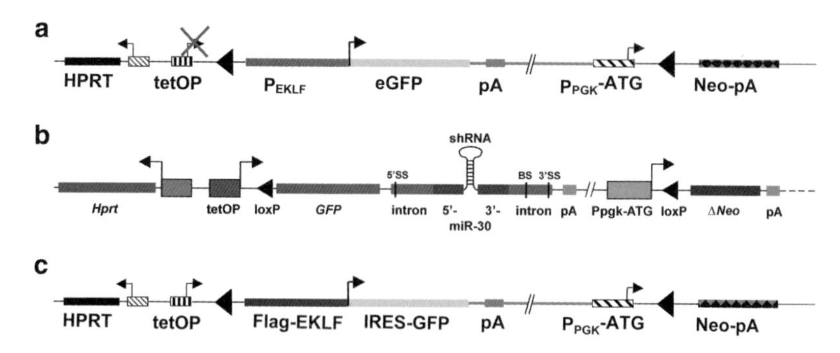

Fig. 2 Modifications of the Kyba/Daley system. The three modifications of the original inducible expression system described by Kyba and Daley (Kyba et al. 2002) used in the experiments described in the presentation are shown. All schematics are shown after their unidirectional, single-copy integration into the single endogenous modified HPRT locus. **A** The EKLF promoter (950 bp; P-EKLF) was placed upstream of the GFP reporter (Lohmann and Bieker 2008). These experiments did not utilize the tet operon but, rather, relied on the EKLF promoter to drive expression of GFP. **B** The miR30 backbone was inserted into an intron in the GFP expression cassette (Lohmann and Bieker 2008). This provides a slot into which any desired shRNA can be inserted; in the present case, they were directed against Smad5. The tet operon is operative and inducible by doxycycline. **C** Flag-tagged EKLF was inserted upstream of an IRES-GFP gene (Frontelo et al. 2007), providing a doxycycline-inducible construct to provide dose-dependent expression of EKLF at any desired time point during embryonic stem (ES) cell differentiation

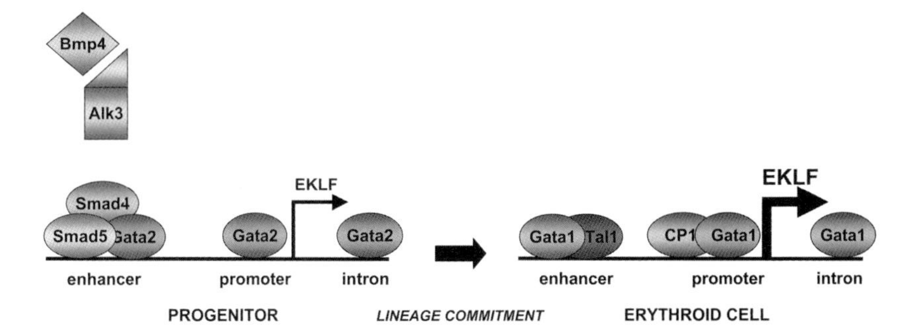

Fig. 3 Model for two-tiered, stage-dependent induction of EKLF during early development. The status of the EKLF promoter is shown before (progenitor) and after (erythroid cell) lineage commitment based on the experiments described in the presentation (Lohmann and Bieker 2008). Multispecies alignments, promoter mutations, ChIP, and shRNA knockdowns suggest that SMAD5 and GATA2 play an initial, necessary role early to generate low levels of EKLF expression, but this switches to a GATA1-dependent, high-level expression pattern after GATA1 protein is induced. Additional proteins (Tal1, CP1, although likely no longer Smad5) may also play a role by binding to other conserved sites

that depend on GATA2 and SMAD5 proteins at early stages, followed by high levels of EKLF transcript after GATA1 protein is produced (Fig. 3).

A prediction from these studies was that selection of EKLF promoter/GFP+ cells should enrich for erythroid progenitors. We therefore established an EKLF promoter/GFP reporter mouse, isolated GFP+ fetal liver cells, and monitored their capacity for colony formation in methylcellulose. Indeed, this selected for erythroid progenitors, but the additional surprise was that this population was also enriched for megakaryocyte progenitors. This was also true when similar assays were performed with EB-derived cells. As a result, there is a line of evidence suggesting that EKLF is positioned by its promoter activity to play an earlier role in hematopoiesis than was originally apparent.

EKLF Gain-of Function and Loss-of-Function Analyses

Although genetic ablation of EKLF led to embryonic lethality due to a profound ß-thalassemia, its expression pattern during embryogenesis (described above) in addition to its presence in some multipotential cell lines suggested other, more subtle functions. In earlier studies, we had used the Kyba/Daley system in its original design for expression of chimeric EKLF proteins (Manwani et al. 2007). Given the enigmatic expression pattern of EKLF cell lines, we used this system (Fig. 2C) to ask whether increased expression of EKLF could alter normal patterns of hematopoiesis during EB differentiation (Frontelo et al. 2007). This powerful system provides a dose-dependent way to increase the target protein at the desired time point and in all the cells within a developing EB. This approach was combined with disaggregation and plating on OP9 stromal cells. By monitoring cell surface marker expression in cells from EBs that had been induced to overexpress EKLF by the addition of doxycycline, we found that megakaryopoiesis (as judged by CD41/CD42d expression) was repressed. This was not a general effect, as c-kit levels were increased and CD71 levels were not affected. At the same time, however, erythropoiesis was accentuated, as Ter119 levels increased during this time, an effect that was visibly apparent by quantitative morphological examination of the cells.

As a complement to these studies, we monitored megakaryocyte status in EKLF-null fetal livers, harvested at E13.5 just prior to lethality. EKLF-null fetal livers contained a higher percentage of megakaryocytes than did the wild type, judged by both CD41/42d and CD41/42b FACS analyses, which also expanded to a greater extent following cell culture.

We then addressed whether these increased numbers of megakaryocytes resulted from an increase in progenitors. Colony assays using total fetal liver cells, lineage-depleted (lin-) cells, or CMP-sorted cells all led to increased numbers of megakaryocyte colonies if they were derived from EKLF-null material. An unanticipated aspect of these assays was that the EKLF-null megakaryocytic colonies were of greater size than those derived from wild-type material. In total, these data provide a second line of evidence that EKLF functions in hematopoiesis prior to erythroid differentiation, as its levels play a critical role in the extent of megakaryopoiesis.

EKLF Expression During Normal Adult Hematopoiesis

Our earlier studies had focused on EKLF expression patterns during normal mammalian embryonic development but had not addressed this issue during normal hematopoietic differentiation that emanates from the hematopoietic stem cell (HSC). We utilized well-established cell surface marker criteria for isolation of specific subpopulations of cells emanating from long-term hematopoietic stem cells and analyzed them for EKLF presence by reverse transcription-polymerase chain reaction (qRT-PCR) (Frontelo et al. 2007). The analyses (Fig. 4) demonstrated that EKLF is expressed at barely detectable levels in hematopoietic stem cells and multipotent progenitors. A clear difference in expression subsequently becomes established, with EKLF absent in common lymphoid progenitors and their B- and T-cell progeny, yet increased in the common myeloid progenitor (CMP). At this point there is another clear demarcation in expression within the CMP progeny, as EKLF levels become higher in the megakaryocyte/erythroid progenitor (MEP) but decline further in the granulocyte/macrophage progenitor (GMP). EKLF expression in the GMP does not develop any further. These data show that there is a gradual restriction in expression of EKLF as hematopoiesis proceeds even though its levels are steadily increasing in more the differentiated cells that express it. Of particular interest, however, the bipotential differentiation of MEPs leads to a dramatic

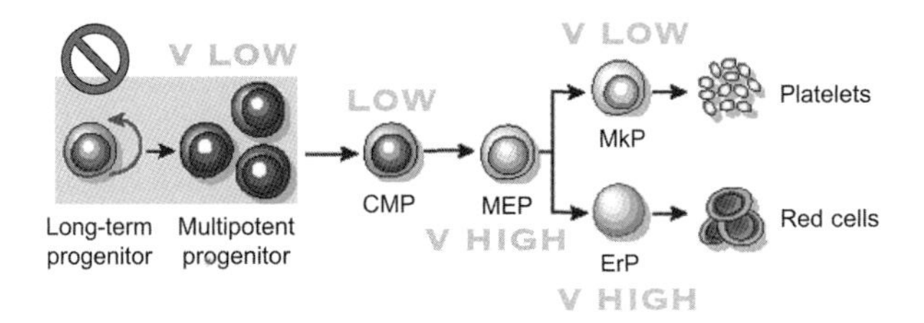

Fig. 4 EKLF expression during normal hematopoiesis. Cell populations from murine bone marrow were sorted and monitored for EKLF expression by a quantitative reverse transcription-polymerase chain reaction, leading to "very low," "low," "high," and "very high" categories based on their relative expression levels. *CMP* = common myeloid progenitor; *MEP* = megakaryocyte erythroid progenitor; *MkP* = megakaryocyte progenitor; *ErP* = erythroid progenitor. Shown are a subset of the total results (Frontelo et al. 2007)

difference in EKLF expression, with erythroid progenitors exhibiting an 80-fold greater level of expression than megakaryocyte progenitors. This demarcates EKLF as having significantly different properties from GATA1, FOG, GFi1b, and SCL—transcription factors whose presence are required for both erythroid and megakaryocytic expansion and differentiation. These data provide a third line of evidence that EKLF is expressed earlier in hematopoiesis and show that it is normally downregulated as MEPs differentiate down the megakaryocytic lineage although being retained at high levels in the erythroid lineage. This also explains the gain-of-function data, which can now be seen as having misregulated the normal shut-off of EKLF in megakaryocytes.

Relevance of EKLF Posttranslational Modifications

EKLF undergoes a range of functionally important posttranslational modifications that encompass phosphorylation (Ouyang et al. 1998), acetylation (Zhang and Bieker 1998), and ubiquitylation (Quadrini and Bieker 2006). Some of these have been shown to alter subsequent protein–protein interactions (Chen and Bieker 2004; Zhang et al. 2001). We noted that mammalian EKLF protein contains a conserved motif near its amino terminus that matches the consensus target site for sumoylation (Siatecka et al. 2007). As a result, we directly tested whether the modification occurs and found that EKLF is sumoylated at a single site within the conserved motif (K74 in the murine sequence), and that PIAS1 plays a critical role in this process. Mutation of this site affects EKLF's repression capability but has no discernible effect on its ability to activate a target promoter. Repression by the wild-type protein can be altered by co-expression of a dominant-negative Ubc9 or the SUMO-specific isopeptidase SSP. EKLF nuclear localization proceeds equivalently irrespective of its sumoylation status. Similar to the other modifications, sumoylated EKLF provides an efficient platform for its protein interactions, in this case with the Mi2ß subunit of the NuRD repression complex, providing a molecular explanation for its effects in vivo.

When considering the functional importance of EKLF sumoylation, we were fortunate to have obtained, by then, the data on EKLF's ability to repress megakaryopoiesis (summarized above). The data strongly suggested to us that they might provide a means to test whether the two observations are related. We did this in two ways. First, we established K562-derived cell lines that contain stable zinc-inducible EKLF constructs, one with wild type and the other with the sumoylation mutant (K74R). K562 cell lines are an erythroleukemic cell line that can be further directed toward erythropoiesis (fetal hemoglobin expression) by the addition of hemin, or redirected toward megakaryopoiesis by addition of the phorbol ester TPA. Importantly, these cells do not express endogenous EKLF. We found that induction of megakaryopoiesis was inhibited by the presence of wild-type EKLF compared to that seen in the presence of K74R EKLF, which exhibited little inhibition of the process. This suggested that the modification might play an important functional role in megakaryocyte repression.

To further support this idea in vivo during normal hematopoiesis, we cloned EKLF downstream of the megakaryocyte-specific platelet factor 4 (PF4) promoter and established transgenic lines from these DNA constructs. Lines that expressed the transgenes in the bone marrow were examined by cellular and molecular analyses. Already at harvest, bone marrow cells from transgenic mice expressing WT-EKLF displayed a threefold decrease in megakaryocyte cellularity in WT-EKLF transgenic bone marrow. This demonstrates that mis-expression of EKLF in the megakaryocyte lineage has a repressive effect on megakaryocyte formation, consistent with our analyses (see above). However, transgenic bone marrow from the K74R-EKLF line revealed no effect of mutant EKLF transgene expression on megakaryocyte cellularity. We also determined whether transgene expression also altered megakaryocyte colony formation and found a trend similar to that seen with the cell assay: WT-EKLF transgenic bone marrow contained fivefold less megakaryocyte colony-forming potential than the nontransgenic control, and the K74R-EKLF transgenic bone marrow was not affected. We concluded that availability of the SUMO modification site in EKLF is absolutely critical for it to exert its normal inhibition of megakaryopoiesis prior to red blood cell onset and provides the fourth line of evidence for its functional importance earlier in hematopoiesis.

Conclusions and Future Perspectives

These four lines of evidence suggesting that EKLF plays a functional role prior to erythropoiesis leads to a testable working model (Fig. 5). The model includes the results of studies (not presented) supporting the idea that Fli1, an ETS-related transcription factor that is also expressed in the MEP but critical for megakaryocyte differentiation, is negatively regulated by EKLF, providing a molecular basis for EKLF's ability to repress megakaryopoiesis (Frontelo et al. 2007).

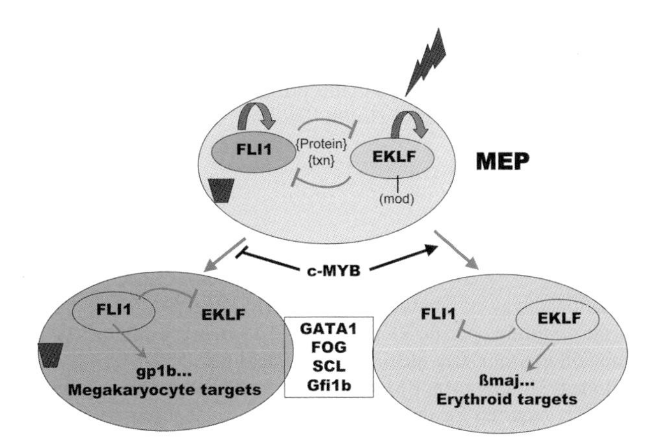

Fig. 5 Model of EKLF's role in erythroid and megakaryocytic decisions. See the text for details.

As a result, we postulate that within the MEP, EKLF and Fli1 can antagonize each other by protein or transcriptional inhibition mechanisms. Stochastic variation in levels, external influences (lightning bolt), or asymmetrically distributed molecules (trapezoid) can enable progeny to attain either an erythroid or a megakaryocyte fate. Also critical is the modification state (mod) of EKLF during these cellular decisions. After cell division, in the erythroid precursor cell EKLF completes the repression of Fli1 while activating downstream red blood cell targets, whereas the converse happens in megakaryocyte precursors as a result of Fli1 suppression of EKLF. Hence, EKLF gain- or loss-of-function leads to divergent effects on the bipotential decision by the MEP. At the same time, and in contrast to the actions of EKLF, other transcription factors (e.g., GATA1, FOG, SCL, Gfi1b) are positively required for both erythroid and megakaryocytic lineages. An interesting exception is c-myb, as hypomorphs have been shown to exhibit effects on MEP bipotential decisions similar to those of EKLF (Mukai et al. 2006).

Recent studies have lent further support for the concepts put forward in this chapter. For example, the role of EKLF in bipotential decisions within the MEP that lead to megakaryocytic repression and erythroid expansion have been supported by directly altering its levels within these progenitor cells and noting their downstream effects (Bouilloux et al. 2008). The importance of single EKLF posttranslational modifications, in this case acetylation of K288, has been shown to be critical for recruiting CBP and the subsequent modification of histone H3, opening the chromatin structure, and transcriptional activation of adult ß-globin (Sengupta et al. 2008). Finally, point mutations in human EKLF have been shown to lead to phenotypic variation of red blood cell antigen expression even in the heterozygous state (Bieker 2008; Singleton et al. 2008). Putting these ideas together leads to the exciting prediction that genetic mutation of EKLF, at sites critical for its protein modification, may in the future be shown to be a causative factor for a specific subset of hematopoietic aberration and disease.

Acknowledgments I particularly wish to recognize the contributions of the present laboratory members Felix Lohmann, Deepa Manwani, Mirka Siatecka, and Li Xue and our collaborator Holger Karsunky to the studies summarized in this presentation. The studies have been supported by funding from the National Institutes of Health (DK46865, DK48721, HL28381).

References

Adelman CA, Chattopadhyay S, Bieker JJ (2002) The BMP/BMPR/Smad pathway directs expression of the erythroid-specific EKLF and GATA1 transcription factors during embryoid body differentiation in serum-free media. Development 129: 539–549.

Armstrong JA, Bieker JJ, Emerson BM (1998) A SWI/SNF-related chromatin remodeling complex, E-RC1, is required for tissue-specific transcriptional regulation by EKLF in vitro. Cell 95: 93–104

Bieker JJ (2001) Kruppel-like factors: Three fingers in many pies. J Biol Chem 276: 34355–34358

Bieker JJ (2008) Blood group antigens reveal their maker. Blood 112: 1554–1555

Bouilloux F, Juban G, Cohet N et al (2008) EKLF restricts megakaryocytic differentiation at the benefit of erythrocytic differentiation. Blood 112: 576–584

Chen X, Bieker JJ (2001) Unanticipated repression function linked to erythroid Kruppel-like factor. Mol Cell Biol 21: 3118–3125

Chen X, Bieker JJ (2004) Stage-specific repression by the EKLF transcriptional activator. Mol Cell Biol 24: 10416–10424

Chen X, Reitman M, Bieker JJ (1998) Chromatin structure and transcriptional control elements of the erythroid Kruppel-like factor (EKLF) gene. J Biol Chem 273: 25031–25040

Crossley M, Tsang AP, Bieker JJ et al (1994) Regulation of the erythroid Kruppel-like factor (EKLF) gene promoter by the erythroid transcription factor GATA-1. J Biol Chem 269: 15440–15444

Donze D, Townes TM, Bieker JJ (1995) Role of Erythroid Krüppel-like Factor (EKLF) in human g- to ß-globin switching. J Biol Chem 270: 1955–1959

Drissen R, Palstra RJ, Gillemans N et al (2004) The active spatial organization of the beta-globin locus requires the transcription factor EKLF. Genes Dev 18: 2485–2490

Feng WC, Southwood CM, Bieker JJ (1994) Analyses of ß-thalassemia mutant DNA interactions with erythroid Krüppel-like factor (EKLF), an erythroid cell-specific transcription factor. J Biol Chem 269: 1493–1500

Frontelo P, Manwani D, Galdass M et al (2007) Novel role for EKLF in megakaryocyte lineage commitment. Blood 110: 3871–3880

Im H, Grass JA, Johnson KD et al (2005) Chromatin domain activation via GATA-1 utilization of a small subset of dispersed GATA motifs within a broad chromosomal region. Proc Natl Acad Sci U S A 102: 17065--17070

Kyba M, Perlingeiro RC, Daley GQ (2002) HoxB4 confers definitive lymphoid-myeloid engraftment potential on embryonic stem cell and yolk sac hematopoietic progenitors. Cell 109: 29-37

Lohmann F, Bieker JJ (2008) Activation of Eklf expression during hematopoiesis by Gata2 and Smad5 prior to erythroid commitment. Development 135: 2071–2082

Manwani D, Galdass M, Bieker JJ (2007) Altered regulation of beta-like globin genes by a redesigned erythroid transcription factor. Exp Hematol 35: 39–47

Miller IJ, Bieker JJ (1993) A novel, erythroid cell-specific murine transcription factor that binds to the CACCC element and is related to the Krüppel family of nuclear proteins. Mol Cell Biol 13: 2776–2786

Mukai HY, Motohashi H, Ohneda O et al (2006) Transgene insertion in proximity to the c-myb gene disrupts erythroid-megakaryocytic lineage bifurcation. Mol Cell Biol 26: 7953–7965

Nuez B, Michalovich D, Bygrave A et al (1995) Defective haematopoiesis in fetal liver resulting from inactivation of the EKLF gene. Nature (London) 375: 316–318

Ouyang L, Chen X, Bieker JJ (1998) Regulation of erythroid Kruppel-like factor (EKLF) transcriptional activity by phosphorylation of a protein kinase casein kinase II site within its interaction domain. J Biol Chem 273: 23019–23025

Perkins AC, Sharpe AH, Orkin SH (1995) Lethal ß-thalassemia in mice lacking the erythroid CACCC-transcription factor EKLF. Nature (London) 375: 318–322

Quadrini KJ, Bieker JJ (2006) EKLF/KLF1 is ubiquitinated in vivo and its stability is regulated by activation domain sequences through the 26S proteasome. FEBS Lett 580: 2285–2293

Sengupta T, Chen K, Milot E et al (2008) Acetylation of EKLF is essential for epigenetic modification and transcriptional activation of the {beta}-globin locus. Mol Cell Biol 28: 6160–6170

Siatecka M, Xue L, Bieker JJ (2007) Sumoylation of EKLF Promotes Transcriptional Repression and Is Involved in Inhibition of Megakaryopoiesis. Mol Cell Biol 27: 8547–8560

Singleton BK, Burton NM, Green C et al (2008) Mutations in EKLF/KLF1 form the molecular basis of the rare blood group In(Lu) phenotype. Blood 112: 2081–2088

Southwood CM, Downs KM, Bieker JJ (1996) Erythroid Kruppel-like Factor (EKLF) exhibits an early and sequentially localized pattern of expression during mammalian erythroid ontogeny. Devel Dyn 206: 248–259

Xue L, Chen X, Chang Y et al (2004) Regulatory elements of the EKLF gene that direct erythroid cell-specific expression during mammalian development. Blood 103: 4078-4083

Zhang W, Bieker JJ (1998) Acetylation and modulation of erythroid Kruppel-like factor (EKLF) activity by interaction with histone acetyltransferases. Proc Natl Acad Sci U S A 95: 9855–9860

Zhang W, Kadam S, Emerson BM et al (2001) Site-specific acetylation by p300 or CREB binding protein regulates erythroid Kruppel-like factor transcriptional activity via its interaction with the SWI-SNF complex. Mol Cell Biol 21: 2413–2422.

Chapter 7
Roles of Krüppel-like Factors in Lymphocytes

Kensuke Takada, Kristin A. Hogquist, and Stephen C. Jameson

Abstract Several family members of Krüppel-like factors (KLFs) are found in lymphocytes, and their expression is tightly regulated during development and differentiation. The related factors KLF2 and KLF4 have been suggested to promote lymphocyte "quiescence" by inducing withdrawal of cells from the cell cycle. Although the physiological role of KLF2 in cell cycle control in lymphocytes is currently unclear, it is potentially due to redundancy between related KLFs. On the other hand, there is growing evidence that individual KLFs regulate migration of lymphocytes (and other cells) during normal homeostasis of the immune system and in inflammatory situations. In addition to KLF2 and KLF4, the roles of KLF10 and KL13 in lymphocytes are briefly discussed.

Introduction

The immune system provides defense against diverse threats, including pathogenic microorganisms and cancer. Immunity depends on a complex process mediated by various cellular and humoral factors. Lymphocytes are especially important in the adaptive immune response, in which antigen-specific responses mediated by T cells and B cells can efficiently eliminate pathogens and tumors and lead to lifelong immunity. During the steady state, naïve lymphocytes with a wide range of specificities are maintained in a resting, "quiescent" state and recirculate between secondary lymphoid organs and the blood, the migration being dictated by various chemokines and adhesion molecules. It is within secondary lymphoid organs that encounters with foreign antigens occur, leading to activation and rapid proliferation of specific T and B cells. Upon successful elimination of the antigen, these specific

K.Takada (✉), K.A.Hogquist, and S.C. Jameson
Department of Laboratory Medicine and Pathology, Center for Immunology,
University of Minnesota, 312 Church Street SE, Minneapolis, MN 55455, USA

R. Nagai et al. (eds.), *The Biology of Krüppel-like Factors*,
DOI: 10.1007/978-4-431-87775-2_7, © Springer 2009

T and B cells are maintained in the memory pool, yielding enhanced responses to the subsequent reencounter with the same antigen. Broadly, T cells are comprised of helper and cytolytic cells, the former having various functions including supporting B cell responses and production of cytokines, and the latter being responsible for killing infected or transformed cells. Activated B cells produce antibodies specific for their target antigen which can, for example, neutralize viruses.

Lymphocyte development, migration, activation, and subsequent differentiation are tightly controlled by a number of transcription factors, including some Krüppel-like factor (KLF) family members (Kuo and Leiden 1999). KLFs are a subfamily of the zinc-finger transcription factors and bind to GC-rich DNA elements through a highly conserved DNA-binding domain containing three $cysteine_2$/$histidine_2$-type zinc finger motifs located at the carboxyl terminus (Haldar et al. 2007; Kaczynski et al. 2003; Suzuki et al. 2005; Turner and Crossley, 1999). The mammalian KLF family is composed of 17 family members (Haldar et al. 2007), and distinct functions within the family can be attributed to differences in the expression profile and the structure of N-terminal domains involved in transcriptional repression and activation (Turner and Crossley 1999). Considerable progress has been made over the past decade on understanding the role of KLFs in regulating lymphocyte "quiescence" and "migration."

Regulated Expression of KLF2 and KLF4 in Lymphocytes

KLF2 and KLF4 are both expressed in T and B cells, and the expression of these KLFs changes dynamically with the differentiation state of the lymphocytes, as discussed below.

In the T cell lineage, KLF2 is expressed in mature T cells developing in the thymus (Carlson et al. 2006; Kuo et al. 1997b; McCaughtry et al. 2007; Mick et al. 2004). KLF2 mRNA is not detected in the abundant immature $CD4^+8^+$ thymocyte stage but is strongly upregulated in mature $CD4^+$ and $CD8^+$ thymocytes and is expressed in peripheral naïve T cells. The factors that induce KLF2 expression during T-cell development are unclear. Studies in lymphocyte and endothelial cell lines have suggested that the MEF2 transcription factor promotes KLF2 expression, and MEF2D activity is induced via activation of the mitogen-activated protein kinase–extracellular signal-regulated kinase (MEK5–ERK5) pathway (Parmar et al. 2006; Sohn et al. 2005). Winoto and colleagues have argued that this pathway may be triggered by either the T-cell receptor (TCR) or interleukin-7 receptor (IL-7R), both of which are critical for T-cell development (Sohn et al. 2005). Studies using $ERK5^{-/-}$ T cells suggest this factor is not essential for KLF2 expression (our unpublished data), although the role of MEF2D in regulation of KLF2 expression warrants further investigation. Recent studies have revealed that the Forkhead Box O protein FOXO-1 can drive expression of KLF2 in T cells (Fabre et al. 2008) and of KLF4 in B cells (see below) (Fruman 2004; Sinclair et al. 2008)) and is under the control of phosphoinositide 3-kinase (PI3K) signaling.

Downregulation of KLF2 is also tightly regulated. In mature T cells, stimulation through the TCR leads to rapid, profound loss of KLF2 (Endrizzi and Jameson 2003; Kuo et al. 1997b; Schober et al. 1999). This is mediated through reduced transcription of the *KLF2* gene (Endrizzi and Jameson 2003; Kuo et al. 1997b) but also appears to involve active degradation of KLF2 protein (Endrizzi and Jameson 2003; Kuo et al. 1997b). Lingrel's group showed that KLF2 protein is targeted for ubiquitin-mediated degradation, involving the WW domain-containing protein 1 (WWP1) E3-ubiquitin ligase (Conkright et al. 2001; Zhang et al. 2004), although it is not clear if this same pathway regulates KLF2 in lymphocytes. Reexpression of KLF2 occurs with the transition from effector to memory phase in vitro and in vivo (Bai et al. 2007; Grayson et al. 2001; Schober et al. 1999). Cytokines can affect KLF2 reexpression, as shown by the ability of IL-7 and IL-15 to induce KLF2 expression in activated T cells, whereas IL-12, IL-4, and high-dose IL-2 all prevent KLF2 reexpression (Bai et al. 2007; Endrizzi and Jameson 2003; Schober et al. 1999). The differential effects of IL-2 and IL-15 can be attributed to the fact that high-dose IL-2 (but not IL-15) induces sustained signaling through PI3K (Cornish et al. 2006; Sinclair et al. 2008), which negatively regulates KLF2 expression (Fruman 2004; Sinclair et al. 2008). Indeed, two downstream targets of the PI3K pathway, the mammalian target of rapamycin (mTOR) and Akt, negatively regulate transcriptional regulation of KLF2 (Fabre et al. 2008; Fruman 2004; Sinclair et al. 2008).

KLF2 regulation in B cells has been analyzed less extensively, but microarray studies indicate that KLF2 is upregulated by pre-B cell receptor (BCR) signals during development but transcriptionally silenced following BCR signals in mature B cells (Glynne et al. 2000; Schuh et al. 2008), suggesting potentially similar general regulation pathways.

KLF4 is highly homologous to KLF2, belonging to the same subclass (Kaczynski et al. 2003), and this factor has been more extensively studied in B cells. During development in the bone marrow, KLF4 is expressed in pre-B-I and small pre-B-II stages at higher levels than in other stages including large pre-B-II cells (van Zelm et al. 2005). Mature naïve B cells express high levels of KLF4 compared to pro-B cells and total pre-B cells (Klaewsongkram et al. 2007). Transcription of the *KLF4* gene is under the direct regulation of FOXO transcription factors (including FOXO1 and FOXO3a) and δEF1, another transcription factor that partially shares the target genes with FOXOs (Yusuf et al. 2008). Upon activation, mRNA and protein are both immediately downregulated (Good and Tangye 2007; Kharas et al. 2007; Klaewsongkram et al. 2007; Yusuf et al. 2008) through the PI3K–Akt signaling pathway (Yusuf et al. 2008). Reexpression occurs in human memory B cells, but this level is evidently lower than that in naïve B cells (Good and Tangye, 2007). In contrast to B cells, in-depth analysis of KLF4 regulation in T cells has not been reported.

Overall, there appear to be striking parallels between general expression patterns and regulation mechanisms for KLF2 and KLF4 in T and B lymphocytes, although current studies do not reveal whether there is truly coordinated regulation of both factors in the two cell types.

Function of KLF2 and KLF4 in Lymphocytes: Quiescence

Cellular quiescence is a state where cells exit the cell cycle from the G_1 into the G_0 phase (Yusuf and Fruman, 2003). Naïve lymphocytes are in a quiescent state until an encounter with specific antigens, and they are characterized by small size, lack of spontaneous proliferation, diminished metabolic rate, and resistance to apoptosis. Quiescence is evidently not a default state but, rather, is actively controlled (Kuo et al. 1997b; Yusuf and Fruman 2003; Yusuf et al. 2008). As discussed above, KLF2 and KLF4 are expressed in resting lymphocytes (naïve and memory cells), but both factors are downregulated rapidly after lymphocyte activation, suggesting that they may be quiescence factors, as previously discussed (Kuo et al. 1997b; Yusuf et al. 2008).

Analysis of KLF2-deficient mice initially suggested a key role in T-cell quiescence. Kuo et al. generated chimeras from *KLF2*$^{-/-}$ embryonic stem (ES) cells in recombinant-activating gene *(RAG) 2*$^{-/-}$ blastocysts (to bypass the embryonic lethality associated with KLF2 deficiency) and observed that *KLF2*$^{-/-}$ T cells developed ostensibly normally in the thymus but were reduced by more than 90% in peripheral lymphoid tissues (Kuo et al. 1997b). The few *KLF2*$^{-/-}$ T cells in peripheral sites had an activated phenotype (CD44^{hi}CD69^{hi}CD62L^{lo}) and were highly sensitive to Fas ligand-induced apoptosis (Kuo et al. 1997b). Based on these observations, a model was suggested that mature T cells undergo spontaneous activation and Fas-mediated death in the absence of KLF2, implying the requirement of KLF2 for T-cell quiescence (Kuo and Leiden 1999; Kuo et al. 1997a). This concept was reinforced by overexpression of KLF2 in the Jurkat T-cell line, which caused a halt in the autonomous proliferation of Jurkat cells, decreased cell size, and reduced metabolic activity (Buckley et al. 2001; Haaland et al. 2005; Shie et al. 2000; Wu and Lingrel 2004; Yusuf et al. 2008; Zhang et al. 2000). An impaired cell cycle was associated with reduced Myc mRNA levels (Buckley et al. 2001; Haaland et al. 2005), and related studies showed induction of p21$^{WAF1/CIP1}$ with KLF2 expression (Buckley et al. 2001; Shie et al. 2000; Wu and Lingrel 2004; Yusuf et al, 2008; Zhang et al. 2000). This expression pattern fits well with the concept that KLF2 expression could lead to cell cycle withdrawal.

Broadly similar functions have been proposed for KLF4 in the B-cell pool: Forced expression of KLF4 in activated B cells by retroviral transduction leads to cell cycle arrest at the G_1 phase (Yusuf et al. 2008). This is accompanied by increased expression of p21$^{WAF1/CIP1}$ and the decreased expression of Myc and cyclin D2, in accordance with the findings in T cells and nonlymphoid cells that KLF2 or KLF4 controls the expression of these cell cycle regulators (Buckley et al. 2001; Shie et al. 2000; Wu and Lingrel 2004; Yusuf et al. 2008; Zhang et al. 2000). Similarly, ectopic expression of KLF4 (or KLF9) in naïve and memory human B cells reduced the number of proliferating cells (Good and Tangye 2007). Concurrently, enforced expression of KLF4 induces apoptosis in B-cell and T-cell leukemia cells (Kharas et al. 2007; Yasunaga et al. 2004). Similarly, in nonlymphoid cells, KLF4 and KLF6 have been shown to induce p21$^{WAF1/CIP1}$ expression, leading to cell cycle arrest, as a key component of the tumor suppressive activity of these KLFs (Narla et al. 2001; Rowland and Peeper 2006; Zhang et al. 2000). On the other hand, *KLF4*'s function in the cell cycle may be context-dependent, as it can act as an oncogene in $p21^{WAF1/CIP1}$-deficient cells (Rowland et al. 2005; Rowland and Peeper 2006). Hence, it may be hazardous to assume that expression levels of these KLFs allow prediction of cell cycle control, especially in cancer.

The most compelling data indicating a role for KLF2 and KLF4 in lymphocyte quiescence have come from overexpression studies. Such experiments suggest that these KLFs are sufficient for restraining cell cycle progression in lymphocytes, but it is less clear whether physiological levels of KLF2 and KLF4 have these roles. As is discussed below, the changes in the T-cell pool observed in $KLF2^{-/-}$ animals can be explained by altered thymic egress and T-cell trafficking rather than compromised quiescence regulation. Indeed, we found that activation, proliferation, and differentiation of KLF2-deficient T cells is quite normal, suggesting that the lack of KLF2 regulation does not lead to drastic changes in cell cycle control (our unpublished data). Likewise, whereas ectopic expression of KLF4 strongly restrains the cell cycle in activated B cells, KLF4 deficiency has minimal effects on B-cell development, survival, or functional reactivity (Klaewsongkram et al. 2007; Yusuf et al. 2008). In one report no difference was observed in cell cycle progression of stimulated WT and $KLF4^{-/-}$ B cells (Yusuf et al. 2008), whereas another study (using the same model system) reported activated $KLF4^{-/-}$ B cells had slightly impaired proliferation after BCR stimulation, showing a mild arrest of the G_1 to S phase transition, and decreased cyclin D2 expression (Klaewsongkram et al. 2007). Such data suggest that KLF4 has a modest role in promoting (rather than restraining) the cell cycle.

Hence current data make it difficult to determine whether physiological levels of KLF2 and KLF4 are relevant to the regulation of quiescence. Given the fact that there are overlapping patterns of expression for these two factors, and that they are highly homologous, there is a strong likelihood for some functional redundancy. Lymphocytes deficient in both factors are required to resolve this issue.

Forced expression of KLF2 and KLF4 has also been reported to moderately induce apoptotic death in T and B cells, respectively (Buckley et al. 2001; Yusuf et al. 2008), although once again it is unclear whether this reflects the role of KLF2 and KLF4 at physiological expression levels.

Control of Immune Cell Migration

Mature thymocytes leave the thymus through the blood and populate the peripheral T-cell pool. Similarly, immature B cells exit the bone marrow and undergo final maturation in the periphery. At steady state, naïve B and T cells continually circulate through secondary lymphoid organs (SLOs), blood, and lymph. These cellular dynamics allow immune cells to sample numerous SLOs in their search for foreign antigens. Such trafficking patterns require recognition of various chemokines and adhesion molecules. For instance, T cells interact with high endothelial venules using CD62L/L-selectin and CCR7 to enter the lymph nodes (von Andrian and Mempel 2003). For entry to the mesenteric lymph nodes and Peyer's patches, β7 integrin is additionally required (von Andrian and Mempel 2003). Egress from lymphoid tissues is also regulated, and the sphingosine-1-phosphate receptor $S1P_1$ plays a critical role in permitting T-cell exit from the thymus (after T-cell development) and from SLOs (during lymphocyte recirculation) (Cyster 2005).

Carlson et al. analyzed *KLF2*$^{-/-}$ fetal liver chimeras (Carlson et al. 2006). Consistent with the finding by Kuo et al. in *KLF2*$^{-/-}$*RAG2*$^{-/-}$ chimeras (Kuo et al. 1997b), there were few T cells in the periphery despite superficially normal thymocyte development. However, in contrast to the previously suggested model in which *KLF2*$^{-/-}$ T cells spontaneously underwent activation-induced apoptosis, *KLF2*$^{-/-}$ thymocytes from fetal liver chimeras survived normally after adoptive transfer and showed an abnormal distribution, being absent in the blood and lymph nodes and accumulating in the spleen. Moreover, intrathymic injection of biotin demonstrated the impaired emigration of *KLF2*$^{-/-}$ thymocytes into the periphery, corresponding to the observation that $CD4^+8^-$ and $CD4^-8^+$ mature thymocytes accumulated in the thymus of *KLF2*$^{-/-}$ fetal liver chimeras. These trafficking abnormalities of *KLF2*$^{-/-}$ T cells were associated with defective expression of CD62L, CCR7, β7 integrin, and $S1P_1$, implying that the expression of these molecules is regulated by KLF2. Indeed, direct activation of promoters of genes encoding $S1P_1$ and CD62L by KLF2 was demonstrated in two studies (Bai et al. 2007; Carlson et al. 2006).

KLF2 has been shown to be reexpressed in the memory stage of T cells (Grayson et al. 2001; Schober et al. 1999). Memory T cells are divided into two major populations depending on their trafficking properties (Sallusto et al. 2004). CD62L and CCR7, whose expressions are affected in KLF2-deficient T cells, are commonly used as markers for this classification, distinguishing $CD62L^-CCR7^-$ effector memory (EM) and $CD62L^+CCR7^+$ central memory (CM) T cells (Carlson et al. 2006; Sallusto et al. 2004). Hence, one might expect a potential role for KLF2 in EM and CM T cell differentiation. $CD4^+$ T cell microarray analysis showed higher expression of KLF2 mRNA in CM cells than in EM cells in both humans (Riou et al. 2007) and the mouse (M.K. Jenkins, personal communication). In contrast, other studies of $CD8^+$ EM and CM T cells did not reveal extensive changes in KLF2 transcripts (Bai et al. 2007; our unpublished observations), although assessment of KLF2 protein expression in these subsets has not been reported.

Sedbza et al. employed conditional gene disruption by Vav-Cre transgene and floxed *KLF2* allele (Sebzda et al. 2008). These investigators replicated the findings reported above but suggested that the phenotype was not due to blocked thymic emigration; rather, it reflected misdirected T-cell trafficking into nonlymphoid organs, such as the liver. This abnormal migration was attributed to the increased expression of multiple chemokine receptors, including CXCR3 and CCR5, with the suggestion that KLF2 might suppress inflammatory chemokine receptor expression in addition to induction of homeostatic homing molecules (Sebzda et al. 2008). However, other studies involving T cell-specific KLF2 deficiency (using CD4-Cre) did not observe trafficking to nonlymphoid sites (our unpublished data).

A potential resolution of this issue arises from the observation that deletion of KLF2 in hematopoietic cells leads to severe alterations in the composition of the T-cell pool, with most T-cell subsets being reduced in number but a population of $CD4^+$ TCRγδ T cells being increased in number and frequency (our unpublished data). Clearly, further studies are required to determine how KLF2 regulates differentiation and trafficking of discreet T-cell subsets.

In the B-cell lineage, the significance of KLF2 for migratory regulation is much less clear. Although $S1P_1$ is also required for naïve B-cell trafficking (Matloubian et al. 2004), *KLF2*$^{-/-}$ naïve B-cell trafficking is not notably compromised (Carlson et al. 2006; Kuo et al. 1997b; our unpublished observations), indicating distinct regulation of $S1P_1$ in naïve B and T cells. On the other hand, the amount of $S1P_1$ in immunoglobulin G (IgG)-secreting plasma cells and plasmablasts determines whether they reside in the splenic red pulp or migrate to bone marrow through the blood (Kabashima et al. 2006). Expression of $S1P_1$ and KLF2 are both higher In the latter population than in the former population, suggesting a possible role of KLF2 in the regulation of plasma cell positioning and differentiation (Kabashima et al. 2006). In addition, pre-B-cell receptor signals strongly upregulate the expression of KLF2 and $S1P_1$, suggesting a potential role of KLF2 in controlling migration during B-cell precursor development (Schuh et al. 2008).

Hence, at present, the significance of KLF2 in B-cell trafficking and the relevance (if any) of KLF4 in T- or B-cell trafficking await further analysis. Given the similar functions of KLF2 and KLF4 in other systems (e.g., their impact on lymphocyte quiescence discussed above and their role in endothelial development as discussed by Lloyd in Chapter 9), we consider it likely that there is some redundancy between these factors.

KLF13 in T Cells

In contrast to the downregulation of KLF2 and KLF4 in activated lymphocytes, KLF13 expression is strongly induced by T-cell activation. KLF13 mRNA levels are similar in naïve and activated T cells (Song et al. 1999), but protein is abundantly detected only during the late stage of T-cell activation. This is due to translational regulation through 5′-untranslated regions of the transcripts that dictate expression of KLF13 protein (Nikolcheva et al. 2002; Song et al. 1999). Translation of KLF13 is dependent on a translational initiation factor, eIF4F, which mediates recruitment of ribosomes to mRNA; its activity is under the control of mitogen-activated protein kinase (MAPK) signaling through p38 and ERK-1/2 and PI3K-mTOR signaling (Nikolcheva et al. 2002). Furthermore, in activated T cells, KLF13 activity is post-translationally downregulated by phosphorylation (Song et al. 1999). Involvement of PRP4, a MAPK family member, has been demonstrated (Huang et al. 2007).

KLF13 was originally identified as a factor that regulates the expression of CCL5/RANTES, a chemokine produced at the late stage of T-cell activation (Song et al. 1999). CCL5 mediates the migration of a wide range of immune cells, including

T cells, monocytes, eosinophils, basophils, and natural killer (NK) cells (Song et al. 1999). The molecular mechanism of CCL5 expression that is driven by KLF13 was investigated in detail by Krensky and colleagues (Ahn et al. 2007; Song et al. 1999, 2002). Reporter gene assays in the Jurkat T-cell line showed that exogenous KLF13 recognizes the CTCCC sequence in the proximal promoter region of the *CCL5* gene and induces its expression (Song et al. 1999, 2002). Expression of CCL5 following activation was suppressed in primary T cells from KLF13-deficient mice and in normal T cells in which KLF13 was knocked down by small interfering RNA (Ahn et al. 2007; Zhou et al. 2007). Binding of KLF13 to the *CCL5* promoter was also demonstrated, indicating a physiological interaction of KLF13 and *CCL5* gene in activated T cells (Ahn et al. 2007). Factors associated with KLF13 for chromatin remodeling were also identified in activated T cells. Soon after activation, MAPK such as Nemo-like kinase binds to KLF13 at the *CCL5* promoter and phosphorylates the near by histones. This enables p300/cyclic AMP response element protein (CBP) and p300/CBP-associated factor to accetylate the histones at the later stage of activation, which is followed by the recruitment of an ATPase involved in chromatin remodeling, Brahma-related gene 1, to the promoter. These events recruit polymerase II to the adjacent TATA box of *CCL5* promoter, leading to transcriptional activation.

In KLF13-deficient mice, abnormalities are observed in multiple stages of lymphocyte development. They include the partial arrest of transition from $CD4^{+}8^{+}$ to $CD4^{+}8^{-}$ thymocytes and from large to small pre-B-II cells, implying that KLF13 regulates lymphocyte differentiation in the downstream of TCR and (pre-) BCR (Outram et al. 2008). However, the target genes of KLF13 in these processes remain to be identified.

KLF13 has also been reported to promote apoptosis. KLF13-deficient mice exhibit enlarged thymi and spleens because of decreased apoptosis of T cells (Zhou et al. 2007). This involves elevated expression of Bcl-X_L, an antiapoptotic factor. KLF13 can bind to the promoter of *Bcl-X_L* and decrease its activity (Zhou et al. 2007). However, KLF13-deficient mice do not show signs of either tumorigenesis or autoimmunity. Effects of KLF13 on lymphocyte cell cycle regulation have not been extensively investigated.

KLF10 in T Cells

KLF10 is induced by transforming growth factor-β (TGF-β) signaling—KLF10's alternative name, TIEG1, stands for TGF-β induced early gene 1—and has recently been found to play a key role in induction of a regulatory T-cell population (inducible Treg). T-cell stimulation in the presence of TGF-β can induce T-cell expression of the transcription factor Foxp3, which dictates a Treg differentiation pathway. KLF10 is a key component in efficient induction of Foxp3 by this mechanism; and to promote Foxp3 expression KLF10 must be monoubiquitinated by the E3 ligase Itch (Venuprasad et al. 2008). The KLF10 knockout shows defective production of inducible Treg. The ability of KLF10 to promote Foxp3 transcription directly is in contrast with the typical activity of this factor as a transcriptional repressor (Kaczynski et al. 2003). A role for KLF10 in regulating differentiation or function of other T-cell subsets has not been reported, but it will be of interest to explore its role in the Th17 T cells. Th17 T cells play an important role in autoimmune diseases and in the control of certain extracellular pathogens (Bettelli et al. 2007; Dong 2008). Like Treg, their differentiation requires TGF-β signals; but in the case of Th17 cells, these are accompanied by signals through IL-6R.

Conclusion

Although several KLF family members are expressed in resting lymphocytes, the closely related factors KLF2 and KLF4 have been best studied. Both factors can induce lymphocyte quiescence, but the physiological significance of this is still unclear and is complicated by likely functional redundancy between KLF2 and KLF4. On the other hand, KLF2 appears to have a unique role in controlling migration of T cells via regulating expression of key trafficking molecules. In contrast, the factors KLF10 and KLF13 are important in postactivation T cells, contributing to functional T-cell differentiation. As in other tissues, the compensatory, combinatorial, or opposing roles of distinct KLF family members expressed in lymphocytes must be taken into consideration to understand the role played by these factors in regulating lymphocyte homeostasis and function.

References

Ahn, Y.T., Huang, B., McPherson, L., Clayberger, C., and Krensky, A.M. (2007). Dynamic interplay of transcriptional machinery and chromatin regulates "late" expression of the chemokine RANTES in T lymphocytes. Molecular and Cellular Biology 27, 253–266.

Bai, A., Hu, H., Yeung, M., and Chen, J. (2007). Kruppel-like factor 2 controls T cell trafficking by activating L-selectin (CD62L) and sphingosine-1-phosphate receptor 1 transcription. J Immunol 178, 7632–7639.

Bettelli, E., Korn, T., and Kuchroo, V.K. (2007). Th17: the third member of the effector T cell trilogy. Current Opinion in Immunology 19, 652–657.

Buckley, A.F., Kuo, C.T., and Leiden, J.M. (2001). Transcription factor LKLF is sufficient to program T cell quiescence via a c-Myc–dependent pathway. Nature Immunology 2, 698–704.

Carlson, C.M., Endrizzi, B.T., Wu, J., Ding, X., Weinreich, M.A., Walsh, E.R., Wani, M.A., Lingrel, J.B., Hogquist, K.A., and Jameson, S.C. (2006). Kruppel-like factor 2 regulates thymocyte and T-cell migration. Nature 442, 299–302.

Conkright, M.D., Wani, M.A., and Lingrel, J.B. (2001). Lung Kruppel-like factor contains an autoinhibitory domain that regulates its transcriptional activation by binding WWP1, an E3 ubiquitin ligase. J Biol Chem 276, 29299–29306.

Cornish, G.H., Sinclair, L.V., and Cantrell, D.A. (2006). Differential regulation of T-cell growth by IL-2 and IL-15. Blood 108, 600–608.

Cyster, J.G. (2005). Chemokines, sphingosine-1-phosphate, and cell migration in secondary lymphoid organs. Annu Rev Immunol 23, 127–159.

Dong, C. (2008). TH17 cells in development: an updated view of their molecular identity and genetic programming. Nature reviews 8, 337–348.

Endrizzi, B.T., and Jameson, S.C. (2003). Differential role for IL-7 in inducing lung Kruppel-like factor (Kruppel-like factor 2) expression by naive versus activated T cells. Int Immunol 15, 1341–1348.

Fabre, S., Carrette, F., Chen, J., Lang, V., Semichon, M., Denoyelle, C., Lazar, V., Cagnard, N., Dubart-Kupperschmitt, A., Mangeney, M., *et al.* (2008). Forkhead Box O1 Regulates L-Selectin and a Network of Human T-Cell Homing Molecules Downstream of Phosphatidylinositol 3-Kinase. J Immunol IN PRESS.

Fruman, D.A. (2004). Phosphoinositide 3-kinase and its targets in B-cell and T-cell signaling. Current opinion in immunology 16, 314–320.

Glynne, R., Ghandour, G., Rayner, J., Mack, D.H., and Goodnow, C.C. (2000). B-lymphocyte quiescence, tolerance and activation as viewed by global gene expression profiling on microarrays. Immunol Rev 176, 216–246.

Good, K.L., and Tangye, S.G. (2007). Decreased expression of Kruppel-like factors in memory B cells induces the rapid response typical of secondary antibody responses. Proc Natl Acad Sci U S A 104, 13420–13425.

Grayson, J.M., Murali-Krishna, K., Altman, J.D., and Ahmed, R. (2001). Gene expression in antigen-specific CD8+ T cells during viral infection. J Immunol 166, 795–799.

Haaland, R.E., Yu, W., and Rice, A.P. (2005). Identification of LKLF-regulated genes in quiescent CD4+ T lymphocytes. Mol Immunol 42, 627–641.

Haldar, S.M., Ibrahim, O.A., and Jain, M.K. (2007). Kruppel-like Factors (KLFs) in muscle biology. J Mol Cell Cardiol 43, 1–10.

Huang, B., Ahn, Y.T., McPherson, L., Clayberger, C., and Krensky, A.M. (2007). Interaction of PRP4 with Kruppel-like factor 13 regulates CCL5 transcription. J Immunol 178, 7081–7087.

Kabashima, K., Haynes, N.M., Xu, Y., Nutt, S.L., Allende, M.L., Proia, R.L., and Cyster, J.G. (2006). Plasma cell S1P1 expression determines secondary lymphoid organ retention versus bone marrow tropism. The Journal of Experimental Medicine 203, 2683–2690.

Kaczynski, J., Cook, T., and Urrutia, R. (2003). Sp1- and Kruppel-like transcription factors. Genome Biol 4, 206.

Kharas, M.G., Yusuf, I., Scarfone, V.M., Yang, V.W., Segre, J.A., Huettner, C.S., and Fruman, D.A. (2007). KLF4 suppresses transformation of pre-B cells by ABL oncogenes. Blood 109, 747–755.

Klaewsongkram, J., Yang, Y., Golech, S., Katz, J., Kaestner, K.H., and Weng, N.P. (2007). Kruppel-like factor 4 regulates B cell number and activation-induced B cell proliferation. J Immunol 179, 4679–4684.

Kuo, C.T., and Leiden, J.M. (1999). Transcriptional regulation of T lymphocyte development and function. Annu Rev Immunol 17, 149–187.

Kuo, C.T., Veselits, M.L., Barton, K.P., Lu, M.M., Clendenin, C., and Leiden, J.M. (1997a). The LKLF transcription factor is required for normal tunica media formation and blood vessel stabilization during murine embryogenesis. Genes Dev 11, 2996–3006.

Kuo, C.T., Veselits, M.L., and Leiden, J.M. (1997b). LKLF: A transcriptional regulator of single-positive T cell quiescence and survival. Science 277, 1986–1990.

Matloubian, M., Lo, C.G., Cinamon, G., Lesneski, M.J., Xu, Y., Brinkmann, V., Allende, M.L., Proia, R.L., and Cyster, J.G. (2004). Lymphocyte egress from thymus and peripheral lymphoid organs is dependent on S1P receptor 1. Nature 427, 355–360.

McCaughtry, T.M., Wilken, M.S., and Hogquist, K.A. (2007). Thymic emigration revisited. The Journal of Experimental Medicine 204, 2513–2520.

Mick, V.E., Starr, T.K., McCaughtry, T.M., McNeil, L.K., and Hogquist, K.A. (2004). The regulated expression of a diverse set of genes during thymocyte positive selection in vivo. J Immunol 173, 5434–5444.

Narla, G., Heath, K.E., Reeves, H.L., Li, D., Giono, L.E., Kimmelman, A.C., Glucksman, M.J., Narla, J., Eng, F.J., Chan, A.M., *et al.* (2001). KLF6, a candidate tumor suppressor gene mutated in prostate cancer. Science 294, 2563–2566.

Nikolcheva, T., Pyronnet, S., Chou, S.Y., Sonenberg, N., Song, A., Clayberger, C., and Krensky, A.M. (2002). A translational rheostat for RFLAT-1 regulates RANTES expression in T lymphocytes. The Journal of Clinical Investigation 110, 119–126.

Outram, S.V., Gordon, A.R., Hager-Theodorides, A.L., Metcalfe, J., Crompton, T., and Kemp, P. (2008). KLF13 influences multiple stages of both B and T cell development. Cell Cycle 7, 2047–2055.

Parmar, K.M., Larman, H.B., Dai, G., Zhang, Y., Wang, E.T., Moorthy, S.N., Kratz, J.R., Lin, Z., Jain, M.K., Gimbrone, M.A., Jr., and Garcia-Cardena, G. (2006). Integration of flow-dependent endothelial phenotypes by Kruppel-like factor 2. The Journal of Clinical Investigation 116, 49–58.

Riou, C., Yassine-Diab, B., Van grevenynghe, J., Somogyi, R., Greller, L.D., Gagnon, D., Gimmig, S., Wilkinson, P., Shi, Y., Cameron, M.J., *et al.* (2007). Convergence of TCR and cytokine signaling leads to FOXO3a phosphorylation and drives the survival of CD4+ central memory T cells. The Journal of Experimental Medicine 204, 79–91.

Rowland, B.D., Bernards, R., and Peeper, D.S. (2005). The KLF4 tumour suppressor is a transcriptional repressor of p53 that acts as a context-dependent oncogene. Nat Cell Biol 7, 1074–1082.

Rowland, B.D., and Peeper, D.S. (2006). KLF4, p21 and context-dependent opposing forces in cancer. Nat Rev Cancer 6, 11–23.

Sallusto, F., Geginat, J., and Lanzavecchia, A. (2004). Central memory and effector memory T cell subsets: function, generation, and maintenance. Annu Rev Immunol 22, 745–763.

Schober, S.L., Kuo, C.T., Schluns, K.S., Lefrancois, L., Leiden, J.M., and Jameson, S.C. (1999). Expression of the transcription factor lung Kruppel-like factor is regulated by cytokines and correlates with survival of memory T cells in vitro and in vivo. J Immunol 163, 3662–3667.

Schuh, W., Meister, S., Herrmann, K., Bradl, H., and Jack, H.M. (2008). Transcriptome analysis in primary B lymphoid precursors following induction of the pre-B cell receptor. Mol Immunol 45, 362–375.

Sebzda, E., Zou, Z., Lee, J.S., Wang, T., and Kahn, M.L. (2008). Transcription factor KLF2 regulates the migration of naive T cells by restricting chemokine receptor expression patterns. Nature Immunology 9, 292–300.

Shie, J.L., Chen, Z.Y., Fu, M., Pestell, R.G., and Tseng, C.C. (2000). Gut-enriched Kruppel-like factor represses cyclin D1 promoter activity through Sp1 motif. Nucleic Acids Res 28, 2969–2976.

Sinclair, L.V., Finlay, D., Feijoo, C., Cornish, G.H., Gray, A., Ager, A., Okkenhaug, K., Hagenbeek, T.J., Spits, H., and Cantrell, D.A. (2008). Phosphatidylinositol-3-OH kinase and nutrient-sensing mTOR pathways control T lymphocyte trafficking. Nature Immunology 9, 513–521.

Sohn, S.J., Li, D., Lee, L.K., and Winoto, A. (2005). Transcriptional regulation of tissue-specific genes by the ERK5 mitogen-activated protein kinase. Molecular and Cellular Biology 25, 8553–8566.

Song, A., Chen, Y.F., Thamatrakoln, K., Storm, T.A., and Krensky, A.M. (1999). RFLAT-1: a new zinc finger transcription factor that activates RANTES gene expression in T lymphocytes. Immunity 10, 93–103.

Song, A., Patel, A., Thamatrakoln, K., Liu, C., Feng, D., Clayberger, C., and Krensky, A.M. (2002). Functional domains and DNA-binding sequences of RFLAT-1/KLF13, a Kruppel-like transcription factor of activated T lymphocytes. J Biol Chem 277, 30055–30065.

Suzuki, T., Aizawa, K., Matsumura, T., and Nagai, R. (2005). Vascular implications of the Kruppel-like family of transcription factors. Arterioscler Thromb Vasc Biol 25, 1135–1141.

Turner, J., and Crossley, M. (1999). Mammalian Kruppel-like transcription factors: more than just a pretty finger. Trends Biochem Sci 24, 236–240.

van Zelm, M.C., van der Burg, M., de Ridder, D., Barendregt, B.H., de Haas, E.F., Reinders, M.J., Lankester, A.C., Revesz, T., Staal, F.J., and van Dongen, J.J. (2005). Ig gene rearrangement steps are initiated in early human precursor B cell subsets and correlate with specific transcription factor expression. J Immunol 175, 5912–5922.

Venuprasad, K., Huang, H., Harada, Y., Elly, C., Subramaniam, M., Spelsberg, T., Su, J., and Liu, Y.C. (2008). The E3 ubiquitin ligase Itch regulates expression of transcription factor Foxp3 and airway inflammation by enhancing the function of transcription factor TIEG1. Nature Immunology 9, 245–253.

von Andrian, U.H., and Mempel, T.R. (2003). Homing and cellular traffic in lymph nodes. Nature Reviews 3, 867–878.

Wu, J., and Lingrel, J.B. (2004). KLF2 inhibits Jurkat T leukemia cell growth via upregulation of cyclin-dependent kinase inhibitor p21WAF1/CIP1. Oncogene 23, 8088–8096.

Yasunaga, J., Taniguchi, Y., Nosaka, K., Yoshida, M., Satou, Y., Sakai, T., Mitsuya, H., and Matsuoka, M. (2004). Identification of aberrantly methylated genes in association with adult T-cell leukemia. Cancer Res 64, 6002–6009.

Yusuf, I., and Fruman, D.A. (2003). Regulation of quiescence in lymphocytes. Trends Immunol 24, 380–386.

Yusuf, I., Kharas, M.G., Chen, J., Peralta, R.Q., Maruniak, A., Sareen, P., Yang, V.W., Kaestner, K.H., and Fruman, D.A. (2008). KLF4 is a FOXO target gene that suppresses B cell proliferation. Int Immunol 20, 671–681.

Zhang, W., Geiman, D.E., Shields, J.M., Dang, D.T., Mahatan, C.S., Kaestner, K.H., Biggs, J.R., Kraft, A.S., and Yang, V.W. (2000). The gut-enriched Kruppel-like factor (Kruppel-like factor 4) mediates the transactivating effect of p53 on the p21WAF1/Cip1 promoter. J Biol Chem 275, 18391–18398.

Zhang, X., Srinivasan, S.V., and Lingrel, J.B. (2004). WWP1-dependent ubiquitination and degradation of the lung Kruppel-like factor, KLF2. Biochem Biophys Res Commun 316, 139–148.

Zhou, M., McPherson, L., Feng, D., Song, A., Dong, C., Lyu, S.C., Zhou, L., Shi, X., Ahn, Y.T., Wang, D., *et al.* (2007). Kruppel-like transcription factor 13 regulates T lymphocyte survival in vivo. J Immunol 178, 5496–5504.

Chapter 8
Krüppel-like Factors in Gastrointestinal Tract Development and Differentiation

Marie-Pier Tétreault and Jonathan P. Katz

Abstract A number of KLF family members are known to play important roles in the regulation of proliferation, differentiation, and development of the gastrointestinal tract. Of these, KLF4 (previously known as GKLF or EZF) and KLF5 (previously known as IKLF or BTEB2) have been the most extensively studied. In this chapter, we review the expression patterns and established functions for KLF family members in the gastrointestinal tract and offer insight into possible future areas of investigation of the KLFs in gastrointestinal differentiation and development.

Development and Differentiation of the Gastrointestinal Tract

In the developing embryo, changes leading to the origins of the organs of the gastrointestinal tract are first detected during the period of gastrulation (Katz and Wu 2004; Lebenthal 1989; Moore and Persaud 1998). The gut tube develops from two invagination events (one at the anterior end and the other at the posterior end) that allow incorporation of the endoderm into the body cavity. These invaginations elongate in the endoderm and fuse in the midline of the embryo to form an elongated tube. Beginning at about the fourth week of human development (from embryonic day 7.5 to 9.5 in mice), the endoderm is internalized as the embryo folds to become the primitive gut (Hogan and Zaret 2002; Wells and Melton 1999). The primitive gut can be divided into three distinct parts during fetal life: foregut, midgut, and hindgut. Each segment of the primitive gut contributes to different components of the gastrointestinal tract. The foregut gives rise to the esophagus, stomach, proximal half of the duodenum, liver, and pancreas. The midgut forms the distal half of the duodenum, as well as the jejunum, ileum, cecum, appendix, ascending colon,

M-P. Tétreault and J.P. Katz (✉)
Division of Gastroenterology, Department of Medicine, University of Pennsylvania School of Medicine, 600 Clinical Research Building, 415 Curie Boulevard, Philadelphia, PA 19104, USA

R. Nagai et al. (eds.), *The Biology of Krüppel-like Factors*,
DOI: 10.1007/978-4-431-87775-2_8, © Springer 2009

and parts of the transverse colon. The hindgut develops into the transverse colon, descending colon, sigmoid colon, and rectum down to the anorectal line. Whereas the gut epithelium originates predominantly from the endoderm, the smooth muscle layers and connective tissue surrounding the epithelium are derived from the mesoderm. Complex interactions occurring between the gut endoderm and different mesenchymal tissues lead to the morphogenesis and differentiation of the esophagus, stomach, small intestine, and colon (Roberts 2000).

Esophageal Development and Differentiation

The esophagus has its origins in the foregut immediately caudal to the primordial pharynx, beginning at around E9.5 in mice and the fourth week of human embryonic development (Katz and Wu 2004; Kaufman 1995; Lebenthal 1989; Moore and Persaud 1998; Que et al. 2006). Initially, a laryngotracheal diverticulum develops from the ventral side of the foregut. As the diverticulum elongates, a tracheoesophageal septum is formed, dividing the trachea and the esophagus by 34–36 days of human gestation. During the seventh and eight weeks of human embryonic development, the esophageal epithelium proliferates and almost completely occludes the lumen. Beginning at 10 weeks, the esophagus recanalizes, eventually forming a columnar ciliated epithelium. By birth, the ciliated epithelium is replaced by a stratified squamous epithelium. This stratified squamous epithelium persists throughout adulthood under normal conditions and consists of three compartments: basal layer, suprabasal (prickle) layer, and superficial layer (Karam 1999). Proliferation occurs in the basal layer, and cells undergo differentiation as they migrate through the suprabasal layer to the superficial layer of the epithelium. The stem cells of the esophagus are located in the basal layer of the epithelium (Seery 2002). In some cases, metaplasia develops in cells of the esophageal epithelium with the formation of a columnar lining similar to that normally found in the intestine. This condition is called intestinal metaplasia or Barrett's esophagus (Fitzgerald 2006). Of note, the esophagus is keratinized in mice but not in humans (Hogan and Zaret 2002).

Gastric Development and Differentiation

Starting around 4 weeks of gestation in the human and approximately E10 in the mouse, the stomach begins to form as a dilatation of the distal foregut (Hogan and Zaret 2002; Katz and Wu 2004; Kaufman,1995; Moore and Persaud 1998). During development the stomach shifts position, and the dorsal wall of the stomach grows faster than the ventral wall, forming the greater and lesser curvatures of the stomach. The stomach continues to rotate until about 8 weeks' gestation and is eventually lined by a columnar epithelium composed of a pit-gland unit with four compartments: foveolus, isthmus, neck, and base. All cells lining the pit-gland unit originate from progenitor cells located in the isthmus, which then migrate inward and/or outward to give rise to several cell types: surface mucus (pit) cells, mucus neck cells, zymogenic (chief) cells, parietal (oxyntic) cells, and enteroendocrine cells (Karam 1999). In the mouse, the proximal stomach (forestomach) is lined by a stratified, keratinized, squamous epithelium, and the distal stomach is lined by a glandular epithelium. Keratinization of the forestomach is first detected at E16.5 (Hogan and Zaret 2002).

Intestinal Development and Differentiation

Up until approximately E14 in the developing mouse and week 8 in humans, the intestine is lined by pseudostratified epithelium (Hogan and Zaret 2002; Traber and Wu 1995). Between E14 and E15 in mice and gestational weeks 9–10 in humans, the intestinal lining undergoes a transition into a single-layered simple columnar epithelium, with the development of finger-like projections called villi as a wave along the craniocaudal axis of the small intestine. This critical transition occurs after less than 25% of gestation in humans but after nearly 75% of gestation in mice (Traber and Wu 1995). Thus, the mouse intestinal epithelium is relatively immature at birth, and changes in gene expression and histogenesis continue over the first 3 weeks of postnatal life in mice (de Santa Barbara et al. 2003). In humans, the crypts begin to form between gestational weeks 10 and 12, and by about gestational week 16 the intestinal epithelium appears similar to that in the adult. In mice, crypts develop from the intervillous zone beginning around E19 and continuing into the second week of life, and the villi continue to lengthen during this time. At the suckling–weaning transition in mice, occurring between postnatal days 18 and 22, further functional differentiation occurs with dramatic changes in gene expression, such that by the middle of the fourth postnatal week the mouse small intestine has attained its adult form (Traber and Silberg 1996).

The adult small intestine contains four cell types: enterocytes, goblet cells, enteroendocrine cells, and Paneth cells (Karam 1999). These cells can be grouped into two functional classes—absorptive (enterocytes) and secretory (goblet cells, enteroendocrine cells, Paneth cells)—with Notch signaling being the key "molecular switch" between the two classes (Yang et al. 2001). The colon generally lacks Paneth cells but contains the other three cell types (enterocytes, goblet cells, enteroendocrine cells). In both the small and large intestine, as in the other luminal gastrointestinal organs, there is a spatial separation of proliferating and differentiating cells. In the small intestine, stem cells and transit-amplifying cells are located in the crypts, with differentiation occurring as cells migrate out of the crypts and along the villi. In the colon, which lacks villi, proliferation is restricted to the lower third of the crypts.

KLFs in Gastrointestinal Development and Differentiation

Among all the KLF family members characterized thus far, KLF4 (previously known as GKLF or EZF) and KLF5 (previously known as IKLF or BTEB2) have been the most extensively studied KLFs in the gastrointestinal tract (McConnell et al. 2007).

KLF4

KLF4 is highly expressed in the differentiated compartments of the gastrointestinal epithelia, including: the suprabasal and superficial layers of the esophagus (Fig.1A); the mid to upper portion of the gastric unit; small intestinal villi (Fig. 1B); and the upper portion of the colonic crypts (Garrett-Sinha et al. 1996; Goldstein et al. 2007;

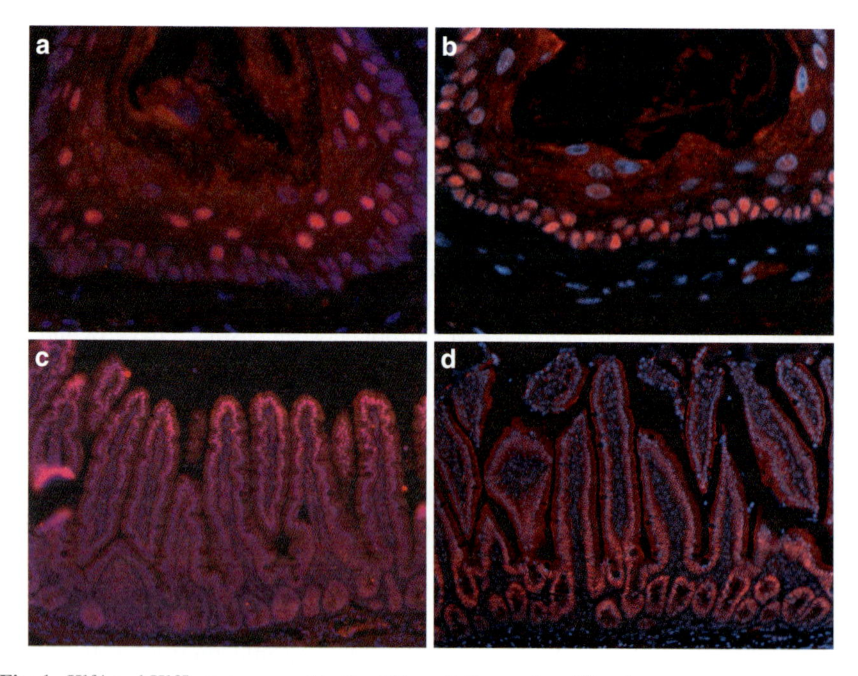

Fig. 1 Klf4 and Klf5 are expressed in the differentiating and proliferative compartments, respectively, in murine gastrointestinal epithelia. In the squamous esophagus, nuclear Klf4 staining (*red*) is seen in cells of the suprabasal layer (**A**), and Klf5 (*red*) is expressed in basal cell nuclei (**B**). In columnar epithelial cells, such as those of the small intestine, Klf4 (*red*) is expressed along the villi (**C**), and Klf5 (*red*) is localized to the crypts (**D**). Similar patterns of expression are seen in proliferative and differentiating regions of the stomach and colon (not shown). Hoechst 33258 (*blue*) is used as a counterstain. **A, B** ×400. **C, D** ×100

Katz et al. 2005; Shields et al. 1996). Expression of the gene encoding *Klf4* is regulated during development, with maximal expression occurring in the late stage of mouse fetal development (Garrett-Sinha et al. 1996; Ton-That et al. 1997). No expression of *Klf4* is detected at E9.5, and the level of expression is very low from E10 to E12. Beginning on E13, the level of the *Klf4* transcript starts to rise, peaking at E17, before decreasing moderately by E19. *Klf4* expression during embryonic development is therefore coincident with the transition in the intestine from squamous cells to columnar epithelium as well as with gut tube morphogenesis (Hogan and Zaret 2002; Traber and Wu 1995).

In 12.5-day embryos, *Klf4* mRNA expression is apparent in epithelial cells of the dorsal surface of the tongue, whereas epithelial cells of the esophagus and colon begin to express *Klf4* transcripts at E15.5 (Garrett-Sinha et al. 1996). In newborn mice, many gastrointestinal epithelia express high levels of *Klf4* mRNA, including the tongue, esophagus, stomach, and colon (Garrett-Sinha et al. 1996; Ton-That et al. 1997). Interestingly, levels of *Klf4* transcripts have been reported to be higher in the colon than in the small intestine in the embryo, newborn, and adult mice. Homozygous null mice for *Klf4* have been generated but die within 15 hours after birth owing to a defect in the barrier function of the skin (Segre et al. 1999). *Klf4* null mice show perturbations of the late-stage differentiation structures of the skin and tongue and abnormal differentiation of goblet cells in the colon (Katz et al. 2002; Segre et al. 1999). Tissue-specific gene ablation of *Klf4* in the glandular epithelium of the stomach results in increased proliferation and altered differentiation of parietal, zymogenic, pit, and mucus neck cells of the gastric epithelia (Katz et al. 2005). Gastric epithelia of these *Klf4* mutant mice are also hypertrophic and display precancerous changes.

Expression of KLF4 in terminally differentiated cells of the gastrointestinal tract suggests an important role for this transcription factor in the switch from cell proliferation to cell differentiation (Ghaleb et al. 2005). In support of this, overexpression of KLF4 in vitro in a human colonic adenocarcinoma cell line (HT-29) results in growth arrest, and DNA synthesis increases following suppression of KLF4 in these cells (Shie et al. 2000). KLF4 also regulates cell proliferation by controlling several genes critical for cell cycle checkpoint control. For example, in the human colon cancer cell line RKO, inducible expression of KLF4 blocks cell cycle progression at the G_1/S transition point (Chen et al. 2001). Furthermore, increased expression of cell cycle checkpoint protein $p21^{WAF1/CIP1}$ is observed in RKO cells following induction of KLF4, and transcriptional profiling reveals that KLF4 induction also leads to increased expression of $p27^{KIP2}$ and to reduced expression of cyclin D1 and CDC2. The protein $p21^{WAF1/CIP1}$ is also a target of Klf4 in the gastric epithelium in vivo (Katz et al. 2005). KLF4 controls p53-dependent G_1/S cell cycle arrest and inhibits the expression of cyclin B1, cyclin D1, and cyclin E following DNA damage in HCT-116 colon cancer cells (Shie et al. 2000; Yoon et al. 2003, 2005). It has also been proposed that KLF4 is a tumor suppressor in gastric and colon cancers (Wei et al. 2005; Zhao et al. 2004). In esophageal cancer cells, KLF4 promotes apoptosis and inhibits invasion (Yang et al. 2005). However, KLF4 is also suggested to be a context-dependent oncogene (Rowland et al. 2005), and nuclear localization of KLF4 in skin and breast is associated with a more aggressive phenotype (Chen et al. 2008; Pandya et al. 2004). Nonetheless, these data generally support a role for KLF4 as a negative regulator of proliferation in gastrointestinal epithelial cells.

In addition to its role in the regulation of cell proliferation, Klf4 regulates a number of genes critical for epithelial differentiation. In esophageal epithelial cells, KLF4 transcriptionally activates keratin 4, a marker of keratinocyte differentiation (Jenkins et al. 1998; Okano et al. 2000). Klf4 also increases expression of another keratinocyte differentiation marker keratin 13 (Goldstein et al. 2007). In newborn colon, loss of *Klf4* results in decreased expression of the goblet cell-specific gene *muc2* and diminished numbers of goblet cells, suggesting that this factor may be critical for goblet cell differentiation (Katz et al. 2002). Klf4 also activates the enterocyte differentiation marker intestinal alkaline phosphatase (Hinnebusch et al. 2004). Thus, in sum, Klf4 appears to play an essential role in epithelial differentiation throughout the gastrointestinal tract.

A number of studies have suggested that KLF4 acts as a tumor suppressor in the gastrointestinal tract. Decreased expression of KLF4 has been observed in dysplastic epithelium of the colon (Shie et al. 2000). Similarly, colonic adenomas and carcinomas from patients with familial adenomatous polyposis have decreased levels of KLF4 expression compared to adjacent normal mucosa (Dang et al. 2000). Reduced levels of *Klf4* mRNA are also observed in the intestine of APC^{Min} mice; and $APC^{Min}/Klf4^{+/-}$ mice develop an increased number of adenomas compared to APC^{Min} mice (Ghaleb et al. 2007). Furthermore, dysregulation of KLF4 gene expression has been reported in a number of human colorectal cell lines (Zhao et al. 2004), gastric cancers (Katz et al. 2005; Wei et al. 2005), and human esophageal squamous cell carcinomas (Ghaleb et al. 2007; Luo et al. 2003; Wang et al. 2002; Wei et al. 2005).

KLF5

Klf5, in contrast to *Klf4*, is expressed predominantly in the proliferative compartments of the gastrointestinal epithelia in adults, including: the basal layers of the esophagus (Fig. 1C); the small intestinal crypts (Fig. 1D); and the lower third of the colonic crypts (Conkright et al. 1999; Goldstein et al. 2007; Ohnishi et al. 2000; Yang et al. 2008). In the mouse embryo, *Klf5* expression is abundant from E7 and is expressed throughout the primitive gut beginning at E10.5 (Ohnishi et al. 2000). Progressively, expression of *Klf5* becomes confined to the crypts of the small intestine by E17.5. Expression is also seen in the epithelium of the tongue at E16.5 and E17.5. Thus, *Klf5* expression remains consistently high throughout development of the gastrointestinal tract.

Klf5 appears to function as a positive regulator of proliferation in non-transformed epithelial cells (Sun et al. 2001). For example, Klf5 expression in IEC-6, IEC-18, and IMCE intestinal epithelial cells increases cell proliferation (Bateman et al. 2004; Chanchevalap et al. 2004), and transgenic expression of *Klf5* in murine esophageal epithelia in vivo results in increased basal cell proliferation (Goldstein et al. 2007). Interestingly, in this model, expression of *Klf5* in the suprabasal layer did not appear to alter keratinocyte proliferation or differentiation, suggesting that context is important for Klf5 function. Other studies suggest that Klf5 may be important in maintaining the transit-amplifying cell state in esophageal epithelia by transcriptionally activating epidermal growth factor receptor (EGFR) and integrin-linked kinase (ILK) to regulate proliferation and migration, respectively (Yang et al. 2007, 2008). Some controversy, however, exists regarding the role of KLF5 in transformation. Although some studies suggest that KLF5 contributes to cellular transformation induced by K-Ras during intestinal tumorigenesis (Nandan et al. 2008), Klf5 appears to inhibit proliferation in some intestinal cancer cell lines; moreover, reduced expression of Klf5 is observed in intestinal adenomas from APC^{Min} mice and familial adenomatous polyposis (FAP) patients (Bateman et al. 2004). Klf5 also inhibits proliferation in esophageal squamous cancer cells (Yang et al. 2005). Mice homozygous for a null mutation in *Klf5* die by E8.5, and heterozygous loss of *Klf5* results in abnormally shaped small intestinal villi (Shindo et al. 2002). In sum, the function of Klf5 in gastrointestinal epithelia appears to be generally pro-proliferative but is also context-dependent.

KLF6

KLF6 (also known as Zf9 or CPBP) was initially cloned from a human placental expression library and is ubiquitously expressed in adult tissues (Koritschoner et al. 1997). During development, *Klf6* is abundantly expressed at E16.5 and E18.5, where it is restricted to the mucosa of the hindgut (Laub et al. 2001). In mice, loss of *Klf6* is lethal to embryos by E12.5 due to markedly reduced hematopoiesis in the yolk sac (Matsumoto et al. 2006). *Klf6*$^{-/-}$ mice also have a poorly defined liver. To date, the early lethality of the *Klf6*$^{-/-}$ mice has precluded analyses of the effects of *Klf6* loss on gastrointestinal development and homeostasis at later stages in the embryo and in the adult. A growing body of evidence implicates KLF6 as a negative regulator of proliferation and/or a potential inducer of differentiation. For example, KLF6 increases expression of the Cdk inhibitor $p21^{Waf1/CIP1}$, which promotes cell cycle arrest in a p53-independent manner (Narla et al. 2001). Moreover, interaction of KLF6 with cyclin D1 has been shown to inhibit proliferation via disruption of the cyclin D1–Cdk4 complex assembly and activity (Benzeno et al. 2004). In human esophageal cells, KLF6 interacts with KLF4 to co-activate the keratin 4 promoter, which is important for epithelial differentiation (Okano et al. 2000). KLF6 has been identified as a putative tumor suppressor in the stomach (Cho et al. 2005) and colon (Cho et al. 2006a, 2006b; Reeves et al. 2004). It is downregulated in Barrett's associated esophageal adenocarcinoma (Peng et al. 2008) and silenced by promoter hypermethylation in esophageal squamous cell cancer (Yamashita et al. 2002).

KLF9

Klf9 (previously known as BTEB) was originally isolated as a positive transcriptional regulator of the CYP1A1 gene promoter (Imataka et al. 1992). Northern blot analyses demonstrated that KLF9 is ubiquitously expressed. In the embryo, *Klf9* is expressed by E8 in the epithelia and smooth muscle layers of the developing gut, and this expression pattern is sustained to E16 (Imataka et al. 1992). To study the function of KLF9 in vivo, homozygous null mice for *Klf9* were generated. *Klf9* null mice are viable but have defects in brain development and in female reproductive capacity (Morita et al. 2003; Simmen et al. 2004). Further analysis of *Klf9*$^{-/-}$ mice revealed significantly reduced levels of proliferation and migration in intestinal epithelial cells of mutant mice (Simmen et al. 2007). These mice also show slight alterations of Paneth and goblet cell differentiation. Whereas these studies suggest that Klf9 is pro-proliferative in vivo, KLF9 expression is decreased in human colorectal cancers by a quantitative real-time polymerase chain reaction (PCR), Western blot analysis, and immunohistochemistry (Kang et al. 2008). More studies are necessary to confirm the function of KLF9 in gastrointestinal homeostasis and carcinogenesis.

KLF13

Klf13 (also known as BTEB3, RFLAT-1, or FKLF2) was identified by screening mouse databases for the presence of C_2H_2 DNA-binding domains (Martin et al. 2000; Scohy et al. 2000). Klf13 is expressed ubiquitously. *Klf13* mRNA is widely expressed at different stages of the developing embryo, and transcripts can be detected by E8 (Martin et al. 2001). Klf13 expression is detected in the muscle and epithelial layers of the developing gut at E11 and E13, but it becomes restricted to the epithelium by E16. The predominant phenotype of the *Klf13*$^{-/-}$ mice is a reduced number of erythrocytes, increased number of lymphoid cells, and a large spleen (Gordon et al. 2008; Zhou et al. 2007). Although homozygous null mutants for *Klf13* are viable, an increase in the number of deaths was observed among mutant mice at 3 weeks of age. No changes in the gastrointestinal tract of these animals have been reported to date.

KLF16

Klf16 (also called BTEB4 or DRRF) was initially described as a zinc finger transcription factor that modulates the activity of the dopamine receptor promoters (Hwang et al. 2001). Using in situ hybridization, *Klf16* can be detected in the intestine at E12, E14, and E16 (D'Souza et al. 2002). In the adult, Klf16 is expressed in a number of tissues, where it is localized to cell nuclei and acts as a transcriptional repressor (Kaczynski et al. 2002).

Conclusion

Krüppel-like factors are known to play a critical role in the differentiation and development of numerous organs and tissues in the body. Among them, KLF4 and KLF5 have been the most extensively characterized in the gastrointestinal tract and are known to play important roles in development and gastrointestinal homeostasis. Although some tissue-restricted KLFs are likely to have little or no role in the gastrointestinal tract in the normal state, the expression and function of these factors during/after gastrointestinal tract injury and cancer have not been well studied. Moreover, numerous other KLFs are expressed ubiquitously or are known to be expressed in the luminal organs of the gastrointestinal tract, but their precise role in differentiation and development is still unclear. In addition, little is known about the interactions among these factors themselves. Finally, given their similar binding sites, some KLFs may be able to replace others when the expression of the latter factors is lost (e.g., through the cellular differentiation process or in cancer and other diseases). Thus, further identifying the roles of the KLFs, their specific interactions, and their ability to antagonize and/or compensate for each other will provide fertile ground for future investigations of KLF function in the gastrointestinal tract.

References

Bateman, N. W., Tan, D., Pestell, R. G., Black, J. D., and Black, A. R. (2004). Intestinal tumor progression is associated with altered function of KLF5. J Biol Chem *279*, 12093–12101.

Benzeno, S., Narla, G., Allina, J., Cheng, G. Z., Reeves, H. L., Banck, M. S., Odin, J. A., Diehl, J. A., Germain, D., and Friedman, S. L. (2004). Cyclin-dependent kinase inhibition by the KLF6 tumor suppressor protein through interaction with cyclin D1. Cancer Res *64*, 3885–3891.

Chanchevalap, S., Nandan, M. O., Merlin, D., and Yang, V. W. (2004). All-trans retinoic acid inhibits proliferation of intestinal epithelial cells by inhibiting expression of the gene encoding Kruppel-like factor 5. FEBS Lett *578*, 99–105.

Chen, X., Johns, D. C., Geiman, D. E., Marban, E., Dang, D. T., Hamlin, G., Sun, R., and Yang, V. W. (2001). Kruppel-like factor 4 (gut-enriched Kruppel-like factor) inhibits cell proliferation by blocking G1/S progression of the cell cycle. J Biol Chem *276*, 30423–30428.

Chen, Y. J., Wu, C. Y., Chang, C. C., Ma, C. J., Li, M. C., and Chen, C. M. (2008). Nuclear Kruppel-like factor 4 expression is associated with human skin squamous cell carcinoma progression and metastasis. Cancer Biol Ther *7*, 777–782.

Cho, Y. G., Choi, B. J., Kim, C. J., Song, J. W., Kim, S. Y., Nam, S. W., Lee, S. H., Yoo, N. J., Lee, J. Y., and Park, W. S. (2006a). Genetic alterations of the KLF6 gene in colorectal cancers. Apmis *114*, 458–464.

Cho, Y. G., Choi, B. J., Song, J. W., Kim, S. Y., Nam, S. W., Lee, S. H., Yoo, N. J., Lee, J. Y., and Park, W. S. (2006b). Aberrant expression of kruppel-like factor 6 protein in colorectal cancers. World J Gastroenterol *12*, 2250–2253.

Cho, Y. G., Kim, C. J., Park, C. H., Yang, Y. M., Kim, S. Y., Nam, S. W., Lee, S. H., Yoo, N. J., Lee, J. Y., and Park, W. S. (2005). Genetic alterations of the KLF6 gene in gastric cancer. Oncogene *24*, 4588–4590.

Conkright, M. D., Wani, M. A., Anderson, K. P., and Lingrel, J. B. (1999). A gene encoding an intestinal-enriched member of the Kruppel-like factor family expressed in intestinal epithelial cells. Nucleic Acids Res *27*, 1263–1270.

D'Souza, U. M., Lammers, C. H., Hwang, C. K., Yajima, S., and Mouradian, M. M. (2002). Developmental expression of the zinc finger transcription factor DRRF (dopamine receptor regulating factor). Mech Dev *110*, 197–201.

Dang, D. T., Bachman, K. E., Mahatan, C. S., Dang, L. H., Giardiello, F. M., and Yang, V. W. (2000). Decreased expression of the *gut-enriched Krüppel-like factor* gene in intestinal adenomas of multiple intestinal neoplasia mice and in colonic adenomas of familial adenomatous polyposis patients. FEBS Lett *476*, 203–207.

de Santa Barbara, P., van den Brink, G. R., and Roberts, D. J. (2003). Development and differentiation of the intestinal epithelium. Cell Mol Life Sci *60*, 1322–1332.

Fitzgerald, R. C. (2006). Molecular basis of Barrett's oesophagus and oesophageal adenocarcinoma. Gut *55*, 1810–1820.

Garrett-Sinha, L. A., Eberspaecher, H., Seldin, M. F., and de Crombrugghe, B. (1996). A gene for a novel zinc-finger protein expressed in differentiated epithelial cells and transiently in certain mesenchymal cells. Journal of Biological Chemistry *271*, 31384–31390.

Ghaleb, A. M., McConnell, B. B., Nandan, M. O., Katz, J. P., Kaestner, K. H., and Yang, V. W. (2007). Haploinsufficiency of Kruppel-like factor 4 promotes adenomatous polyposis coli dependent intestinal tumorigenesis. Cancer Res *67*, 7147–7154.

Ghaleb, A. M., Nandan, M. O., Chanchevalap, S., Dalton, W. B., Hisamuddin, I. M., and Yang, V. W. (2005). Kruppel-like factors 4 and 5: the yin and yang regulators of cellular proliferation. Cell Res *15*, 92–96.

Goldstein, B. G., Chao, H. H., Yang, Y., Yermolina, Y. A., Tobias, J. W., and Katz, J. P. (2007). Overexpression of Kruppel-like factor 5 in esophageal epithelia in vivo leads to increased proliferation in basal but not suprabasal cells. Am J Physiol Gastrointest Liver Physiol *292*, G1784–1792.

Gordon, A. R., Outram, S. V., Keramatipour, M., Goddard, C. A., Colledge, W. H., Metcalfe, J. C., Hager-Theodorides, A. L., Crompton, T., and Kemp, P. R. (2008). Splenomegaly and modified erythropoiesis in KLF13–/– mice. J Biol Chem *283*, 11897–11904.

Hinnebusch, B. F., Siddique, A., Henderson, J. W., Malo, M. S., Zhang, W., Athaide, C. P., Abedrapo, M. A., Chen, X., Yang, V. W., and Hodin, R. A. (2004). Enterocyte differentiation marker *intestinal alkaline phosphatase* is a target gene of the gut-enriched Krüppel-like factor. Am J Physiol Gastrointest Liver Physiol *286*, G23–30.

Hogan, B. L. M., and Zaret, K. S. (2002). Mouse Development: Patterning, Morphogenesis, and Organogenesis (San Diego: Academic Press).

Hwang, C. K., D'Souza, U. M., Eisch, A. J., Yajima, S., Lammers, C. H., Yang, Y., Lee, S. H., Kim, Y. M., Nestler, E. J., and Mouradian, M. M. (2001). Dopamine receptor regulating factor, DRRF: a zinc finger transcription factor. Proc Natl Acad Sci U S A *98*, 7558–7563.

Imataka, H., Sogawa, K., Yasumoto, K., Kikuchi, Y., Sasano, K., Kobayashi, A., Hayami, M., and Fujii-Kuriyama, Y. (1992). Two regulatory proteins that bind to the basic transcription element (BTE), a GC box sequence in the promoter region of the rat P-4501A1 gene. Embo J *11*, 3663–3671.

Jenkins, T. D., Opitz, O. G., Okano, J., and Rustgi, A. K. (1998). Transactivation of the human keratin 4 and Epstein-Barr virus ED-L2 promoters by gut-enriched Kruppel-like factor. Journal of Biological Chemistry *273*, 10747–10754.

Kaczynski, J. A., Conley, A. A., Fernandez Zapico, M., Delgado, S. M., Zhang, J. S., and Urrutia, R. (2002). Functional analysis of basic transcription element (BTE)-binding protein (BTEB) 3 and BTEB4, a novel Sp1-like protein, reveals a subfamily of transcriptional repressors for the BTE site of the cytochrome P4501A1 gene promoter. Biochem J *366*, 873–882.

Kang, L., Lu, B., Xu, J., Hu, H., and Lai, M. (2008). Downregulation of Kruppel-like factor 9 in human colorectal cancer. Pathol Int *58*, 334–338.
Karam, S. M. (1999). Lineage commitment and maturation of epithelial cells in the gut. Front Biosci *4*, D286–298.
Katz, J. P., Perreault, N., Goldstein, B. G., Actman, L., McNally, S. R., Silberg, D. G., Furth, E. E., and Kaestner, K. H. (2005). Loss of KLF4 in mice causes altered proliferation and differentiation and precancerous changes in the adult stomach. Gastroenterology *128*, 935–945.
Katz, J. P., Perreault, N., Goldstein, B. G., Lee, C. S., Labosky, P. A., Yang, V. W., and Kaestner, K. H. (2002). The zinc-finger transcription factor KLF4 is required for terminal differentiation of goblet cells in the colon. Development *129*, 2619–2628.
Katz, J. P., and Wu, G. D. (2004). Abnormalities of gastrointestinal organogenesis, Vol 2 (Philadelphia: Mosby).
Kaufman, M. M. (1995). The Atlas of Mouse Development, Revised edn (Academic Press: San Diego).
Koritschoner, N. P., Bocco, J. L., Panzetta-Dutari, G. M., Dumur, C. I., Flury, A., and Patrito, L. C. (1997). A novel human zinc finger protein that interacts with the core promoter element of a TATA box-less gene. J Biol Chem *272*, 9573–9580.
Laub, F., Aldabe, R., Ramirez, F., and Friedman, S. (2001). Embryonic expression of Kruppel-like factor 6 in neural and non-neural tissues. Mech Dev *106*, 167–170.
Lebenthal, E. (1989). Human gastrointestinal development (New York: Raven Press).
Luo, A., Kong, J., Hu, G., Liew, C. C., Xiong, M., Wang, X., Ji, J., Wang, T., Zhi, H., Wu, M., and Liu, Z. (2003). Discovery of Ca^{2+}-relevant and differentiation-associated genes downregulated in esophageal squamous cell carcinoma using cDNA microarray. Oncogene.
Martin, K. M., Cooper, W. N., Metcalfe, J. C., and Kemp, P. R. (2000). Mouse BTEB3, a new member of the basic transcription element binding protein (BTEB) family, activates expression from GC-rich minimal promoter regions. Biochem J *345 Pt 3*, 529-533.
Martin, K. M., Metcalfe, J. C., and Kemp, P. R. (2001). Expression of KLF9 and KLF13 in mouse development. Mech Dev *103*, 149–151.
Matsumoto, N., Kubo, A., Liu, H., Akita, K., Laub, F., Ramirez, F., Keller, G., and Friedman, S. L. (2006). Developmental regulation of yolk sac hematopoiesis by Kruppel-like factor 6. Blood *107*, 1357–1365.
McConnell, B. B., Ghaleb, A. M., Nandan, M. O., and Yang, V. W. (2007). The diverse functions of Kruppel-like factors 4 and 5 in epithelial biology and pathobiology. Bioessays *29*, 549–557.
Moore, K. L., and Persaud, T. V. N. (1998). The developing human : clinically oriented embryology, 6th edn (Philadelphia: Saunders).
Morita, M., Kobayashi, A., Yamashita, T., Shimanuki, T., Nakajima, O., Takahashi, S., Ikegami, S., Inokuchi, K., Yamashita, K., Yamamoto, M., and Fujii-Kuriyama, Y. (2003). Functional analysis of basic transcription element binding protein by gene targeting technology. Mol Cell Biol *23*, 2489–2500.
Nandan, M. O., McConnell, B. B., Ghaleb, A. M., Bialkowska, A. B., Sheng, H., Shao, J., Babbin, B. A., Robine, S., and Yang, V. W. (2008). Kruppel-like factor 5 mediates cellular transformation during oncogenic KRAS-induced intestinal tumorigenesis. Gastroenterology *134*, 120–130.
Narla, G., Heath, K. E., Reeves, H. L., Li, D., Giono, L. E., Kimmelman, A. C., Glucksman, M. J., Narla, J., Eng, F. J., Chan, A. M., et al. (2001). KLF6, a candidate tumor suppressor gene mutated in prostate cancer. Science *294*, 2563–2566.
Ohnishi, S., Laub, F., Matsumoto, N., Asaka, M., Ramirez, F., Yoshida, T., and Terada, M. (2000). Developmental expression of the mouse gene coding for the Kruppel-like transcription factor KLF5. Dev Dyn *217*, 421–429.
Okano, J., Opitz, O. G., Nakagawa, H., Jenkins, T. D., Friedman, S. L., and Rustgi, A. K. (2000). The Krüppel-like transcriptional factors Zf9 and GKLF coactivate the human *keratin 4* promoter and physically interact. FEBS Lett *473*, 95–100.

Pandya, A. Y., Talley, L. I., Frost, A. R., Fitzgerald, T. J., Trivedi, V., Chakravarthy, M., Chhieng, D. C., Grizzle, W. E., Engler, J. A., Krontiras, H., et al. (2004). Nuclear localization of KLF4 is associated with an aggressive phenotype in early-stage breast cancer. Clin Cancer Res *10*, 2709–2719.

Peng, D., Sheta, E. A., Powell, S. M., Moskaluk, C. A., Washington, K., Goldknopf, I. L., and El-Rifai, W. (2008). Alterations in Barrett's-related adenocarcinomas: a proteomic approach. Int J Cancer *122*, 1303–1310.

Que, J., Choi, M., Ziel, J. W., Klingensmith, J., and Hogan, B. L. (2006). Morphogenesis of the trachea and esophagus: current players and new roles for noggin and Bmps. Differentiation *74*, 422–437.

Reeves, H. L., Narla, G., Ogunbiyi, O., Haq, A. L., Katz, A., Benzeno, S., Hod, E., Harpaz, N., Goldberg, S., Tal-Kremer, S., Eng, F. J., Arthur, M. J., Martignetti, J. A., and Friedman, S. L. (2004). Krüppel-like factor 6 (KLF6) is a tumor-suppressor gene frequently inactivated in colorectal cancer. Gastroenterology *126*, 1090–1103.

Roberts, D. J. (2000). Molecular mechanisms of development of the gastrointestinal tract. Dev Dyn *219*, 109–120.

Rowland, B. D., Bernards, R., and Peeper, D. S. (2005). The KLF4 tumour suppressor is a transcriptional repressor of p53 that acts as a context-dependent oncogene. Nat Cell Biol *7*, 1074–1082.

Scohy, S., Gabant, P., Van Reeth, T., Hertveldt, V., Dreze, P. L., Van Vooren, P., Riviere, M., Szpirer, J., and Szpirer, C. (2000). Identification of KLF13 and KLF14 (SP6), novel members of the SP/XKLF transcription factor family. Genomics *70*, 93–101.

Seery, J. P. (2002). Stem cells of the oesophageal epithelium. J Cell Sci *115*, 1783–1789.

Segre, J. A., Bauer, C., and Fuchs, E. (1999). KLF4 is a transcription factor required for establishing the barrier function of the skin. Nat Genet *22*, 356–360.

Shie, J. L., Chen, Z. Y., O'Brien, M. J., Pestell, R. G., Lee, M. E., and Tseng, C. C. (2000). Role of gut-enriched Kruppel-like factor in colonic cell growth and differentiation. Am J Physiol Gastrointest Liver Physiol *279*, G806–814.

Shields, J. M., Christy, R. J., and Yang, V. W. (1996). Identification and characterization of a gene encoding a gut-enriched Krüppel-like factor expressed during growth arrest. Journal of Biological Chemistry *271*, 20009–20017.

Shindo, T., Manabe, I., Fukushima, Y., Tobe, K., Aizawa, K., Miyamoto, S., Kawai-Kowase, K., Moriyama, N., Imai, Y., Kawakami, H., et al. (2002). Kruppel-like zinc-finger transcription factor KLF5/BTEB2 is a target for angiotensin II signaling and an essential regulator of cardiovascular remodeling. Nat Med *8*, 856–863.

Simmen, F. A., Xiao, R., Velarde, M. C., Nicholson, R. D., Bowman, M. T., Fujii-Kuriyama, Y., Oh, S. P., and Simmen, R. C. (2007). Dysregulation of intestinal crypt cell proliferation and villus cell migration in mice lacking Kruppel-like factor 9. Am J Physiol Gastrointest Liver Physiol *292*, G1757–1769.

Simmen, R. C., Eason, R. R., McQuown, J. R., Linz, A. L., Kang, T. J., Chatman, L., Jr., Till, S. R., Fujii-Kuriyama, Y., Simmen, F. A., and Oh, S. P. (2004). Subfertility, uterine hypoplasia, and partial progesterone resistance in mice lacking the Kruppel-like factor 9/basic transcription element-binding protein-1 (Bteb1) gene. J Biol Chem *279*, 29286–29294.

Sun, R., Chen, X., and Yang, V. W. (2001). Intestinal-enriched Kruppel-like factor (Kruppel-like factor 5) is a positive regulator of cellular proliferation. J Biol Chem *276*, 6897–6900.

Ton-That, H., Kaestner, K. H., Shields, J. M., Mahatanankoon, C. S., and Yang, V. W. (1997). Expression of the gut-enriched Kruppel-like factor gene during development and intestinal tumorigenesis. FEBS Letters *419*, 239–243.

Traber, P. G., and Silberg, D. G. (1996). Intestine-specific gene transcription. Annu Rev Physiol *58*, 275–297.

Traber, P. G., and Wu, G. D. (1995). Intestinal Development and Differentiation (Philadelphia: Lippencott-Raven).

Wang, N., Liu, Z. H., Ding, F., Wang, X. Q., Zhou, C. N., and Wu, M. (2002). Down-regulation of gut-enriched Krüppel-like factor expression in esophageal cancer. World J Gastroenterol *8*, 966–970.

Wei, D., Gong, W., Kanai, M., Schlunk, C., Wang, L., Yao, J. C., Wu, T. T., Huang, S., and Xie, K. (2005). Drastic down-regulation of Krüppel-like factor 4 expression is critical in human gastric cancer development and progression. Cancer Res *65*, 2746–2754.

Wells, J. M., and Melton, D. A. (1999). Vertebrate endoderm development. Annu Rev Cell Dev Biol *15*, 393–410.

Yamashita, K., Upadhyay, S., Osada, M., Hoque, M. O., Xiao, Y., Mori, M., Sato, F., Meltzer, S. J., and Sidransky, D. (2002). Pharmacologic unmasking of epigenetically silenced tumor suppressor genes in esophageal squamous cell carcinoma. Cancer Cell *2*, 485–495.

Yang, Q., Bermingham, N. A., Finegold, M. J., and Zoghbi, H. Y. (2001). Requirement of Math1 for secretory cell lineage commitment in the mouse intestine. Science *294*, 2155–2158.

Yang, Y., Goldstein, B. G., Chao, H. H., and Katz, J. P. (2005). KLF4 and KLF5 regulate proliferation, apoptosis and invasion in esophageal cancer cells. Cancer Biol Ther *4*, 1216–1221.

Yang, Y., Goldstein, B. G., Nakagawa, H., and Katz, J. P. (2007). Kruppel-like factor 5 activates MEK/ERK signaling via EGFR in primary squamous epithelial cells. Faseb J *21*, 543–550.

Yang, Y., Tetreault, M. P., Yermolina, Y. A., Goldstein, B. G., and Katz, J. P. (2008). Kruppel-like factor 5 controls keratinocyte migration via the integrin-linked kinase. J Biol Chem *283*, 18812–18820.

Yoon, H. S., Chen, X., and Yang, V. W. (2003). Krüppel-like factor 4 mediates p53-dependent G1/S cell cycle arrest in response to DNA damage. J Biol Chem *278*, 2101–2105.

Yoon, H. S., Ghaleb, A. M., Nandan, M. O., Hisamuddin, I. M., Dalton, W. B., and Yang, V. W. (2005). Krüppel-like factor 4 prevents centrosome amplification following gamma-irradiation-induced DNA damage. Oncogene.

Zhao, W., Hisamuddin, I. M., Nandan, M. O., Babbin, B. A., Lamb, N. E., and Yang, V. W. (2004). Identification of *Krüppel-like factor 4* as a potential tumor suppressor gene in colorectal cancer. Oncogene *23*, 395–402.

Zhou, M., McPherson, L., Feng, D., Song, A., Dong, C., Lyu, S. C., Zhou, L., Shi, X., Ahn, Y. T., Wang, D., et al. (2007). Kruppel-like transcription factor 13 regulates T lymphocyte survival in vivo. J Immunol *178*, 5496–5504.

Chapter 9
Gene Interactions Between Krüppel-like Factors in Development

Joyce A. Lloyd

Abstract Krüppel-like factors (KLFs) are transcription factors involved in differentiation and development. EKLF (or KLF1), KLF2, and KLF4 belong to a subclass of KLFs that are similar in regard to their zinc finger DNA-binding domains. EKLF knockout (KO) mouse embryos die between embryonic day 14.5 (E14.5) and E16.5 due to anemia. KLF2 KO embryos die between E12.5 and E14.5 and exhibit heart failure and hemorrhaging. KLF4 KO mice die perinatally owing to a loss of skin-barrier function. EKLF is expressed in erythroid cells, but KLF2 and KLF4 are expressed in multiple tissues. Our laboratory has analyzed compound mutant embryos for EKLF and KLF2 and for KLF2 and KLF4. The double KO embryos have more severe erythroid and/or cardiovascular defects and die earlier than do single knockouts. This indicates that there are interactions between these pairs of KLF genes during development. The phenotypes of the compound mutants and the possible mechanisms for KLF gene interactions are discussed.

Introduction

Krüppel-like factors (KLFs) are a family of DNA-binding proteins with sequence homology to the *Drosophila* transcription factor, Krüppel. KLFs have three C2/H2 zinc finger domains and share conserved residues located primarily within these domains (Bieker 2001; Philipsen and Suske 1999). Erythroid Krüppel-like factor (EKLF or KLF1) was the first of 17 KLFs to be identified in the mouse and humans (Miller and Bieker 1993). It is expressed specifically in erythroid cells and positively regulates the adult β-globin gene (Donze et al. 1995; Miller and Bieker 1993). $EKLF^{-/-}$ mice develop fatal anemia during definitive (fetal liver) erythropoiesis due to a defect in the maturation of red blood cells and die by embryonic day 16.5 (E16.5) (Coghill et al. 2001; Nuez et al. 1995; Perkins et al. 1995). Several other

J.A. Lloyd (✉)
Department of Human and Molecular Genetics and Massey Cancer Center, Virginia Commonwealth University, 401 College Street, Richmond, VA 23298, USA

R. Nagai et al. (eds.), *The Biology of Krüppel-like Factors*,
DOI: 10.1007/978-4-431-87775-2_9, © Springer 2009

members of the KLF family, including KLF2 (lung Krüppel-like factor, LKLF), are also expressed in erythroid cells (Basu et al. 2004, 2005; Crossley et al. 1996; Matsumoto et al. 2006).

Based on phylogenetic analyses, the zinc finger domains of EKLF, KLF2, and KLF4 are 90% similar (Bieker 2001; Kaczynski et al. 2003; Philipsen and Suske 1999; Zhang et al. 2005). $KLF2^{-/-}$ mice die between E12.5 and E14.5 owing to heart failure and severe hemorrhaging caused by defects in vascular endothelial cells and in the stabilization of immature vessels by recruited smooth muscle cells (Kuo et al. 1997a; Wani et al. 1998). Prior to E12.5, $KLF2^{-/-}$ embryos reportedly have normal vasculogenesis and angiogenesis (Kuo et al. 1997a; Wani et al. 1998).

KLF2 also plays an important role in hematopoietic cell biology. We reported that KLF2 is essential for embryonic yolk sac (primitive) erythropoiesis and positively regulates the embryonic β-like globin genes in vivo. E10.5 $KLF2^{-/-}$ primitive erythroid cells have abnormal morphology (Basu et al. 2005). KLF2 also regulates T-cell activation. Deficiency of KLF2 leads to a decrease in the peripheral T-cell pool (Kuo et al. 1997b) owing to defective thymocyte emigration (Carlson et al. 2006). Overexpression of KLF2 in mice inhibits proinflammatory activation of peripheral blood monocytes (Das et al. 2006). The fact that EKLF and KLF2 are both important in erythroid cell development led to our hypothesis that they may have compensatory functions.

KLF4 (gut Krüppel-like factor, GKLF) is essential for normal skin barrier function, and KLF4-null mice die perinatally (Segre et al. 1999). In addition, KLF4 is important in monocytes and macrophages (Alder et al. 2008; Feinberg et al. 2005), endothelial cells responding to inflammation and shear stress (Hamik et al. 2007), and vascular smooth muscle cells responding to injury (Liu et al. 2005). It is interesting to note that KLF2 and KLF4 are not only highly related structurally but are both important in the cardiovascular system.

We have shown that simultaneous ablation of the EKLF and KLF2 genes has a severely negative impact on both primitive erythropoiesis and endothelial cell development in vivo. This was the first report in knockout mice showing that KLF family members have compensatory roles in development. More recently, our laboratory has established that KLF2 and KLF4 have overlapping roles in cardiovascular development, suggesting that interactions between KLF genes may be a more general phenomenon.

There is additional evidence in the literature from experiments not involving knockout mice that KLF genes have compensatory roles. The simultaneous depletion of KLF2, KLF4, and KLF5 by knockdown in embryonic stem (ES) cells causes the cells to differentiate, indicating that these three KLF genes are required for ES cell self-renewal (Jiang et al. 2008). ES cells do not differentiate when any pair of these KLF genes is knocked down, indicating that there is functional redundancy of the genes. There is also evidence that KLF4 and KLF5 have contrasting effects on transcriptional regulation and cell proliferation in other cell types (reviewed in Ghaleb et al. 2005; Suzuki et al. 2005).

It is important to define what is meant by gene interactions in the context of this review. We consider there to be a gene interaction if, in compound mutant mice, there is a unique phenotype(s) not present in either single gene knockout and/or there is a phenotype that occurs earlier in development than in mice with a single gene mutation. One example of a gene interaction would be if double-mutant embryos die earlier in development than either single-KO embryo.

Gene Interactions Between Transcription Factor Family Members

Several pairs of closely related transcription factors are known to partially or fully compensate functionally for each other during development in double-KO mouse models. A list of examples, which is representative but not exhaustive, is shown in Table 1. In some cases, the compound mutants phenotypes were apparent when one gene was null and the other had a heterozygous mutation. In most cases, the related, interacting genes have overlapping expression patterns in the embryo, which has implications for predicting the mechanism by which they interact (see Figure 1, below). Multiple pairs of Hox genes have compensatory roles in developmental mouse models and provide many classic examples of gene interactions beyond those listed here. For instance, both Hoxa11 and Hoxd11, and Hoxa13 and Hoxd13, interact with each other in pattern formation of the limbs (Boulet and Capecchi 2004; Fromental-Ramain et al. 1996).

Although a number of examples are shown in Table 1, for the purpose of this review we focus on gene interactions that affect the development of the hematopoietic and cardiovascular systems. EKLF and KLF2 gene interactions are discussed in detail below. Mice lacking both GATA-1 and GATA-2 have virtually no embryonic erythroid cells, whereas the single-KO models of these genes show only modest effects

Table 1 Examples of genetic interactions between transcription factor family members during development in mouse knockout models

Factor family	Compound gene knockout models	Tissue(s)/system(s) affected	References
Hox	Hoxa11/Hoxd11	Limbs	Boulet & Capecchi 2004; Fromental-Ramain et al. 1996
	Hoxa13/Hoxd13	Limbs	
KLF	EKLF/KLF2	Erythroid	Basu et al. 2007
GATA	GATA-1/GATA-2	Erythroid	Fujiwara et al. 2004
Ets	PU.1/Spi-B	Lymphoid	Garrett-Sinha et al. 1999
Forkhead	Foxc1/Foxc2	Heart	Seo & Kume 2006
Pax	Pax1/Pax9	Vertebral column	Peters et al. 1999
Myogenic bHLH	MRF/MyoD	Muscle	Rawls et al. 1998
Dlx	Dlx5/Dlx6	Limbs	Hsu et al. 2006
Tead	Tead1/Tead2	Notochord	Sawada et al. 2008
Cdx	Cdx1/Cdx2	Vertebral column	van den Akker et al. 2002
TCF/LEF	Lef/Tcf1	Limbs, neural tube	Galceran et al. 1999
	Tcf4/Tcf1	Gastrointestinal tract	Gregorieff et al. 2004

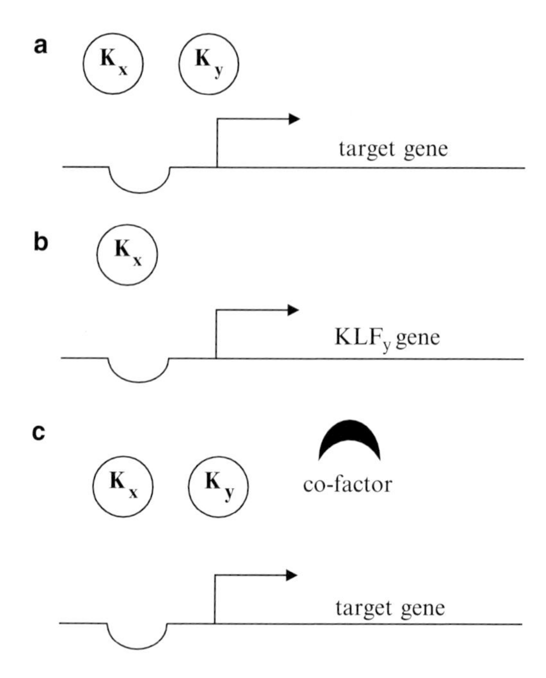

Fig. 1 Possible models for interactions between KLF genes. The *circles* represent KLF proteins, where K_x and K_y are different KLFs. The *lines* represent KLF target genes, and the semicircle is a KLF protein-binding site in the promoter. The *right-facing arrow* indicates the transcription start site. The three models are not mutually exclusive. **A** Two different KLF proteins can bind to and regulate the same target gene. **B** One KLF protein controls the transcription of another KLF gene. **C** Two different KLF proteins can bind the same cofactor, which is required for target gene transcription.

on primitive erythropoiesis (Fujiwara et al. 2004). Compound PU.1/Spi-B mutants have more extensive B-cell defects than do single gene mutants (Garrett-Sinha et al. 1999). $Foxc1^{-/-}Foxc2^{-/-}$ embryos have a wide spectrum of cardiac abnormalities, including anomalies of the outflow tract, whereas single Foxc mutants have no obvious cardiac defects at E9.0 (Seo and Kume 2006). These findings emphasize the importance of exploring the overlapping roles of EKLF, KLF2, and KLF4.

EKLF and KLF2 in Erythropoiesis

Transcription factors controlling the developmental regulation of the globin genes are a topic of intensive investigation. Using KLF2 KO mice, we have shown that KLF2 positively regulates murine (Ey- and βh1-) embryonic β-like globin genes in the embryonic yolk sac, the site of primitive erythropoiesis (Basu et al. 2005). Using $KLF2^{-/-}$ embryos with a human β-globin locus transgene, we showed that the human (ε-) embryonic gene is similarly controlled by KLF2. However, expression of these genes is diminished by only about 50% in $KLF2^{-/-}$ mice, suggesting that an additional factor(s) is involved. We speculated that EKLF and KLF2 have compensatory roles in primitive erythropoiesis and globin gene regulation.

It was initially reported that EKLF affects only adult—not embryonic or fetal—globin gene expression. Interestingly, however, EKLF is expressed early in mouse and chicken embryonic development, as early as the primitive streak stage, followed by expression in the yolk sac blood islands where primitive erythroid cells develop (Chervenak et al. 2006; Southwood et al. 1996). EKLF also plays an essential role in hemoglobin metabolism and membrane stability in primitive erythroid cells (Drissen et al. 2005; Hodge et al. 2006). Moreover, in EKLF$^{-/-}$ embryos that carry a YAC transgene with the entire human β-globin locus, the human embryonic ε-globin gene is downregulated (Tanimoto et al. 2000). Recently, ChIP analyses have shown that EKLF binds to the CACCC boxes in the Ey- and βh1-globin gene promoters in primitive erythroid cells (Zhou et al. 2006). These studies established the important role of EKLF in primitive erythropoiesis.

To investigate potential KLF gene interactions, EKLF/KLF2 double-mutant embryos were analyzed (Basu et al. 2007). EKLF$^{-/-}$KLF2$^{-/-}$ mice are anemic at E10.5 and die before E11.5, whereas single-KO EKLF$^{-/-}$ or KLF2$^{-/-}$ embryos are grossly normal beyond E11.5. At E10.5, Ey- and βh1-globin mRNA is greatly reduced in EKLF$^{-/-}$KLF2$^{-/-}$ embryos, compared to EKLF$^{-/-}$ or KLF2$^{-/-}$ embryos, which is consistent with the observed anemia. Light and electron microscopic analyses of E9.5 EKLF$^{-/-}$KLF2$^{-/-}$ yolk sacs indicate that erythroid cell precursors are morphologically more abnormal than in either single-KO embryos. Cytospins of E9.5 EKLF$^{-/-}$KLF2$^{-/-}$ blood show that the erythroid cells are markedly more irregularly shaped than either single-KO cells, suggesting possible membrane abnormalities.

A unique phenotype involving abnormal morphology of endothelial cells is observed only in the E10.5 EKLF/KLF2 double-KO yolk sacs, not in single-KO embryos. This was a surprising finding because there is no evidence that EKLF is expressed in endothelial cells. EKLF and KLF2 may have coordinate roles in the hemangioblast, a common progenitor to erythroid and endothelial cells. Alternatively, correct signaling between erythroid and endothelial cells may be controlled by EKLF and KLF2 and is required for normal endothelial cell development in the embryonic yolk sac.

The data indicate that EKLF and KLF2 have redundant functions in embryonic β-like globin gene expression, primitive erythropoiesis, and endothelial cell development. There is residual expression of both Ey- and bh1-globin mRNA in EKLF/KLF2 double-mutant embryos. This leaves open the possibility that yet another KLF can partially compensate for the simultaneous loss of EKLF and KLF2.

EKLF and KLF2 in Cardiovascular Development

We have recently discovered that E9.5 KLF2$^{-/-}$ embryos have a defect in the atrioventricular (AV) endocardial cushions (J.A. Lloyd, unpublished work, 2008). The AV cushions develop into the AV valves of the heart. In the KLF2$^{-/-}$ embryos, there is an accumulation of additional AV endocardial cushion cells compared to wild-type (WT). Although this phenotype has not previously been observed in KLF2 KO mice, there is a high level of expression of KLF2 mRNA in normal AV cushion cells (Lee et al. 2006), suggesting that KLF2 plays a role in these cells. At the electron microscopic level, endocardial cells in the KLF2$^{-/-}$ AV cushions show cytoplasmic projections extending from their apical surfaces, which are not found in the wild-type. Interestingly, although EKLF is reportedly expressed only in erythroid cells, EKLF/KLF2 double-KO embryos may have a more severe defect of the AV canal than do KLF2$^{-/-}$ embryos. This indicates that interactions between these genes likely occur in the AV cushions as well as in the yolk sac. At the electron microscopic level, EKLF/KLF2 KO embryos appear to have a more convoluted AV canal structure than do KLF2$^{-/-}$ embryos. These data may indicate that the endocardial cushion cells in the mutant embryos are improperly proliferating and/or are not proceeding through normal endothelial-to-mesenchymal transformation (EMT) and/or are not able to invade the surrounding cardiac jelly. These processes are required for valve development. Further studies are needed to determine which of these processes may be affected.

KLF2 and KLF4 in Cardiovascular Development

KLF2$^{-/-}$ embryos die by E14.5 and have a complex vascular phenotype involving hemorrhaging, heart failure, and a lack of mesenchymal cell recruitment to nascent vessels (Kuo et al. 1997a; Lee et al. 2006; Wani et al. 1998; Wu et al. 2008). KLF4$^{-/-}$ mice have no reported vascular phenotype, although KLF4 does have an established role in the cardiovascular system. KLF4 is induced in endothelial cells by shear stress (Hamik et al. 2007). In mouse models of vascular injury, KLF4 is upregulated and plays a role in phenotypic switching of vascular smooth muscle cells (Liu et al. 2005).

Our laboratory has recently shown that KLF2$^{-/-}$KLF4$^{-/-}$ mice die before E10.5 and can exhibit gross hemorrhaging (J.A. Lloyd, unpublished work, 2008), although E10.5 single KO embryos are grossly normal. Developmental abnormalities were observed in the primary head vein and carotid artery of KLF2$^{-/-}$KLF4$^{+/-}$ embryos. These abnormalities include a substantial decrease in the number of mesenchymal cells surrounding the vessels compared to WT. Additionally, in KLF2$^{-/-}$KLF4$^{+/-}$ vessels, the endothelial layer can lack structural integrity, and erythroid cells can be found in the tissue surrounding the breached vessels. Therefore, as with EKLF and KLF2, the KLF2 and KLF4 genes appear to interact. In E10.5 embryos with mutations in both the KLF2 and KLF4 genes, there are defects in blood vessel integrity that have not been described for either single mutant. Further studies are clearly needed to determine the mechanism of action of KLF2 and KLF4 gene interactions in cardiovascular development.

Conclusions: Possible Mechanism(s) for KLF Gene Interactions

The mechanism(s) by which KLF gene interactions occur is not known. Figure 1 details several possible models, which are not mutually exclusive. Perhaps the simplest model is that multiple KLF proteins with highly similar DNA-binding domains could bind to the same regulatory element in a target gene (Fig. 1A). The factor most likely to bind may depend on the relative levels of the two KLFs (K_x and K_y) in a particular cell type. Although there is no direct evidence in the erythroid or cardiovascular systems, it is plausible that the EKLF/KLF2/KLF4 subfamily of factors has redundant activity because their DNA-binding domains are so similar that they regulate common target genes. There is evidence for such regulatory circuits in ES cells (Jiang et al. 2008). Chromatin immunoprecipitation and subsequent microarray assays (ChIP-on-chip) were utilized to identify in vivo target genes for KLF2, KLF4, and KLF5 in view of their coordinate roles in ES cell self-renewal. These experiments showed, for example, that these three KLFs directly bind the Nanog gene. Nanog is an important factor for sustaining stem cell pluripotency. Electrophoretic mobility shift assay (EMSA) experiments indicate that the KLFs bind a distal enhancer region of the Nanog gene via a CCC<u>CACCC</u> motif. Co-transfection and luciferase assays showed that Nanog enhancer activity is reduced if the CACCC nucleotides are mutated. Furthermore, the ChIP-on-chip data also suggest that KLF2, KLF4, and KLF5 share many other common gene targets in ES cells.

A second possible mechanism for KLF gene interactions is that the level of expression of one KLF gene is controlled by another KLF (K_x controls the KLF_y gene in Fig. 1B). If a simple compensation model is invoked, one would expect that $KLF2^{-/-}$ erythroid cells, for example, would have increased expression of EKLF. In fact, however, we observe the opposite result, a modest decrease in EKLF mRNA in $KLF2^{-/-}$ embryonic erythroid cells and a modest decrease of KLF2 mRNA in $EKLF^{-/-}$ embryonic erythroid cells (J.A. Lloyd, unpublished work, 2008). This argues against the model in Fig. 1B for gene interactions between EKLF and KLF2 in erythroid cells. However, there is ample evidence that KLF proteins regulate KLF genes. From the ChIP-on-chip experiments in ES cells described above, it was revealed that KLF2, KLF4, and KLF5 regulate each other as well as themselves, thus forming a regulatory loop (Jiang et al. 2008). KLF4 and KLF5 both regulate the KLF4 gene in transient transfection assays, albeit with opposing effects (Dang et al. 2002). Furthermore, EKLF directly activates the KLF3 gene in erythroid cells (Funnell et al. 2007).

Finally, it is possible that two KLF proteins (K_x and K_y) could interact with a common cofactor required for target gene transcription, resulting in apparent gene interactions (Fig. 1C). If either KLF is available, transcription occurs; but if neither is present, the target gene is not transcribed. Although different KLF proteins are similar to each other mainly in their zinc finger DNA-binding domains, and not throughout the entire protein, the model in Figure 1C is still viable.

EKLF is known to interact with multiple chromatin remodeling proteins (Zhang and Bieker 1998), and extensive sequence homology would likely not be required for other KLF proteins to interact with these same cofactors. For example, EKLF and KLF2 can both interact with cyclic AMP response element-binding protein (CBP/p300), and this cofactor increases the ability of the KLFs to trans-activate reporter genes (SenBanerjee et al. 2004; Zhang and Bieker 1998).

A caveat for any of these proposed mechanisms for KLF gene interactions is that the two KLF proteins must be co-expressed in the same cell type. More complex models must been invoked if, for example, EKLF is solely expressed in erythroid and not endothelial cells and yet the EKLF and KLF2 genes interact in generating an endothelial phenotype. Likewise, KLF2 is thought to be expressed mainly in endothelial cells and KLF4 mainly in mesenchymal cells in the developing cardiovascular system. However, it is possible that in knockout models, the cell-type specificity of KLF expression is altered, resulting in the observed gene interactions. Future work is clearly needed to address the mechanisms by which KLF genes interact.

Acknowledgments I thank the current and past members of the laboratory, who have contributed both intellectually and experimentally to our understanding of KLF gene interactions. These hard-working young scientists include Priyadarshi Basu, Tina Lung, Megan Smith, Sean Fox, Mohua Basu, Latasha Redmond, Aditi Chiplunkar, Yousef Alhashem, and Christopher Pang. Our work would not be possible without the expertise, guidance, and contributions of Dr. Jack Haar. I also thank Dr. Gordon Ginder for many helpful discussions and for his critical review of our work.

References

Alder JK, Georgantas RW III, Hildreth RL et al (2008) Krüppel-like factor 4 is essential for inflammatory monocyte differentiation in vivo. J Immunol 180:5645–5652

Basu P, Lung TK, Lemsaddek W et al (2007) EKLF and KLF2 have compensatory roles in embryonic beta-globin gene expression and primitive erythropoiesis. Blood 110:3417–3425

Basu P, Morris PE, Haar JL et al (2005) KLF2 is essential for primitive erythropoiesis and regulates the human and murine embryonic beta-like globin genes. Blood 106:2566–2571

Basu P, Sargent TG, Redmond LC et al (2004) Evolutionary conservation of KLF transcription factors and functional conservation of human gamma-globin gene regulation in chicken. Genomics 84:311–319

Bieker JJ (2001) Krüppel-like factors: three fingers in many pies. J Biol Chem 276: 34355–34358

Boulet AM and Capecchi MR (2004) Multiple roles of Hoxa11 and Hoxd11 in the formation of the mammalian forelimb zeugopod. Development 131:299–309

Carlson CM, Endrizzi BT, Wu J et al (2006) Krüppel-like factor 2 regulates thymocyte and T-cell migration. Nature 442:299–302

Chervenak AP, Basu P, Shin M et al (2006) Identification, characterization, and expression pattern of the chicken EKLF gene. Dev Dyn 235:1933–1940

Coghill E, Eccleston S, Fox V et al (2001) Erythroid Krüppel-like factor (EKLF) coordinates erythroid cell proliferation and hemoglobinization in cell lines derived from EKLF null mice. Blood 97:1861–1868

Crossley M, Whitelaw E, Perkins A et al (1996) Isolation and characterization of the cDNA encoding BKLF/TEF-2, a major CACCC-box-binding protein in erythroid cells and selected other cells. Mol Cell Biol 16:1695–1705

Dang DT, Zhao W, Mahatan CS et al (2002) Opposing effects of Krüppel-like factor 4 (gut-enriched Krüppel-like factor) and Krüppel-like factor 5 (intestinal-enriched Krüppel-like factor) on the promoter of the Krüppel-like factor 4 gene. Nucleic Acids Res 30:2736–2741
Das H, Kumar A, Lin Z et al (2006) Krüppel-like factor 2 (KLF2) regulates proinflammatory activation of monocytes. Proc Natl Acad Sci U S A 103:6653–6658
Donze D, Townes TM, and Bieker JJ (1995) Role of erythroid Krüppel-like factor in human gamma- to beta-globin gene switching. J Biol Chem 270:1955–1959
Drissen R, von Lindern M Kolbus A et al (2005) The erythroid phenotype of EKLF-null mice: defects in hemoglobin metabolism and membrane stability. Mol Cell Biol 25:5205–5214
Feinberg MW, Cao Z, Wara AK et al (2005) Krüppel-like factor 4 is a mediator of proinflammatory signaling in macrophages. J Biol Chem 280:38247–38258
Fromental-Ramain C, Warot X, Messadecq N et al (1996) Hoxa-13 and Hoxd-13 play a crucial role in the patterning of the limb autopod. Development 122:2997–3011
Fujiwara Y, Chang AN, Williams AM, and Orkin SH (2004) Functional overlap of GATA-1 and GATA-2 in primitive hematopoietic development. Blood 103:583–585
Funnell AP, Maloney CA, Thompson LJ et al (2007) Erythroid Krüppel-like factor directly activates the basic Krüppel-like factor gene in erythroid cells. Mol Cell Biol 27:2777–2790
Galceran J, Farinas I, Depew MJ et al (1999) Wnt3a-/–like phenotype and limb deficiency in Lef1(-/-)Tcf1(-/-) mice. Genes Dev 13:709–717
Garrett-Sinha LA, Su GH, Rao S et al (1999) PU.1 and Spi-B are required for normal B cell receptor-mediated signal transduction. Immunity 10:399–408
Ghaleb AM, Nandan MO, Chanchevalap S et al (2005) Krüppel-like factors 4 and 5: the yin and yang regulators of cellular proliferation. Cell Res 15:92–96
Gregorieff A, Grosschedl R, and Clevers H (2004) Hindgut defects and transformation of the gastro-intestinal tract in Tcf4(-/-)/Tcf1(-/-) embryos. EMBO J 23:1825–1833
Hamik A, Lin Z, Kumar A et al (2007) Krüppel-like factor 4 regulates endothelial inflammation. J Biol Chem 282:13769–13779
Hodge D, Coghill E, Keys J et al (2006) A global role for EKLF in definitive and primitive erythropoiesis. Blood 107:3359–3370
Hsu SH, Noamani B, Abernethy DE et al (2006) Dlx5- and Dlx6-mediated chondrogenesis: Differential domain requirements for a conserved function. Mech Dev 123:819–830
Jiang J, Chan YS, Loh YH et al (2008) A core Klf circuitry regulates self-renewal of embryonic stem cells. Nat Cell Biol 10:353–360
Kaczynski J, Cook T, and Urrutia R (2003) Sp1- and Krüppel-like transcription factors. Genome Biol. doi: 10.1186/gb-2003-4-2-206
Kuo CT, Veselits ML, Barton KP et al (1997a) The LKLF transcription factor is required for normal tunica media formation and blood vessel stabilization during murine embryogenesis. Genes Dev 11:2996–3006
Kuo CT, Veselits ML, and Leiden JM (1997b) LKLF: A transcriptional regulator of single-positive T cell quiescence and survival. Science 277:1986–1990
Lee JS, Yu Q, Shin JT et al (2006) Klf2 is an essential regulator of vascular hemodynamic forces in vivo. Dev Cell 11:845–857
Liu Y, Sinha S, McDonald OG et al (2005) Krüppel-like factor 4 abrogates myocardin-induced activation of smooth muscle gene expression. J Biol Chem 280:9719–9727
Matsumoto N, Kubo A, Liu H et al (2006) Developmental regulation of yolk sac hematopoiesis by Krüppel-like factor 6. Blood 107:1357–1365
Miller IJ and Bieker JJ (1993) A novel, erythroid cell-specific murine transcription factor that binds to the CACCC element and is related to the Krüppel family of nuclear proteins. Mol Cell Biol 13:2776–2786
Nuez B, Michalovich, D, Bygrave, A et al (1995) Defective haematopoiesis in fetal liver resulting from inactivation of the EKLF gene. Nature 375:316–318
Perkins AC, Sharpe AH, and Orkin SH (1995) Lethal beta-thalassaemia in mice lacking the erythroid CACCC-transcription factor EKLF. Nature 375:318–322
Peters H, Wilm B, Sakai N et al (1999) Pax1 and Pax9 synergistically regulate vertebral column development. Development 126:5399–5408

Philipsen S and Suske G (1999) A tale of three fingers: the family of mammalian Sp/XKLF transcription factors. Nucleic Acids Res 27:2991–3000
Rawls A, Valdez MR, Zhang W et al (1998) Overlapping functions of the myogenic bHLH genes MRF4 and MyoD revealed in double mutant mice. Development 125:2349–2358
Sawada A, Kiyonari H, Ukita K et al (2008) Redundant roles of Tead1 and Tead2 in notochord development and the regulation of cell proliferation and survival. Mol Cell Biol 28:3177–3189
Segre JA, Bauer C, and Fuchs E (1999) Klf4 is a transcription factor required for establishing the barrier function of the skin. Nat Genet 22:356–360
SenBanerjee S, Lin Z, Atkins GB et al (2004) KLF2 Is a novel transcriptional regulator of endothelial proinflammatory activation. J Exp Med 199:1305–1315
Seo S and Kume T (2006) Forkhead transcription factors, Foxc1 and Foxc2, are required for the morphogenesis of the cardiac outflow tract. Dev Biol 296:421–436
Southwood CM, Downs KM, and Bieker JJ (1996) Erythroid Krüppel-like factor exhibits an early and sequentially localized pattern of expression during mammalian erythroid ontogeny. Dev Dyn 206:248–259
Suzuki T, Aizawa K, Matsumura T, and Nagai R (2005) Vascular implications of the Krüppel-like family of transcription factors. Arterioscler Thromb Vasc Biol 25:1135–1141
Tanimoto K, Liu Q, Grosveld F et al (2000) Context-dependent EKLF responsiveness defines the developmental specificity of the human epsilon-globin gene in erythroid cells of YAC transgenic mice. Genes Dev 14:2778–2794
van den Akker E, Forlani S, Chawengsaksophak K et al (2002) Cdx1 and Cdx2 have overlapping functions in anteroposterior patterning and posterior axis elongation. Development 129:2181–2193
Wani MA, Means RTJ, and Lingrel JB (1998) Loss of LKLF function results in embryonic lethality in mice. Transgenic Res 7:229–238
Wu J, Bohanan CS, Neumann JC, and Lingrel JB (2008) KLF2 transcription factor modulates blood vessel maturation through smooth muscle cell migration. J Biol Chem 283:3942–3950
Zhang P, Basu P, Redmond LC et al (2005) A functional screen for Krüppel-like factors that regulate the human gamma-globin gene through the CACCC promoter element. Blood Cells, Molecules and Diseases 35:227–235
Zhang W and Bieker JJ (1998) Acetylation and modulation of erythroid Krüppel-like factor (EKLF) activity by interaction with histone acetyltransferases. Proc Natl Acad Sci U S A 95:9855–9860
Zhou D, Pawlik KM, Ren J et al (2006) Differential binding of erythroid Krüppel-like factor to embryonic/fetal globin gene promoters during development. J Biol Chem 281:16052–16057

Chapter 10
Krüppel-like Factors in Stem Cell Biology

Masatsugu Ema, Satoru Takahashi, and Yoshiaki Fujii-Kuriyama

Abstract Embryonic stem (ES) cells are derived from the blastocyst and have the potential to give rise to derivatives of each germ layer. Induced pluripotent stem (iPS) cells can be derived from lineage-restricted cells, such as fibroblasts and lymphocytes, by forced expression of specific transcription factors. iPS cells are transcriptionally and epigenetically similar to ES cells. Although recent studies indicate that Krüppel-like factors (KLFs) are essential for both maintenance of ES cell self-renewal and reprogramming of somatic cells into a pluripotent state, the molecular mechanism of these processes remains unknown. Thus, understanding the molecular mechanism of ES cell self-renewal and somatic cell reprogramming by Klfs is important for the efficient generation of patient-specific pluripotent stem cells and for the development of regenerative medicine.

Introduction

Pluripotency is defined as the ability to give rise to all of the types of cell that exist in an adult organism. Murine embryonic stem (ES) cells are pluripotent cells derived from the inner cell mass (ICM) of the blastocyst and can be maintained indefinitely in a self-renewing state (Evans and Kaufman 1981; Martin 1981). The potential of ES cells to differentiate into various tissues should find use in the development of regenerative medicine. However, the molecular mechanisms underlying the regulation of both pluripotency and proliferation of ES cells are not fully understood. Rodent and human somatic cells can be reprogrammed into induced pluripotent

M. Ema (✉) and S. Takahashi
Department of Anatomy and Embryology, Institute of Basic Medical Sciences, Graduate School of Comprehensive Human Sciences, University of Tsukuba, Tennoudai 1-1-1, Tsukuba 305-8577, Japan

Y. Fujii-Kuriyama
SORST, Japan Science and Technology Agency, 4-1-8 Honcho, Kawaguchi 332-0012, Japan
TARA (Tsukuba Advanced Research Alliance) Center, University of Tsukuba, Tennoudai 1-1-1, Tsukuba 305-8577, Japan

R. Nagai et al. (eds.), *The Biology of Krüppel-like Factors*,
DOI 10.1007/978-4-431-87775-2_10,

stem (iPS) cells by forced expression of a specific set of transcription factors. As iPS cells are similar to ES cells, they may represent a valuable alternative resource for transplantation. Here, we discuss the role of Krüppel-like factors (KLFS) in the self-renewal of ES cells and in somatic cell reprogramming.

Control of Self-renewal of ES Cells by KLFs

Murine ES cells can be maintained indefinitely in a self-renewing pluripotent state. Many previous studies have revealed that the pluripotency of ES cells is maintained by leukemia inhibitory factor (LIF) (Smith et al. 1988; Williams et al. 1988), bone morphogenetic protein (BMP) (Ying et al. 2003), Wnt (Sato et al. 2004), STAT3 (Matsuda et al. 1999; Niwa et al. 1998), Oct3/4 (Nichols et al. 1998; Niwa et al. 2000), Nanog (Chambers et al. 2003; Mitsui et al. 2003), Sox2 (Avilion et al. 2003), and c-Myc (Cartwright et al. 2005). In addition to these soluble factors and transcription factors, Li and coworkers found that KLF4 is able to maintain mouse ES cells in an undifferentiated state in the absence of LIF (Li et al. 2005).

A recent report indicated that Klf family members have overlapping functions during hematopoietic development (Basu et al. 2007). Consistent with this idea, Jiang and coworkers showed that there are overlapping functions among Klfs in ES cells (Jiang et al. 2008). They showed that Klf2, Klf4, and Klf5 are expressed abundantly in undifferentiated ES cells, whereas expression of these factors is markedly decreased following differentiation. Furthermore, triple knockdown of Klf2, Klf4, and Klf5 resulted in defective self-renewal of mouse ES cells (Jiang et al. 2008), demonstrating that these three Klfs have similar functions in the process of ES cell self-renewal.

Originally, Klf5/BTEB2 was identified as a transcription factor that binds to a basic transcription element (BTE) in the promoter of *Cyp1a1* and was presumed to regulate transcription from this promoter (Imataka 1991; Sogawa et al. 1992). To investigate the physiological roles of *Klf5*, we generated *Klf5* knockout (KO) mice and found that loss of *Klf5* results in early embryonic lethality, which is consistent with a previous report (Shindo et al. 2002). Detail analysis indicated that loss of *Klf5* results in defective implantation, perhaps due to reduced Cdx2 expression (Ema et al. 2008) (Fig.1A). *Klf5* is also expressed in cells of the ICM, and loss of *Klf5* results in failure of ICM development and generation of ES cells (Ema et al. 2008) (Fig. 1B–E). We also showed that *Klf5* is essential for normal self-renewal of ES cells (Ema et al. 2008). *Klf5* KO ES cells show increased expression of differentiation-related marker genes, such as *Fgf5* and *Brachyury*, and exhibit frequent spontaneous differentiation, whereas *Klf5* overexpression suppresses the expression of these marker genes and is able to maintain ES cells in a pluripotent state even in the absence of LIF (Ema et al. 2008). Our data also demonstrated that *Klf5* regulates ES cell proliferation through G_1 progression. Intriguingly, our results indicate that *Klf4* has a function similar to that of *Klf5* in suppressing differentiation of ES cells (Ema et al. 2008) (Fig. 2A).

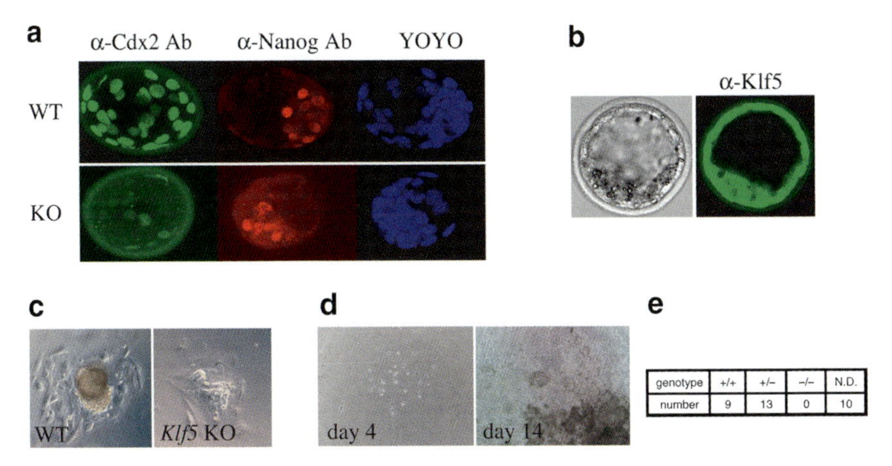

Fig. 1 *Klf5* is an indispensable Klf family member for derivation of embryonic stem (ES) cells from the inner cell mass (ICM). **a** Analysis of Cdx2 and Nanog protein expressions in *Klf5* knockout (*KO*) blastocysts. **b** Immunohistochemical analysis of Klf5 demonstrates ubiquitous expression in the blastocyst. **c** Phase contrast images of outgrowth cultures from WT (*WT*) and *Klf5* KO blastocysts. **d** Highly proliferative ICM cells isolated from wild-type blastocysts. The isolated ICM cells were cultured in ES cell medium for 2 weeks on a gelatin-coated dish and then were genotyped. **e** Summary of immunosurgery experiments. A total of 22 ES cell lines were established and were genotyped. There was no *Klf5* KO ES cell line. The *N.D.* group is that in which the genotype could not be determined

On the other hand, *Klf4*-overexpressing *Klf5* KO ES cells exhibited reduced cell proliferation, a result that is in sharp contrast to that found following reexpression of *Klf5* (Fig. 2A). This observation is consistent with previous reports showing that Klf4 and Klf5 exert opposing effects on cell proliferation (Ghaleb et al. 2005). Taken together, *Klf4* and *Klf5* both appear to suppress differentiation but give rise to opposing effects upon cellular proliferation.

Genes That Act Upstream and Downstream of KLFs to Mediate ES Cell Self-renewal

Klfs are essential for the normal self-renewal of ES cells (Ema et al. 2008; Jiang et al. 2008). We have shown that genes involved in the Tcl1-Akt1 signaling pathway transmit signals downstream of Klf5 to mediate cell proliferation (Ema et al. 2008) (Fig. 2B). Consistent with this idea, Jiang and coworkers used ChIP-on-ChIP studies to show that Klf2, Klf4, and Klf5 could bind to a genomic region localized in close proximity to the *Fgf5*, *Tcl1*, and *Nanog* genes (Jiang et al. 2008). Thus, our studies, together with those of Jiang et al. imply that KLF5 regulates these downstream genes directly. It is of note that loss of KLF5 alters the expression of *Fgf5* and *Tcl1* but not *Nanog*, whereas synergistic knockdown of KLF2 and KLF4

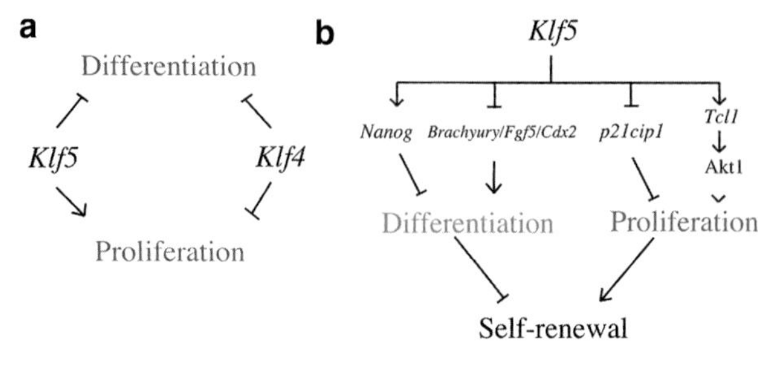

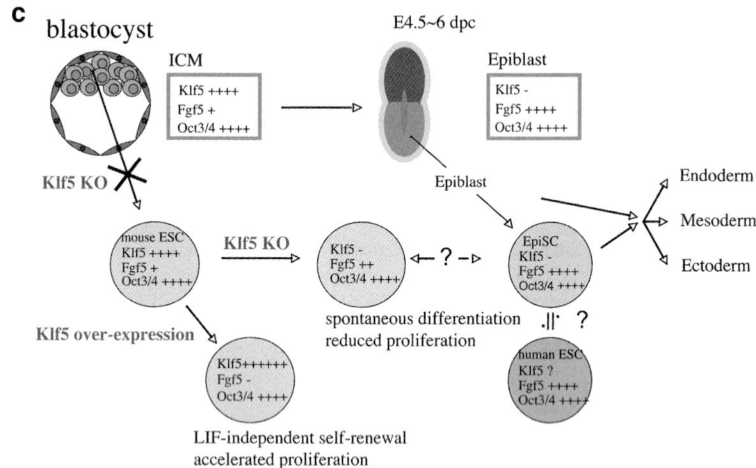

Fig. 2 **a** *Klf4* has redundant and nonoverlapping function with *Klf5*. **b** Model of how *Klf5* regulates self-renewal of ES cells. Klf5 represses the expression of *p21cip1* and activates Akt1 phosphorylation through Tcl1 expression resulting in a shortened G_1 phase, thereby contributing to proliferation. Klf5 also represses differentiation-related genes such as *Brachyury, Cdx2, Fgf5,* and others, and it activates *Nanog*. Thus, Klf5 plays a role in the self-renewal of mouse ES cells. **c** Summary of Klf5 roles in stem cell functions

in *Klf5* KO ES cells abrogates *Nanog* expression, indicating that the dependence of downstream gene expression on Klf factors is gene-specific. Kim et al. examined the target promoters of Oct3/4, Sox2, Klf4, c-Myc, and five other transcription factors by ChIP-on-ChIP analysis and found that a significant number of these promoters are occupied by these transcription factors (Kim et al. 2008).

As Jiang and coworkers reported, Klf2, Klf4, and Klf5 are expressed in undifferentiated ES cells; and expression of these factors decreases rapidly following the induction of differentiation. Consistent with this, we reported that Klf5 is expressed in the ICM of blastocysts, whereas its expression is decreased and undetectable in the epiblast (Ema et al. 2008). However, the factors controlling this large decrease in Klf5 expression remain unknown. Sato and coworkers showed that Wnt is essential for mouse and human ES self-renewal (Sato et al. 2004). Interestingly, Klf5 is induced by Wnt signaling in some cancerous cells (Ziemer et al. 2001), but the relation between Wnt and Klf5 expression in ES cells is not clear.

Reprogramming of Somatic Cells into Pluripotent Stem Cells by KLF4 and Other Klf Family Members

Takahashi and Yamanaka reported that rodent somatic cells could be reprogrammed into pluripotent ES-like cells by forced expression of four genes: *c-Myc, Oct3/4, Sox2*, *KLF4* (Takahashi and Yamanaka 2006). They were also able to reprogram human somatic cells into a pluripotent state following forced expression of the same combination of factors (Takahashi et al. 2007). In addition, Yu and coworkers showed that forced expression of OCT3/4, SOX2, NANOG, and LIN28 was able to reprogram a somatic cell into a pluripotent state (Yu et al. 2007). Subsequently, other groups showed that the same combination of transcription factors was able to reprogram human somatic cells into pluripotent cells (Maherali et al. 2007; Park et al. 2007; Wernig et al. 2007). The molecular mechanism of the reprogramming process is still poorly defined, although recent reports indicate that the gene-expression signature of the somatic cell changes in a stepwise manner that leads to the expression of pluripotency-related genes (Brambrink et al. 2008; Stadtfeld et al. 2008). Other members of the Klf family (e.g., *Klf1, Klf2, Klf5*) can substitute for *Klf4* function to induce pluripotency (Nakagawa et al. 2007). These studies suggest that Klf family members may have similar functions in somatic cell reprogramming, as has been shown during hematopoietic cell development or ES cell self-renewal (Basu et al. 2007; Ema et al. 2008; Jiang et al. 2008), although there is a significant difference in the efficiency of reprogramming by individual family members. Future work may allow the development of small molecules that can substitute for the forced expression of the reprogramming factors so as to avoid the use of virus-derived vectors in the generation of pluripotent stem cells.

Potential Functions of Klf5 During Embryogenesis and in Adults

Our studies have demonstrated that *Klf5* is a crucial factor for the derivation of ES cells and/or ICM development (Ema et al. 2008). In contrast, the roles of *Klf2* and KLF4 in these processes seem to be dispensable because *Klf2* KO mice show defects in hematovascular development and *Klf4* KO mice show defects in goblet cell differentiation and skin barrier formation (Katz et al. 2002; Segre et al. 1999), although further studies are needed for clarification.

Klf5 is expressed in trophectoderm and trophoblast stem (TS) cells. In fact, the lack of *Klf5* in embryos resulted in failure of implantation due to defective trophectoderm development. Thus, Klf5 may be important for TS cell functions as well as ICM development.

Triple knockdown of *Klf2, Klf4,* and *Klf5* in ES cells resulted in increased expression of *Fgf5* and *Brachyury* and decreased expression of *Nanog* and *Tcl1* (Ema et al. 2008; Jiang et al. 2008). This gene expression signature is similar to that of EpiSC, which is a pluripotent cell line derived from the epiblast

(Brons et al. 2007; Tesar et al. 2007). Interestingly, *Klf5* KO ES cells also show increased expression of *Fgf5* and *Brachyury* and decreased expression of *Tcl1,* although this alteration was not as dramatic as that observed in the triple knockdown mutant, and *Nanog* expression was unchanged (Ema et al. 2008) (Fig. 2C). Furthermore, *Klf5* KO ES cells can differentiate at an earlier developmental time point than WT ES cells, suggesting that *Klf5* KO ES cells may be primed for differentiation toward an EpiSC cell-like phenotype. Consistent with this idea, *Klf5* is expressed abundantly in ICM cells at the blastocyst stage but is not detectable in the epiblast of embryos at 5.5 and 6.5 dpc, implying that reduced *Klf5* expression may be required for differentiation of ICM cells into epiblast cells (Fig. 2C).

Taken together, several results demonstrate that *Klf* function is required for ES cell derivation from ICM cells and normal self-renewal of ES cells. Future studies should determine the molecular mechanism of how *Klf* regulates blastocyst development, represses expression of differentiation-inducing genes, and enhances progression of the cell cycle in ES cells. Such studies should increase our understanding of the mechanism underlying the maintenance of the pluripotent and proliferative state of ES cells as well as the mechanisms of the process of reprogramming a somatic cell into a pluripotent ES cell-like state.

References

Avilion, A.A., Nicolis, S.K., Pevny, L.H., Perez, L., Vivian, N., and Lovell-Badge, R. (2003). Multipotent cell lineages in early mouse development depend on SOX2 function. Genes Dev. *17*, 126 – 140

Basu, P., Lung, T.K., Lemsaddek, W., Sargent, T.G., Williams, D.C. Jr, Basu, M., Redmond, L.C., Lingrel, J.B., Haar, J.L., and Lloyd, J.A. (2007). EKLF and KLF2 have compensatory roles in embryonic ß-globin gene expression and primitive erythropoiesis. Blood *110*, 3417 – 3425

Brambrink, T., Foreman, R., Welstead, G. G., Lengner, C. J., Wernig, M., Suh, H., and Jaenisch, R. (2008). Sequential Expression of Pluripotency Markers during Direct Reprogramming of Mouse Somatic Cells. Cell Stem Cell *2*, 151–159

Brons, I. G., Smithers, L. E., Trotter, M. W., Rugg-Gunn, P., Sun, B., Chuva de Sousa Lopes, S. M., Howlett, S. K., Clarkson, A., Ahrlund-Richter, L., Pedersen, R. A., and Vallier, L. (2007). Derivation of pluripotent epiblast stem cells from mammalian embryos. Nature *448*, 191–195.

Cartwright, P., McLean, C., Sheppard, A., Rivett, D., Jones, D., and Dalton, S. (2005). LIF/STAT3 controls ES cell self-renewal and pluripotency by a Myc-dependent mechanism. Development *132*, 885–896.

Chambers, I., Colby, D., Robertson, M., Nichols, J., Lee, S., Tweedie, S. and Smith, A. (2003). Functional expression cloning of Nanog, a pluripotency sustaining factor in embryonic stem cells. Cell *113*, 643–655.

Ema, M., Mori, D., Niwa, H., Hasegawa, Y., Yamanaka, Y., Hitoshi, S., Mimura, J., Kawabe, Y., Hosoya, T., Morita, M., Shimosato, D., Uchida, K., Suzuki, N., Yanagisawa, J., Sogawa, K., Rossant, J., Yamamoto, M., Takahashi, S., and Fujii-Kuriyama, Y. (2008) Krüppel-like factor 5 is essential for blastocyst development and the normal self-renewal of mouse ESCs. Cell Stem Cell. *3*. 555–567.

Evans, M. J. and Kaufman, M. (1981). Establishment in culture of pluripotential cells from mouse embryos. Nature *292,* 154–156.

Ghaleb, A. M., Nandan, M. O., Chanchevalap, S., Dalton, W. B., Hisamuddin, I. M., and Yang, V. W. (2005). Kruppel-like factors 4 and 5: the yin and yang regulators of cellular proliferation. Cell Res. *15,* 92–96

Imataka, H., Sogawa, K., Yasumoto, K., Kikuchi, Y., Sasano, K., Kobayashi, A., Hayami, M., and Fujii-Kuriyama. Y. (1992). Two regulatory proteins that bind to the basic transcription element (BTE), a GC box sequence in the promoter region of the rat P-4501A1 gene. EMBO J *11,* 3663–3671.

Jiang, J., Chan, Y.S., Loh, Y.H., Cai, J., Tong, G.Q., Lim, C.A., Robson, P., Zhong, S., and Ng, H.H. (2008). A core KLF circuitry regulates self-renewal of embryonic stem cells. Nat. Cell Biol. *10,* 353–360.

Katz, J. P., Perreault, N., Goldstein, B. G., Lee, C. S., Labosky, P. A., Yang, V. W., and Kaestner, K. H. (2002). The zinc-finger transcription factor KLF4 is required for terminal differentiation of goblet cells in the colon. Development *129,* 2619–2628

Kim, J., Chu, J., Shen, X., Wang, J., and Orkin, S. H. (2008). An Extended Transcriptional Network for Pluripotency of Embryonic Stem Cells. Cell *132,* 1049–1061

Li, Y., McClintick, J., Edenberg, H.J., Yoder, M.C., and Chan, R.J., (2005). Murine embryonic stem cell differentiation is promoted by SOCS-3 and inhibited by the zinc finger transcription factor KLF4. Blood *105,* 635–637.

Maherali, N., Sridharan, R., Xie, W., Utikal, J., Eminli, S., Arnold, K., Stadtfeld, M., Yachechko, R., Tchieu, J., Jaenisch, R., Plath, K., and Hochedlinger, K. (2007). Directly reprogrammed fibroblasts show global epigenetic remodeling and widespread tissue contribution. Cell Stem Cell. *1,* 55–70.

Martin, G. R. (1981). Isolation of a pluripotent cell line from early mouse embryos cultured in medium conditioned by teratocarcinoma stem cells. Proc. Natl Acad. Sci. USA *78,* 7634–7638.

Matsuda, T., Nakamura, T., Nakao, K., Arai, T., Katsuki, M., Heike, T. and Yokota, T. (1999). STAT3 activation is sufficient to maintain an undifferentiated state of mouse embryonic stem cells. *EMBO J.* 18, 4261–4269.

Mitsui, K., Tokuzawa, Y., Itoh, H., Segawa, K., Murakami, M., Takahashi, K., Maruyama, M., Maeda, M. and Yamanaka, S. (2003). The homeoprotein Nanog is required for maintenance of pluripotency in mouse epiblast and ES cells. Cell *113,*631–642.

Nakagawa, M., Koyanagi, M., Tanabe, K., Takahashi, K., Ichisaka, T., Aoi, T., Okita, K., Mochiduki, Y., Takizawa, N., and Yamanaka, S. (2008). Generation of induced pluripotent stem cells without Myc from mouse and human fibroblasts. Nat. Biotechnol. *26,* 101–106.

Nichols, J., Zevnik, B., Anastassiadis, K., Niwa, H., Klewe-Nebenius, D., Chambers, I., Scholer, H., and Smith, A. (1998). Formation of pluripotent stem cells in the mammalian embryo depends on the POU transcription factor Oct4. Cell *95,* 379–391.

Niwa H, Burdon T, Chambers I, and Smith A. (1998). Self-renewal of pluripotent embryonic stem cells is mediated via activation of STAT3. Genes Dev. *12,* 2048–2060.

Niwa, H., Miyazaki, J., and Smith, A. G. (2000). Quantitative expression of Oct3/4 defines differentiation, dedifferentiation or self-renewal of ES cells. Nature Genet. *24,* 372–376.

Park, I.H., Zhao, R., West, J.A., Yabuuchi, A., Huo, H., Ince, T.A., Lerou, P.H., Lensch, M.W., and Daley, G.Q. (2007). Reprogramming of human somatic cells to pluripotency with defined factors. Nature *451,* 141–146.

Sato, N., Meijer, L., Skaltsounis, L., Greengard, P. and Brivanlou, A. H. (2004). Maintenance of pluripotency in human and mouse embryonic stem cells through activation of Wnt signaling by a pharmacological GSK-3-specific inhibitor. Nat. Med. *10,* 55–63.

Segre, J.A., Bauer, C., and Fuchs, E. (1999). KLF4 is a transcription factor required for establishing the barrier function of the skin, Nat. Genet *22,* 356–360

Shindo, T., Manabe, I., Fukushima, Y., Tobe, K., Aizawa, K., Miyamoto, S., Kawai-Kowase, K., Moriyama, N., Imai, Y., Kawakami, H., Nishimatsu, H., Ishikawa, T., Suzuki, T., Morita, H., Maemura, K., Sata, M., Hirata, Y., Komukai, M., Kagechika, H., Kadowaki, T., Kurabayashi, M., and Nagai, R. (2002). Kruppel-like zinc-finger transcription factor KLF5/BTEB2 is a target for angiotensin II signaling and an essential regulator of cardiovascular remodeling. Nat. Med. *8,* 856–863.

Smith, A. G., Heath, J. K., Donaldson, D. D., Wong, G. G., Moreau, J., Stahl, M., and Rogers, D. (1988). Inhibition of pluripotential embryonic stem cell differentiation by purified polypeptides. Nature *336*,688–690.

Sogawa, K., Imataka, H., Yamasaki, Y., Kusume, H., Abe, H., and Fujii-Kuriyama. Y. (1993). cDNA cloning and transcriptional properties of a novel GC box-binding protein, BTEB2. Nucleic Acids Res. *21,* 1527–1532.

Stadtfeld, M., Maherali, N., Breault, D. T., and Hochedlinger, K. (2008). Defining Molecular Cornerstones during Fibroblast to iPS Cell Reprogramming in Mouse. Cell Stem Cell *2*, 230–240

Takahashi, K., and Yamanaka, S. (2006). Induction of pluripotent stem cells from mouse embryonic and adult fibroblast cultures by defined factors. Cell *126*, 663–676.

Takahashi, K., Tanabe, K., Ohnuki, M., Narita, M., Ichisaka, T., Tomoda, K., and Yamanaka, S. (2007). Induction of pluripotent stem cells from adult human fibroblasts by defined factors. Cell *131*, 861–372.

Tesar, P.J., Chenoweth, J.G., Brook, F.A., Davies, T.J., Evans, E.P., Mack, D.L., Gardner, R.L., McKay, R.D. (2007). New cell lines from mouse epiblast share defining features with human embryonic stem cells. Nature *448*, 196–199.

Wernig, M., Meissner, A., Foreman, R., Brambrink, T., Ku, M., Hochedlinger, K., Bernstein, B.E., and Jaenisch, R. (2007). In vitro reprogramming of fibroblasts into a pluripotent ES-cell-like state. Nature *448*, 318–324.

Williams, R. L., Hilton, D. J., Pease, S., Willson, T. A., Stewart, C. L., Gearing, D. P., Wagner, E. F., Metcalf, D., Nicola, N. A., and Gough, N. M. (1988). Myeloid leukaemia inhibitory factor maintains the developmental potential of embryonic stem cells. Nature *336,* 684-687.

Ying, Q.-L., Nichols, J., Chambers, I. and Smith, A. (2003). BMP induction of Id proteins suppresses differentiation and sustains embryonic stem cell self-renewal in collaboration with STAT3. Cell *115*, 281–292.

Yu, J., Vodyanik, M.A., Smuga-Otto, K., Antosiewicz-Bourget, J., Frane, J.L.,Tian, S., Nie, J., Jonsdottir, G.A., Ruotti, V., Stewart, R., Slukvin, I. I., and Thomson, J. A. (2007). Induced pluripotent stem cell lines derived from human somatic cells. Science *318*, 1917–1920.

Ziemer, L. T., Pennica, D., and Levine, A. J. (2001). Identification of a murine homolog of the human BTEB2 transcriptional factor as a beta-catenin-independent Wnt-1 responsive gene. Mol. Cell. Biol. *21,* 562–574.

Part 4
Krüppel-like Factors in Organ Function and Disease

Chapter 11
Krüppel-like Factors and the Liver

Goutham Narla and Scott L. Friedman

Abstract The Krüppel-like zinc finger transcription factor family encodes a family of proteins that currently includes at least 15 members regulating remarkably diverse processes, including cell growth, signal transduction, and differentiation. KLF genes are highly conserved evolutionarily with homologues expressed in zebrafish and *Xenopus*. The role of the Krüppel-like factor (KLF) family in the liver is equally diverse, with roles in the regulation of adipogenesis, gluconeogenesis, apoptosis, cell cycle progression, cellular differentiation, energy homeostasis, oxidative stress, and hepatic stellate cell activation. Of the KLFs studied to date in the liver, KLF6 is the best characterized. This chapter therefore highlights the functional diversity of KLF6 in the liver with the understanding that it serves as a template for the study of other KLFs in the liver. Specifically, we focus on the *KLF6* gene and its cloning and identification from activated hepatic stellate cells, the importance of alternative splicing in the regulation of *KLF6* gene function, the contribution of the KLF family to liver development and injury, and finally, dysregulation of KLF6 function in disease through several important mechanisms including alternative splicing, decreased expression, loss of heterozygosity, and somatic mutation.

G. Narla
Department of Genetics and Genomic Sciences, Mount Sinai School of Medicine, New York, NY 10029, USA

G. Narla and S.L. Friedman (✉)
Department of Medicine, Mount Sinai School of Medicine, 1425 Madison Avenue, New York, NY 10029, USA
e-mail: Scott.Friedman@mssm.edu

R. Nagai et al. (eds.), *The Biology of Krüppel-like Factors*,
DOI 10.1007/978-4-431-87775-2_11, © Springer 2009

Introduction

KLF Transcription Factor Family

The Krüppel-like factor (KLF) family of transcription factors is named after the developmental regulator Krüppel from *Drosophila melanogaster,* inactivation of which causes a crippled phenotype in flies (Gloor et al. 1950). To date, this family of transcription factors includes at least 24 members, including both Sp1-like (Sp1–Sp8) and KLF-like (KLF1–KLF16) factors. They are phylogenetically organized into three general subgroups based on structural and functional features and can act as either activators or repressors of their target genes depending on the cellular environment. As a family, they regulate remarkably diverse processes, including cell growth, signal transduction, cellular differentiation, and key metabolic pathways/processes including adipogenesis and gluconeogenesis (Dang et al. 2000; Huber et al. 2001; Philipsen and Suske 1999; Turner and Crossley 1999).

The KLF family of transcription factors is characterized by a unique modular structure that defines members of this gene family. All KLF members have a highly conserved C-terminal 81-amino-acid C_2H_2 zinc finger DNA-binding domain that can interact with either "GC-box" or "CACC-box" DNA motifs in responsive promoters. In addition, the zinc finger motifs may regulate protein–protein interactions that modulate DNA-binding specificity (Li et al. 2005; Song et al. 2003; Zhang et al. 2002). The amino-terminal domains of the KLF proteins are highly variable and are believed to provide functional identity through the recruitment and regulation of specific protein–protein interactions (Song et al. 2003; Zhang et al. 2002). Also, these divergent N-terminal regions may include specific activation and/or repression domains that facilitate interactions with various co-activators and co-repressors, respectively, and account for a remarkably broad range of biological activities. The role of this gene family in the liver is best characterized and exemplified by the *KLF6* gene.

KLF6 Gene

KLF6 was originally cloned and independently isolated from three tissues: hepatic mesenchymal stellate cells (Kim et al. 1998), placental cells (Koritschoner et al. 1997), and peripheral blood lymphocytes from a chronic B-cell lymphocytic leukemia (B-CLL) patient (El Rouby et al. 1996). KLF6 can regulate a remarkably diverse range of cellular processes including growth, differentiation (Matsumoto et al. 2006), adhesion, and endothelial motility. This is accomplished through the regulation of gene expression, including placental glycoprotein (Koritschoner et al. 1997), human immunodeficiency virus-1 long terminal repeat (HIV-1 LTR) (Zhao et al. 2000), keratins 4 and 12 (Chiambaretta et al. 2002; Okano et al. 2000), transforming growth factor-β1 (TGF-β1), types I and II TGF-β receptors (Kojima et al. 2000), matrix metallopeptidase-9 (MMP-9) (Das et al. 2006), p21 (Narla et al. 2001), and E-cadherin (DiFeo et al. 2006). KLF6 has also been implicated in the regulation of various metabolic processes, including adipogenesis (Inuzuka et al. 1999; Li et al. 2005).

KLFs and Alternative Splicing

Alternative mRNA splicing is a key molecular event that generates protein diversity and is critical in the regulation of gene function. Through this process the exons of primary transcripts (pre-mRNAs) from a single gene can be spliced in different arrangements to produce structurally and functionally distinct mRNA and protein variants. Thus, a single gene can generate multiple protein products that can display different, even antagonist, biological functions (Shin and Manley 2004). Much of the generation of functional diversity in higher organisms is attributed to alternative splicing (Mercatante et al. 2002). The role of dysregulated alternative splicing in disease progression is now recognized in a range of human diseases (Cáceres and Kornblihtt 2002; Faustino and Cooper 2003; Garcia-Blanco et al. 2004). Many cancer-associated genes are alternatively spliced, and expression of these genes leads to the production of multiple splice variants with antagonist functions (Mercatante et al. 2002). As the prototypical member of the KLF family, recent evidence from our group and others have identified an important functional role for alternative splicing in the regulation of *KLF6* gene function. In one of the largest, multiinstitutional association studies in prostate cancer, KLF6 was found to be aberrantly spliced, with a single nucleotide polymorphism (SNP) in intron 1 of the *KLF6* gene increasing alternative splicing of the gene (Narla et al. 2005a). KLF6 alternative splicing results in the generation of three biologically active splice isoforms, KLF6-SV1, KLF6-SV2, and KLF6-SV3, in both normal and cancerous tissue (Narla et al. 2005b) (Fig. 1). These variants lack either regions of the KLF6 activation domain and/or the DNA-binding domain. In particular, KLF6-SV1 lacks all three zinc finger DNA-binding domains but retains most of the KLF6 N-terminal activation domain (Narla et al. 2005a). This splice variant is particularly important, as it is significantly upregulated in many cancers (Camacho-Vanegas et al. 2007; DiFeo et al. 2008; Narla et al. 2005a, 2005b; Teixeira et al. 2007). Increased KLF6-SV1 expression is dependent on oncogenic Ras/PI3-K/Akt-signaling thereby altering the relative ratio of KLF6 to KLF6-SV1 in hepatocellular carcinoma (Yea et al. 2008). More studies to determine the presence of alternative spliced variants of other KLFs is warranted. This work could provide important insights into an additional mechanism of functional regulation of this gene family in the liver.

KLFs and Liver Development

KLFs play key roles in several developmental pathways based on studies in both lower species (Oates et al. 2001) and mammals. All *Klf* homozygous knockout (KO) mice generated to date have a lethal phenotype. These models have revealed roles of KLFs in blood vessel stability (KLF2) (Kuo et al. 1997), β-globin synthesis during erythropoiesis (KLF1) (Perkins et al. 1995), and epithelial barrier integrity (KLF4) (Segre et al. 1999). Studies using chimeric mice derived from *Rag2*$^{-/-}$ and *Klf2*$^{-/-}$ animals have also indicated a role of *Klf2* in T-cell activation (Kuo et al. 1997). These developmental activities are ascribed to both those KLFs that are tissue-restricted, as well as those that are ubiquitously expressed. For example, *Klf4* and *Klf6* are both widely expressed, yet their developmental expression is restricted (Blanchon et al. 2001; Fischer et al. 2001; Laub et al. 2001b; Ton-That et al. 1997). Importantly, even heterozygous *Klf* gene-deleted mice may reveal an abnormal phenotype, as in the case of *Klf5*$^{+/-}$ animals, which have a defect in arterial wall remodeling when stressed (Shindo et al. 2002).

Among the subfamilies of KLFs, the *Klf6* gene is most closely related to *Klf7*, and they share a common progenitor in *Drosophila*, the *Luna* gene, whose inactivation using siRNA leads to defects in organogenesis and terminal differentiation (De Graeve et al. 2003). Despite its ubiquitous expression in adult tissues, developmental expression of *Klf6* is somewhat restricted and distinct from *Klf7*. Whereas *Klf7* mRNA is primarily expressed in neuronal tissue, *Klf6* transcripts are also found in several nonneural sites, including hindgut, heart, lung, kidney, and limb buds (Laub et al. 2001a,b).

Based on these findings, we analyzed the role of *Klf6* in mouse development. *Klf6*$^{-/-}$ mice die by E 12.5 and are characterized by markedly reduced hematopoietic differentiation in yolk sacs. Complementing this approach, we employed a mouse embryonic stem (ES) cell differentiation system to assess the capacity of murine *Klf6*$^{-/-}$ ES cells to differentiate in vitro (Gouon-Evans et al., 2006; Zhao et al., 2006). The differentiated *Klf6*$^{+/+}$ and *Klf6*$^{+/-}$ ES cell-derived cultures expressed similar levels of endoderm markers as well as early (e.g., α-fetoprotein, transthyretin) and late (albumin) hepatic mRNAs. Cytoplasmic albumin expression was also made apparent by immunocytochemistry analysis. Differentiation was accompanied by a biphasic increase in *Klf6* mRNA on day 2 and again after day 4. In contrast to *Klf*$^{+/+-}$ ES cells, *Klf6*$^{-/-}$ ES cells displayed significantly reduced expression of endoderm markers and do not express hepatocyte markers (α-fetoprotein, transthyretin, albumin). Furthermore, forced expression of KLF6 at specific time points using a tet-inducible system increasea endoderm and hepatic marker expression. These findings indicate that KLF6 regulates endoderm formation and hepatic differentiation in a temporally specific manner in ES cells. However, it is unclear whether these defects are restricted to liver or affect all endoderm-derived tissues (e.g., pancreas).

No studies reported to date implicate other KLFs in liver development. However, this area requires much more study, particularly in view of the essential role of KLF2 in stem cell reprogramming (Takahashi et al. 2006).

KLFs and Liver Injury

Hepatic fibrosis is the liver's wound-healing response to injury of any type and can lead to cirrhosis, characterized by scar accumulation and nodule formation. Hepatic stellate cells are the principal source of extracellular matrix (ECM) in hepatic fibrosis and play a central role in the injury response by undergoing activation, which connotes the transition from a quiescent vitamin A-rich cell to a contractile, proliferative, fibrogenic cell type (Friedman 2008).

Our previous efforts to understand the molecular basis of stellate cell activation utilized subtraction hybridization to clone a novel zinc finger transcription factor, KLF6 (initially called Zf9), which is induced as an immediate–early gene in hepatic stellate cells during liver injury in vivo (Ratziu et al. 1998). Subsequent studies have broadened KLF6's roles in injury to include growth responses of vascular endothelial cells (Botella et al. 2002) and hepatocytes among others.

The discovery that *KLF6*, a growth-suppressive gene, is rapidly induced when stellate cells undergo a proliferative burst, presented a paradox that has now been resolved with the discovery that KLF6 can undergo alternative splicing in rodents and humans to shorter, dominant negative isoforms (KLF6sv1, KLF6sv2, KLF6sv3) that lack all or part of the DNA-binding domain (Narla et al. 2005a) (Fig. 1).

Nonalcoholic fatty liver disease (NAFLD) represents the manifestation of the metabolic syndrome in the liver. It encompasses the entire spectrum of liver disease from steatohepatitis to frank cirrhosis. The prevalence of NAFLD is growing rapidly, and it is now the most common cause of chronic liver disease in Western countries. The cardiovascular health implications of the metabolic syndrome are significant, with the risk of progressive liver disease, cirrhosis, and liver cancer set to have a further major impact. Despite its high prevalence, however, less than one-fourth of subjects with NAFLD ever progress beyond steatosis to develop significant fibrosis or liver cancer. The reasons for these differences in individual susceptibility to progressive disease are unclear; and although several candidate genes have been studied, as yet no genetic associations with advanced NAFLD have been replicated in large studies. Recent evidence from our group suggests an important role for KLF6 in NAFLD. Specifically, KLF6 expression is increased in association with increased steatosis, inflammation, and fibrosis in NAFLD livers. In addition, a common polymorphism *KLF6*-IVS1-27G>A that promotes increased alternative splicing of the *KLF6* gene (Narla et al. 2005a) was significantly associated significantly with mild NAFLD (Miele et al. 2008).

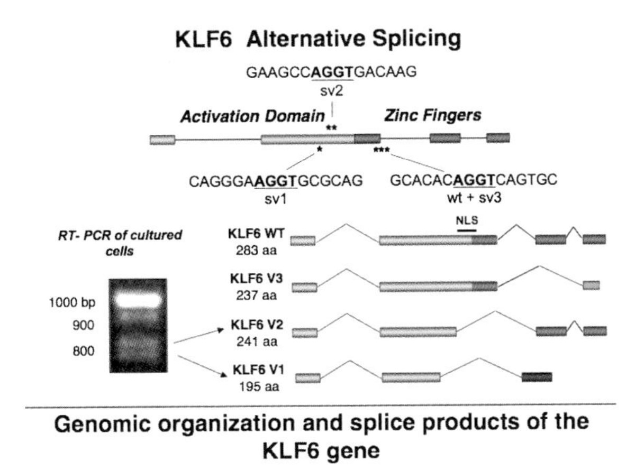

Fig. 1. Structure of KLF6 and its splice variants. Genomic organization, cryptic splice site sequences, and mRNA structures for wild-type and KLF6 splice variants. *NLS* = nuclear localization signal. Activation domain is *light blue*. Zinc fingers are *dark blue*. Novel amino acids are *green* and *pink*. *RT-PCR* = reverse transcription-polymerase chain reaction. (Adapted from Difeo A, Martignetti J, Narla G. The role of KLF6 and its splice variants in cancer therapy. Drug Resistance Updates, 12(1–2) 11–7, 2009

KLF6 and Hepatocellular Carcinoma

Hepatocellular carcinoma (HCC), an often fatal neoplasm in patients with chronic liver disease, is the fifth most common cancer worldwide and the third leading cause of cancer death. As with most cancers, hepatic carcinogenesis is characterized by loss of differentiation and deregulated growth of malignant cells. Much is still unknown about HCC pathogenesis, and many of the specific molecular mechanisms underlying its development and progression have yet to be identified. Recent evidence suggests that KLF6 is a bona fide tumor-suppressor gene inactivated in HCC by a number of mechanisms. Since the initial discovery of its role in prostate cancer, mounting evidence from our laboratory and others has highlighted a variety of KLF6 *inactivating* mechanisms relevant to tumor growth and spread in HCC. These mechanisms include the following.

- Loss of heterozygosity (LOH) and somatic mutation (Kremer-Tal et al. 2007; Tal-Kremer et al. 2004; Wang et al. 2004)
- LOH and/or decreased expression (Kremer-Tal et al. 2007; Wang et al. 2004)
- Transcriptional silencing through promoter hypermethylation (Hirasawa et al. 2007)
- Dysregulated alternative splicing (Yea et al. 2008)

These data confirm that KLF6 plays a general role in the development and progression of human HCC. Additional studies of other KLFs, including KLF4, KLF5, and KLF7 are warranted in HCC given their structural and functional similarity to the *KLF6* tumor suppressor gene to determine their role in disease development and progression.

Conclusions and Future Directions

The Krüppel-like zinc finger transcription factor family encodes a diverse family of transcription factors that regulate a range of physiological and pathological processes. The role of the KLF family in the liver is equally diverse, with functions being identified in the regulation of adipogenesis (Banerjee et al. 2003; Birsoy et al. 2008; Kawamura et al. 2005; Li et al. 2005; Oishi et al. 2005; Sue et al. 2008; Wu et al. 2005), gluconeogenesis (Gray et al. 2007), apoptosis (Sirach et al. 2007), cell cycle progression (Narla et al. 2007; Tal-Kremer et al. 2004), cellular differentiation (Kremer-Tal et al. 2007), energy homeostasis (Hashmi et al. 2008; Teshigawara et al. 2005), oxidative stress (Starkel et al. 2003), and hepatic stellate cell activation (Friedman 2006; Kojima et al. 2000; Mann and Smart 2002). KLF6 was originally cloned from the activated hepatic stellate cells and serves as the prototypical example of the role of the KLF gene family in both normal liver physiology and disease states including NAFLD, cirrhosis, and HCC. Further studies of additional members of this family of transcription factors is warranted as the role of many of its members in the liver have yet to be identified. In addition, many of the studies of the KLF6 gene highlight and illustrate a remarkable functional diversity in mechanisms of gene regulation and inactivation in cancer and disease development. The liver, with its complex structure and diverse functions provides the ideal model to uncover new roles for this extraordinary family of transcription factors.

References

Banerjee SS, Feinberg MW, Wantanabe M, Gray S, Haspel RL, Denkinger DJ, Kawahara R, Hauner H, Jain MK (2003) The Kruppel-like factor KLF2 inhibits peroxisome proliferator-activated receptor-gamma expression and adipogenesis. J Biol Chem. 278:2581–2584.
Birsoy K, Chen Z, Friedman J (2008) Transcriptional regulation of adipogenesis by KLF4. Cell Metab. 7:339–347.
Blanchon L, Bocco JL, Gallot D, et al (2001) Co-localization of KLF6 and KLF4 with pregnancy-specific glycoproteins during human placenta development. Mech Dev. 105:185–189.
Botella LM, Sanchez-Elsner T, Sanz-Rodriguez F, et al (2002) Transcriptional activation of endoglin and transforming growth factor-beta signaling components by cooperative interaction between Sp1 and KLF6: their potential role in the response to vascular injury. Blood 100: 4001-4010.
Cáceres JF, Kornblihtt AR (2002) Alternative splicing: multiple control mechanisms and involvement in human disease. Trends Genet. 18: 186–193.
Camacho-Vanegas O, Narla G, Teixeira MS, et al (2007) Functional inactivation of the KLF6 tumor suppressor gene by loss of heterozygosity and increased alternative splicing in glioblastoma. Int. J. Cancer 121: 1390–1395.
Chiambaretta F, Blanchon L, Rabier B, et al (2002) Regulation of corneal keratin-12 gene expression by the human Kruppel-like transcription factor 6. Invest. Ophthalmol. Vis. Sci. 43: 3422-3429.
Dang DT, Pevsner J, Yang VW (2000) The biology of the mammalian Krüppel-like family of transcription factors. Int. J. Biochem. Cell Biol. 32: 1103–1121.
Das A, Fernandez-Zapico, ME, Cao S, et al (2006) Disruption of an SP2/KLF6 repression complex by SHP is required for farnesoid X receptor-induced endothelial cell migration. J. Biol. Chem. 281: 39105–39113.
De Graeve F, Smaldone S, Laub F, et al (2003) Identification of the Drosophila progenitor of mammalian Kruppel-like factors 6 and 7 and a determinant of fly development. Gene.314:55–62.
DiFeo A, Feld L, Rodriguez E, et al (2008) A functional role for KLF6-SV1 in lung adenocarcinoma prognosis and chemotherapy response. Cancer Res. 68: 965–970.
DiFeo A, Narla G, Camacho-Vanegas O, et al (2006a) E-cadherin is a novel transcriptional target of the KLF6 tumor suppressor. Oncogene 25: 6026–6031.
El Rouby S, Newcomb EW (1996) Identification of Bcd, a novel proto-oncogene expressed in B-cells. Oncogene. 13(12):2623–2630.
Faustino NA, Cooper TA (2003) Pre-mRNA splicing and human disease. Genes Dev. 17: 419-437.
Fischer EA, Verpont MC, Garrett-Sinha LA, et al (2001) Klf6 is a zinc finger protein expressed in a cell-specific manner during kidney development. J Am Soc Nephrol. 12:726–735.
Friedman SL (2006) Transcrptional regulation of stellate cell activation. J Gastroenterol Hepatol, 21: S79–83.
Friedman SL (2008) Hepatic Stellate Cells – Protean, Multifunctional, and Enigmatic Cells of the Liver. Physiological Reviews. 88:125–172.
Garcia-Blanco MA, Baraniak AP, Lasda EL (2004) Alternative splicing in disease and therapy. Nat. Biotechnol. 22: 535–546.
Gloor H, (1950) Arch. Julius Klaus-Stift. 25: 38–44.
Gouon-Evans V, Boussemart L, Gadue P, et al (2006) BMP-4 is required for hepatic specification of mouse embryonic stem cell-derived definitive endoderm. Nat Biotechnol. 24:1402–1411.
Gray S, Wang B, Orihuela Y, et al (2007) Regulation of gluconeogenesis by Kruppel-like factor 15. Cell Metab 5:305–312.

Hashmi S, Ji Q, Zhang J, Parhar RS, Huang CH, Brey C, Gaugler R (2008) A Kruppel-like factor in *Caenorhabditis elegans* with essential roles in fat regulation, cell death, and phagocytosis. DNA Cell Biol. 27:545–551.

Hirasawa Y, Arai M, Imazeki F, et al (2007) Methylation status of genes upegulated by demethylating agent 5-aza-2′-deoxycytidine in hepatocellular carcinoma. Oncology 71:77–85.

Huber TL, Perkins AC, Deconinck AE et al (2001) Neptune, A Kruppel-like transcription factor that participates in primitive erythropoiesis in *Xenopus*. Curr Biol 11:1456–1461.

Inuzuka H, Nanbu-Wakao R, Masuho Y et al (1999) Differential regulation of immediate early gene expression in preadipocyte cells through multiple signaling pathways. Biochem. Biophys. Res. Commun. 265: 664–668.

Kawamura Y, Tanaka Y, Kawamori R, et al (2006). Overexpression of Kruppel-like factor 7 regulates adipocytokine gene expressions in human adipocytes and inhibits glucose-induced insulin secretion in pancreatic beta-cell-line. Mol Endocrinol. 20: 844–56.

Kim Y, Ratziu V, Choi SG, et al (1998) Transcriptional activation of transforming growth factor beta1 and its receptors by the Kruppel-like factor Zf9/core promoter-binding protein and Sp1. Potential mechanisms for autocrine fibrogenesis in response to injury. J. Biol. Chem. 273: 33750–33758.

Kojima, S, Hayashi S, Shimokado K, et al (2000) Transcriptional activation of urokinase by the Kruppel-like factor Zf9/COPEB activates latent TGF-beta1 in vascular endothelial cells. Blood 95: 1309–1316.

Koritschoner NP, Bocco JL, Panzetta-Dutari GM, et al (1997) A novel human zinc finger protein that interacts with the core promoter element of a TATA box-less gene. J. Biol. Chem. 272: 9573–9580.

Kremer-Tal S, Narla, G, Chen Y, et al (2007) Downregulation of KLF6 is an early event in hepatocarcinogenesis, and stimulates proliferation while reducing differentiation. J. Hepatol. 46: 645–654.

Kuo CT, Veselits ML, Barton KP, et al (1997) The LKLF transcription factor is required for normal tunica media formation and blood vessel stabilization during murine embryogenesis. Genes Dev. 11:2996–3006.

Laub F, Aldabe R, Ramirez F, et al (2001a) Embryonic expression of Kruppel-like factor 6 in neural and non-neural tissues. Mech Dev.106:167–170.

Laub F, Aldabe R, Friedrich V, et al (2001b) Developmental expression of mouse Kruppel-like trnascription factor KLF7 suggests a potential role in neurogenesis. Dev Biol. 233:305–318.

Li D, Yea S, Dolios G, et al (2005) Kruppel-like factor-6 promotes preadipocyte differentiation through histone deacetylase 3-dependent repression of DLK1. J. Biol. Chem. 280: 26941-26952.

Mann DA, Smart DE (2002) Transcriptional regulation of hepatic stellate cell activation. Gut 50:891–896.

Matsumoto, N., Kubo, A., Liu, H., Akita, K., Laub, F., et al., 2006. Developmental regulation of yolk sac hematopoiesis by Kruppel-like factor 6. Blood 107, 1357–1365.

Mercatante, DR, Mohler JL, Kole R (2002) Cellular response to an antisense-mediated shift of Bcl-x pre-mRNA splicing and antineoplastic agents. J. Biol. Chem. 277: 49374–49382.

Miele L, Beale G, Patman G et al (2008) The Kruppel-like factor 6 genotype is associated with fibrosis in nonalcoholic fatty liver disease. 135: 282–291.

Narla G, DiFeo A, Reeves HL et al (2005b) A germline DNA polymorphism associated with increased prostate cancer risk enhances alternative splicing of the KLF6 tumor suppressor gene. Cancer Res. 65: 1213–1222.

Narla G, Kremer-Tal S, Matsumoto N, et al (2007). In vivo regulation of p21 by the Kruppel-like factor 6 tumor suppressor gene in mouse liver and human hepatocellular carcinoma. Oncogene. 26: 4428–4434.

Narla G, DiFeo A, Yao S, et al (2005a) Targeted inhibition of the KLF6 splice variant, KLF6 SV1, suppresses prostate cancer cell growth and spread. Cancer Res. 65: 5761–5768.

Narla G, Heath KE, Reeves HL, et al (2001) KLF6, a candidate tumor suppressor gene mutated in prostate cancer. Science 294: 2563–2566.

Oates, AC, Pratt, SJ, Vail, YI, et al (2001) The zebrafish klf gene family. Blood 98: 1792–1801.
Oishi Y, Manabe I, Tobe K et al (2005) Kruppel-like transcription factor KLF5 is a key regulator of adipocyte differentiation. Cell Metab 1:27–39.
Okano J, Opitz OG, Nakagawa H, et al (2000) The Kruppel-like transcriptional factors Zf9 and GKLF coactivate the human keratin 4 promoter and physically interact. FEBS Lett. 473: 95–100.
Perkins AC, Sharpe AH, Orkin SH (1995) Lethal beta-thalassaemia in mice lacking the erythroid CACC-transcription factor EKLF. Nature. 375:318–322.
Philipsen S, Suske G (1999) A tale of three fingers: the family of mammalian Sp/XKLF transcription factors. Nucl. Acids Res. 27: 2991–3000.
Ratziu V, Lalazar A, Wong L, et al (1998) Zf9, a Kruppel-like transcription factor up-regulated in vivo during early hepatic fibrosis. Proc Natl Acad Sci U S A. 95:9500–9505.
Segre JA, Bauer C, Fuchs E (1999) Klf4 is a transcription factor required for establishing the barrier function of the skin. Nat Genet. 22:356–360.
Shin C, Manley JL (2004) Cell signaling and the control of pre-mRNA splicing. Nat. Rev. Mol. Cell. Biol. 5: 727–738.
Shindo T, Manabe I, Fukushima Y, et al (2002) Kruppel-like zinc-finger transcription factor KLF5/BTEB2 is a target for angiotensin II signaling and an essential regulator of cardiovascular remodeling. Nat Med. 8:856–63.
Sirach, E, Bureau, C, Peron JM, et al (2007) KLF6 transcription factor protects hepatocellular carcinoma-derived cells from apoptosis. Cell Death Differ. 14: 1202–1210.
Song CZ, Keller K, Chen Y, et al (2003) Functional interplay between CBP and PCAF in acetylation and regulation of transcription factor KLF13 activity. J. Mol. Biol. 329: 207–215.
Starkel P, Sempoux C, Leclercq I et al (2003) Oxidative stress, KLF6 and transforming growth factor-beta upregulation differentiate non-alcoholic steatohepatitis progressing to fibrosis from uncomplicated steatosis in rats. J Hepatol 39:538–46.
Sue N, Jack BH, Eaton SA et al (2008) Targeted disruption of the basic Kruppel-like factor gene (KLF3) reveals a role in adipogenesis. Mol Cell Biol. 28: 3967–78.
Takahashi K, Yamanaka S (2006) Induction of pluripotent stem cells from mouse embryonic and adult fibroblast cultures by defined factors. Cell;126:663–676.
Tal-Kremer S, Reeves HL, Narla, G, et al (2004) Frequent inactivation of the tumor suppressor Kruppel-like Factor 6 (KLF6) in hepatocellular carcinoma. Hepatology 40: 1047–1052.
Teixeira MS, Camacho-Vanegas O, Fernandez Y, et al (2007) KLF6 allelic loss is associated with tumor recurrence and markedly decreased survival in head and neck squamous cell carcinoma. Int. J. Cancer 121: 1976–1983.
Teshigawara K, Ogawa W, Mori T, et al (2005) Role of Kruppel-like factor 15 in PEPCK gene expression in the liver. Biochem Biophys Res Commun 327: 920–926.
Ton-That H, Kaestner KH, Shields JM, et al (1997) Expression of the gut-enriched Kruppel-like factor gene during development and intestinal tumorigenesis. FEBS Lett. 419:239–243.
Turner J, Crossley M (1999) Mammalian Krüppel-like transcription factors: more than just a pretty finger. Trends Biochem. Sci. 24: 236–240.
Wang S, Chen X, Zhang W, et al (2004) KLF6mRNA expression in primary hepatocellular carcinoma. J. Huazhong Univ. Sci. Technolog. Med. Sci. 24: 585–587.
Wu J, Srinivasan S, Neumann JC, et al (2005) The KLF2 transcription factor does not affect the formation of preadipocytes but inhibits their differentiation into adipocytes. Biochemistry 44: 11098–11105.
Yea S, Narla G, Zhao X, et al (2008) Ras promotes growth by alternative splicing-mediated inactivation of the KLF6 tumor suppressor in hepatocellular carcinoma. Gastroenterology 134: 1521–1531.
Zhang D, Zhang XL, Michel FJ, et al (2002) Direct interaction of the Kruppel-like family (KLF) member, BTEB1, and PR mediates progesterone-responsive gene expression in endometrial epithelial cells. Endocrinology 143: 62–73.

Zhang W, Shields JM, Sogawa K, et al (1998) The gut-enriched Kruppel-like factor suppresses the activity of the CYP1A1 promoter in an Sp1-dependent fashion. J. Biol. Chem. 273: 17917–17925.

Zhao JL, Austen KF, Lam BK (2000) Cell-specific transcription of leukotriene C(4) synthase involves a Kruppel-like transcription factor and Sp1. J. Biol. Chem. 275: 8903-8910.

Zhao X, Matsumoto N, Gouon-Evans V, et al (2006) The role of Krüppel-like factor 6 (KLF6) in differentiation of embryonic stem cells into hepatocytes. Hepatology, abstract.

Chapter 12
Role of Krüppel-like Factor 15 in Adipocytes

Wataru Ogawa, Hiroshi Sakaue, and Masato Kasuga

Abstract Krüppel-like factor 15 (KLF15) has been implicated in energy metabolism in various tissues including muscle, heart, liver, and adipose tissue. The expression of KLF15 is induced by the synergistic action of CCAAT/enhancer-binding protein β (C/EBPβ) and C/EBPδ during the differentiation of preadipocytes into adipocytes. The time course of KLF15 expression during this process is similar to that for C/EBPα, and these two proteins appear to promote the differentiation program in a cooperative manner through induction of the peroxisome proliferator-activated receptor γ (PPARγ) gene and other adipocyte-specific genes. A combination of microarray-based chromatin immunoprecipitation and gene expression analyses identified six genes whose promoters bound KLF15 and whose expression was either increased or decreased by forced expression of KLF15 in 3T3-L1 adipocytes. The gene for adrenomedullin, a vasodilatory hormone implicated in the pathogenesis of obesity-induced hypertension and insulin resistance, was one of these genes whose expression appears to be regulated by KLF15. KLF15 may thus also control the function of mature adipocytes through regulation of such genes.

Introduction

Krüppel-like factor 15 (KLF15) was first identified as a protein that binds to the promoter of the gene for CLC-K1, a kidney-specific CLC chloride channel (Uchida et al. 2000); KLF15 was thus formerly designated kidney KLF (KKLF).

W. Ogawa and M. Kasuga
Division of Diabetes, Metabolism, and Endocrinology, Department of Internal Medicine, Kobe University Graduate School of Medicine, Kobe 650-0017, Japan

H. Sakaue
Department of Nutrition and Metabolism, University of Tokushima Graduate School, Institute of Health Biosciences, Tokushima 770-8503, Japan

M. Kasuga (✉)
Research Institute, International Medical Center of Japan, Tokyo 162-8655, Japan

R. Nagai et al. (eds.), *The Biology of Krüppel-like Factors*,
DOI 10.1007/978-4-431-87775-2_12, © Springer 2009

KLF15 was subsequently shown to regulate the expression of various genes whose products contribute to energy metabolism. For example, KLF15 regulates transcription of the gene for GLUT4 (Gray et al. 2002), an insulin-sensitive glucose transporter expressed in muscle and adipocytes. It is also implicated in the fasting-induced expression in skeletal muscle of the gene for acetyl-coenzyme A (CoA) synthetase 2 (Yamamoto et al. 2004), an enzyme of the mitochondrial matrix that provides acetyl-CoA for the citric acid cycle. Moreover, KLF15 has been shown to control glucose production by the liver through regulation of genes for enzymes involved in either gluconeogenesis (Teshigawara et al. 2005) or amino acid catabolism (Gray et al. 2007). It also appears to influence energy metabolism in the whole body through regulation of the differentiation of adipocytes as well as through control of the expression of various adipocyte-specific genes in mature adipocytes. In this chapter, we first summarize the functions of KLF15 and other members of the KLF family in differentiation of preadipocytes into adipocytes and then address the possible role of KLF15 in mature adipocytes.

Roles of KLF15 and Other KLF Family Proteins in Adipocyte Differentiation

The amount of adipose tissue in the body is an important determinant of energy homeostasis in living animals and is altered in various physiological or pathological conditions (Rosen and Spiegelman 2006). An increase in adipose tissue mass can result from an increase in cell size, cell number, or both (Rosen and Spiegelman 2000). The number of adipocytes is thought to increase as a result of the proliferation of preadipocytes and their subsequent differentiation into mature adipocytes. Such differentiation of preadipocytes is characterized by marked changes in the pattern of gene expression that are achieved by the sequential induction of various transcription factors. Preadipocytes exposed to hormonal inducers of differentiation thus manifest an early, transient increase in expression of the transcription factors CCAAT/enhancer–binding protein β (C/EBPβ) and C/EBPδ, which in turn promote a subsequent increase in the expression of C/EBPα and peroxisome proliferator-activated receptor γ (PPARγ) (Rangwala and Lazar 2000; Rosen and Spiegelman 2006) (Fig. 1). C/EBPα and PPARγ are thought to act synergistically in the transcriptional activation of a variety of adipocyte-specific genes, with each also reciprocally activating the expression of the other (Rangwala and Lazar 2000; Rosen and Spiegelman 2006).

Certain members of the KLF family of proteins have been implicated in the differentiation of preadipocytes. The differentiation of mouse 3T3-L1 preadipocytes into mature adipocytes in response to various hormonal inducers is a widely studied model of adipogenesis. Whereas 3T3-L1 preadipocytes express KLF2 and KLF3 at high levels, the amounts of these two proteins decrease rapidly after exposure of the cells to hormonal inducers of differentiation (Banerjee et al. 2003; Sue et al. 2008) (Fig. 2). Overexpression of KLF2 or KLF3 in 3T3-L1 preadipocytes has been shown

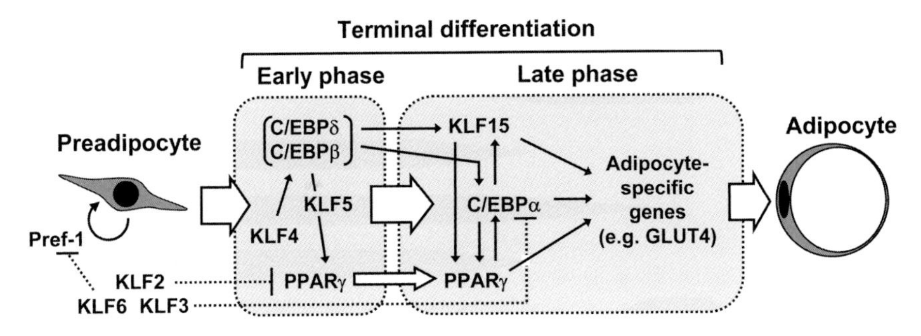

Fig. 1 Model for transcriptional control of adipocyte differentiation. Various members of the Krüppel-like factor (KLF) family of transcription factors contribute to transcriptional control of the differentiation of preadipocytes into adipocytes

to inhibit the differentiation of these cells by preventing the expression of PPARγ and C/EBPα, respectively (Banerjee et al. 2003; Sue et al. 2008). These observations suggest that the downregulation of these KLF proteins is important for differentiation of preadipocytes into adipocytes. Indeed, differentiation of embryonic fibroblasts derived from mice lacking KLF3 into adipocytes is enhanced compared with that for the corresponding wild-type cells (Sue et al. 2008). Preadipocyte factor-1 (Pref-1), also known as delta-like 1 (Dlk1), is a transmembrane protein that inhibits differentiation of preadipocytes. KLF6, which is transiently induced at an early phase of differentiation, inhibits the expression of Pref-1 and thereby promotes the differentiation of preadipocytes (Li et al. 2005) (Figs. 1, 2). The expression of KLF4 is also increased at an early phase of preadipocyte differentiation (within 1 hour after exposure to hormonal inducers), and this protein appears to contribute to differentiation by stimulating the expression of C/EBPβ and C/EBPδ (Birsoy et al. 2008).

KLF15 is also implicated in the differentiation of preadipocytes. Expression of KLF15 is markedly increased at a relatively late phase of the differentiation process (Mori et al. 2005) (Fig. 3A). Given that forced expression of C/EBPβ or C/EBPδ in NIH 3T3 fibroblasts results in a synergistic increase in the amount of KLF15 mRNA (Mori et al. 2005), the former two proteins likely trigger expression of KLF15 during the differentiation of preadipocytes (Fig. 2). Forced expression of KLF15 in NIH 3T3 fibroblasts or C2C12 myoblasts directs these nonadipocyte cell lines into the adipocyte lineage (Mori et al. 2005). Inhibition of the function of KLF15 in 3T3-L1 preadipocytes, by expression of a dominant negative mutant or by RNA interference, was found to attenuate expression of PPARγ as well as adipocytic differentiation (Mori et al. 2005), indicating that KLF15 is essential for the differentiation process in these cells. The time course of KLF15 expression during differentiation is similar to that for C/EBPα. KLF15 and C/EBPα stimulate, in an additive manner, both adipocytic differentiation of nonadipocyte cell lines as well as the activity of the PPARγ gene promoter (Mori et al. 2005), suggesting that these two proteins coordinately regulate gene transcription associated with the terminal differentiation of adipocytes.

Although a dominant negative mutant of KLF15 was shown to inhibit expression of PPARγ during the late phase of adipocytic differentiation in 3T3-L1 cells, it did not

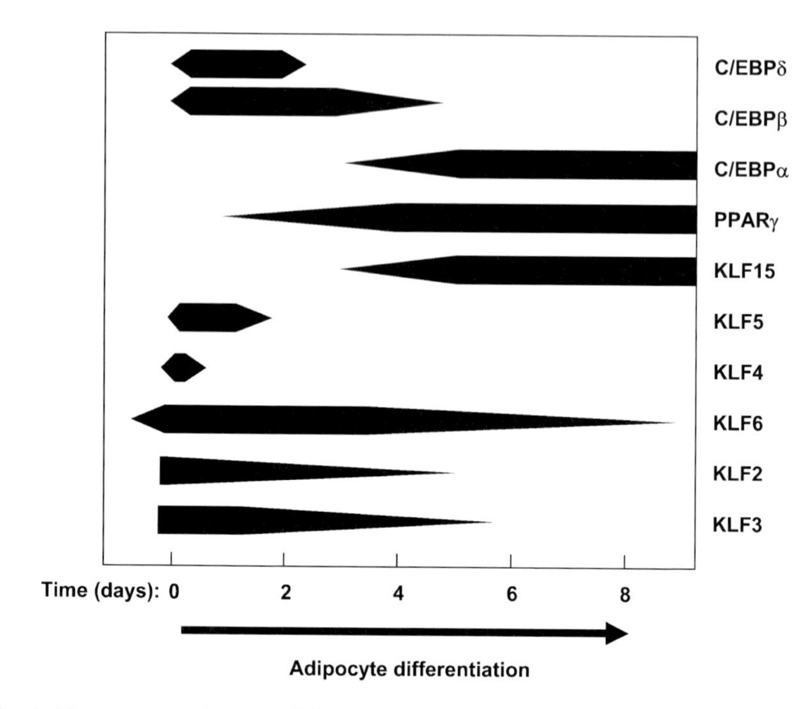

Fig. 2 Time course and extent of the expression of various transcription factors involved in regulation of the differentiation of 3T3-L1 preadipocytes into adipocytes in response to hormonal inducers. *IgG* = immunoglobulin G; *ChIP-chip* = chromatin immunoprecipitation with a promoter oligonucleotide microarray

affect early adipogenesis (on day 2) before the onset of KLF15 expression (Mori et al. 2005). These observations suggest that whereas KLF15 contributes to maintenance of the expression of PPARγ at a high level during the late phase of differentiation the early induction of PPARγ is achieved by a mechanism independent of KLF15. Both C/EBPβ and C/EBPδ bind directly to the promoter of the PPARγ gene and stimulate expression of PPARγ (Rangwala and Lazar 2000). Moreover, KLF5, whose expression is induced at an early phase of differentiation, upregulates the expression of PPARγ, and loss of function of KLF5 was shown to inhibit adipocyte differentiation (Oishi et al. 2005). It is thus likely that KLF5 contributes to the induction of PPARγ in coordination with C/EBPβ and C/EBPδ at an early phase of differentiation (Fig. 2).

Function of KLF15 in Mature Adipocytes

Given that KLF15 is expressed at a high level in mature adipocytes, it likely contributes not only to the differentiation of preadipocytes but also to maintenance of the function of mature adipocytes. Forced expression of KLF15 in mature adipocytes increases the expression of GLUT4 (Gray et al. 2002), a marker protein for mature adipocytes. This effect of KLF15 is likely attributable to its direct activation of the promoter of the GLUT4 gene (Gray et al. 2002), suggesting that KLF15 contributes to the function of mature adipocytes by directly regulating the expression of adipocyte-specific genes.

To characterize further the function of KLF15 in mature adipocytes, we attempted to identify novel target genes of this transcription factor in adipocytes by a combination of chromatin immunoprecipitation with a promoter oligonucleotide microarray (ChIP-chip) and analysis of gene expression with an oligonucleotide expression microarray (Nagare et al. 2009). We performed ChIP-chip analysis with immunoprecipitates prepared from fully differentiated 3T3-L1 adipocytes with antibodies to KLF15 or control immunoglobulin. We found that among ~6000 genes on the microarray the promoter regions of 132 genes showed a reproducibly significant difference in hybridization signal between the two samples (Fig. 3) (Nagare et al. 2009). We next profiled genes whose level of expression changed in association with adenovirus-mediated overexpression of KLF15 in 3T3-L1 adipocytes with the use of an oligonucleotide expression microarray. This analysis revealed that forced expression of KLF15 in the mature adipocytes was accompanied by an increase in expression of 337 genes and a decrease in that of 274 genes (Fig. 3) (Nagare et al. 2009). Comparison of the ChIP-chip data with the expression microarray data resulted in identification of six genes whose promoters bound KLF15 and whose expression was either increased (*Slc16a9*, *Cdk9*, *P4ha2*, *Klf3*) or decreased (*Aprt, Adm*) by forced expression of KLF15 (Nagare et al. 2009).

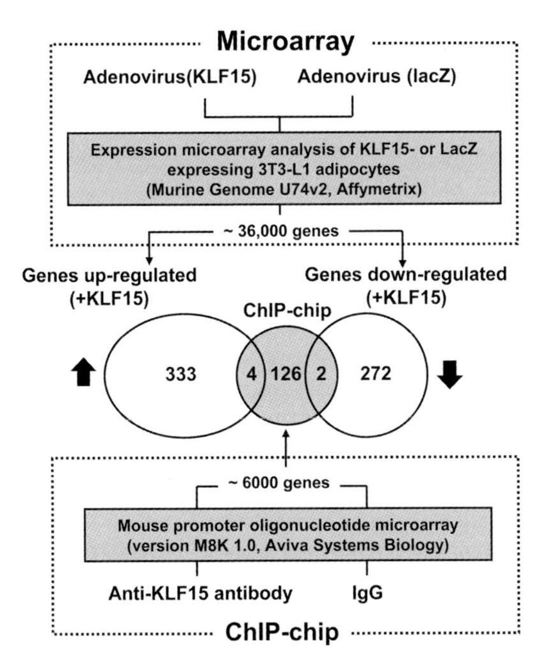

Fig. 3 Identification of target genes of KLF15. Strategy for the identification of target genes of KLF15 and a Venn diagram of the numbers of genes whose promoters were found to bind KLF15 by ChIP-chip analysis and whose expression was found to be increased or decreased in response to KLF15 overexpression by expression microarray analysis in 3T3-L1 adipocytes (Nagare et al. 2009)

Slc16a9 encodes a transporter for monocarboxylic acids, which are important metabolites of carbohydrates and fatty acids (Halestrap and Price 1999). The protein encoded by *Cdk9* is a member of the cyclin-dependent kinase (CDK) family and has been shown to participate in adipocyte differentiation through direct interaction with and phosphorylation of PPARγ (Iankova et al. 2006). *P4ha2* encodes prolyl 4-hydroxylase, which contributes to the synthesis of collagen and to oxygen homeostasis (Myllyharju 2008), the latter of which has been shown to influence the function of adipocytes (Trayhurn et al. 2008). The protein encoded by *Klf3* (KLF3), as mentioned above, appears to regulate adipogenesis by inhibiting the expression of C/EBPα (Sue et al. 2008). *Aprt* encodes adenine phosphoribosyltransferase, an enzyme involved in purine nucleotide metabolism (Delbarre et al. 1974).

Adrenomedullin, the protein encoded by *Adm*, is a potent vasodilatory hormone that was originally identified in pheochromocytoma cells (Kitamura et al. 1993). Mature adipocytes were subsequently shown to be a major source of adrenomedullin in the body (Harmancey et al. 2007; Nambu et al. 2005). We found that the expression of adrenomedullin in 3T3-L1 adipocytes was increased as a result of KLF15 depletion by RNA interference and that KLF15 inhibits the activity of the adrenomedullin gene promoter by directly binding to the most proximal CACCC element (Nagare et al. 2009), confirming the notion that KLF15 is a negative regulator of the adrenomedullin gene. Adipocyte-derived adrenomedullin is thought to protect against the development of hypertension, insulin resistance, and the complications of these conditions in obese subjects (Paulmyer-Lacroix et al. 2006). Expression of KLF15 is decreased in adipose tissue of mice with diet-induced or genetic (*ob/ob*) obesity (H.S. and M.K., unpublished observations), animals in which the expression of adrenomedullin in adipose tissue is increased (Harmancey et al. 2007). It is thus possible that obesity-induced downregulation of KLF15 in adipose tissue is related to the pathogenesis of obesity-induced health disorders induced by a decrease in adrenomedullin production.

References

Banerjee SS, Feinberg MW, Watanabe M et al (2003) The Kruppel-like factor KLF2 inhibits peroxisome proliferator-activated receptor-gamma expression and adipogenesis. J Biol Chem 278:2581-2584

Birsoy K, Chen Z, Friedman J (2008) Transcriptional regulation of adipogenesis by KLF4. Cell Metab 7:339-347

Delbarre F, Aucher C, Amor B et al (1974) Gout with adenine phosphoribosyltransferase deficiency. Biomedicine 21:82-85

Gray S, Feinberg MW, Hull S et al (2002) The Krüppel-like factor KLF15 regulates the insulin-sensitive glucose transporter GLUT4. J Biol Chem 277:34322-34328

Gray S, Wang B, Orihuela Y et al (2007) Regulation of gluconeogenesis by Kruppel-like factor 15. Cell Metab 5:305-312

Halestrap AP, Price NT (1999) The proton-linked monocarboxylate transporter (MCT) family: structure, function and regulation. Biochem J 343:281-299

Harmancey R, Senard JM, Rouet P et al (2007) Adrenomedullin inhibits adipogenesis under transcriptional control of insulin. Diabetes 56:553-563

Iankova I, Petersen RK, Annicotte JS et al (2006) Peroxisome proliferator-activated receptor gamma recruits the positive transcription elongation factor b complex to activate transcription and promote adipogenesis. Mol Endocrinol 20:1494-1505

Kitamura K, Kangawa K, Kawamoto M et al (1993) Adrenomedullin: a novel hypotensive peptide isolated from human pheochromocytoma. Biochem Biophys Res Commun 192:553-560

Li D, Yea S, Li S et al (2005) Kruppel-like factor-6 promotes preadipocyte differentiation through histone deacetylase 3-dependent repression of DLK1. J Biol Chem 280:26941-26952

Mori T, Sakaue H, Iguchi H et al (2005) Role of Kruppel-like factor 15 (KLF15) in transcriptional regulation of adipogenesis. J Biol Chem 280:12867-12875

Myllyharju J (2008) Prolyl 4-hydroxylases, key enzymes in the synthesis of collagens and regulation of the response to hypoxia, and their roles as treatment targets. Ann Med 40:402-417

Nagare T, Sakaue H, Takashima M et al (2009) The Kruppel-like factor KLF15 inhibits transcription of the adrenomedullin gene in adipocytes. Biochem Biophys Res Commun 379:98-103

Nambu T, Arai H, Komatsu Y et al (2005) Expression of the adrenomedullin gene in adipose tissue. Regul Pept 132:17-22

Oishi Y, Manabe I, Tobe K et al (2005) Kruppel-like transcription factor KLF5 is a key regulator of adipocyte differentiation. Cell Metab 1:27-39

Paulmyer-Lacroix O, Desbriere R, Poggi M et al (2006) Expression of adrenomedullin in adipose tissue of lean and obese women. Eur J Endocrinol 155:177-185

Rangwala SM, Lazar MA (2000) Transcriptional control of adipogenesis. Annu Rev Nutr 20:535-559

Rosen ED, Spiegelman BM (2000) Molecular regulation of adipogenesis. Annu Rev Cell Dev Biol 16:145-171

Rosen ED, Spiegelman BM (2006) Adipocytes as regulators of energy balance and glucose homeostasis. Nature 444:847-853

Sue N, Jack BH, Eaton SA et al (2008) Targeted disruption of the basic Kruppel-like factor gene (Klf3) reveals a role in adipogenesis. Mol Cell Biol 28:3967-3978

Teshigawara K, Ogawa W, Mori T et al (2005) Role of Kruppel-like factor 15 in PEPCK gene expression in the liver. Biochem Biophys Res Commun 327:920-926

Trayhurn P, Wang B, Wood IS (2008) Hypoxia in adipose tissue: a basis for the dysregulation of tissue function in obesity? Br J Nutr 100:227-235

Uchida S, Tanaka Y, Ito H et al (2000) Transcriptional regulation of the CLC-K1 promoter by myc-associated zinc finger protein and kidney-enriched Krüppel-like factor, a novel zinc finger repressor. Mol Cell Biol 20:7319-7331

Yamamoto J, Ikeda Y, Iguchi H et al (2004) Kruppel-like factor KLF15 contributes to fasting-induced transcriptional activation of mitochondrial acetyl-CoA synthetase gene AceCS2. J Biol Chem 279:16954-16962

Chapter 13
Krüppel-like Factors in the Heart

Daiji Kawanami, Saptarsi M. Haldar, and Mukesh K. Jain

Abstract Despite the development of numerous therapies, heart disease is a major source of morbidity, mortality, and economic burden to society worldwide. A better understanding of the molecular underpinnings that lead to heart failure are likely to facilitate the development of novel therapies. The Krüppel-like factor (KLF) family of zinc finger transcription factors play important roles in modulating cellular functions in a broad range of mammalian cell types, and accumulating evidence demonstrates important roles of these factors in cardiovascular biology. This chapter describes our current understanding of the role of the KLF gene family in cardiac biology and the potential for these factors to serve as therapeutic targets.

Introduction

Heart disease is a major cause of morbidity and mortality worldwide (Jain and Ridker 2005) and a better understanding of the molecular mechanisms underlying its pathogenesis is extremely valuable from both scientific and therapeutic standpoints. Heart failure is a condition that results from a broad array of insults that impair the pump function of the heart, including ischemia, valvular disease, hypertension, and diabetes. Accumulating evidence provides that these stressors trigger a complex series of signaling cascades that can alter gene programs in both cardiac myocytes and interstitial tissues (Braunwald 2008). Ultimately, these alterations in cell signaling and gene expression can lead to pathological remodeling of the heart, which is characterized by hypertrophic enlargement of myocytes, interstitial fibrosis, electrophysiological abnormalities, contractile dysfunction, altered calcium homeostasis, and metabolic derangements. However, the precise molecular mechanisms by which these alterations in gene expression occur remain incompletely understood.

D. Kawanami, S.M. Haldar, and M.K. Jain (✉)
Case Cardiovascular Research Institute, Case Western Reserve University School of Medicine, University Hospitals Harrington-McLaughlin Heart & Vascular Institute, 2103 Cornell Road, Room 4-522, Cleveland, OH 44106, USA

R. Nagai et al. (eds.), *The Biology of Krüppel-like Factors*,
DOI 10.1007/978-4-431-87775-2_13, © Springer 2009

There has been intense interest in the cytosolic and nuclear signaling pathways that control cardiac development and remodeling (Heineke and Molkentin 2006), and a number of key transcription factors to critical regulators of these processes have been shown (Adhikari et al. 2006; Akazawa and Komuro 2003; Epstein and Parmacek 2005; Finck and Kelly 2006; Oettgen 2006; Perry and Soreq 2002; Puigserver and Spiegelman 2003). Studies from our group and others have recently demonstrated a critical role for the Krüppel-like factor (KLF) family of zinc finger transcription factors in cardiac biology (Feinberg et al. 2004; Perry and Soreq 2002; Suzuki et al. 2005; Wei et al. 2006). This chapter focuses on the emerging role of the KLF family of transcription factors in the heart with an emphasis on the pathobiology of heart failure.

Overview of KLFs in the Heart

Although there has been an explosion of studies on KLFs in a broad variety of tissues and disease states, the number of reports describing the role of KLFs in the heart are few (Haldar et al. 2007). The published reports describing a role for KLFs in the heart are as follows: (1) KLF5 in cardiac fibroblasts (Shindo et al. 2002); (2) KLF15 in postnatal cardiomyocyte (Fisch et al. 2007) and cardiac fibroblast (Wang et al. 2008) biology; (3) KLF13 in the developing vertebrate heart (Lavallee et al. 2006); and (4) a brief report describing the cardiac phenotype of KLF10 knockout mice (Rajamannan et al. 2007). There are also two of reports profiling the expression of various KLFs in cultured cardiomyocytes in response to endothelin-1 (ET-1), oxidative stress and cytokines (Clerk et al. 2006; Cullingford et al. 2008). These published findings of KLFs in cardiac biology are summarized in Table 1. The remainder of this chapter is divided into subsections organized by each KLF family member.

KLF5 in the Heart

KLF5 (also known as BTEB2/IKLF) was cloned as a novel GC box-binding protein from a human placenta cDNA library and originally identified as a positive regulator of *SMemb*, a gene induced in activated smooth muscle cells (Sogawa et al. 1993; Watanabe et al. 1999). Elegant work from the laboratory of Ryozo Nagai has delineated the importance of this factor in cardiac and vascular biology (Shindo et al. 2002). In the heart, KLF5 is expressed primarily in cardiac fibroblasts and serves as a critical effector of angiotensin II signaling in these cells. Angiotensin II stimulation induces KLF5 expression in primary cardiac fibroblasts. Moreover, the angiotensin II mediated induction of platelet-derived growth factor-A (PDGF-A) is dependent on the recruitment of KLF5 to the PDGF-A promoter.

To further understand the role of KLF5 in cardiovascular biology, KLF5 was targeted systemically in the mouse germline (Shindo et al. 2002). KLF5 homozygous-null

Table 1 Function and regulation of KLFs in the heart

Krüppel-like factor	Function/regulation/observation
KLF2	• Expressed in cultured cardiomyocytes • Induced by ET-1 and hydrogen peroxide • Downregulated by TNF-α and IL-1β
KLF3	• Expressed in cultured cardiomyocytes • Downregulated by ET-1
KLF4	• Expressed in cultured cardiomyocytes • Induced by ET-1 and hydrogen peroxide
KLF5	• Expressed in cardiac fibroblasts • Expressed in cultured cardiomyocytes • Induced by angiotensin II • Induced by ET-1 and hydrogen peroxide • Regulates expression of PDGF-A and TGF-β • Reduced hypertrophic remodeling is seen in response to ngiotensin II infusion in KLF5 haplo-insufficient mice • Interacts with retinoic acid receptor-α
KLF6	• Expressed in cultured cardiomyocytes • Induced by ET-1 and hydrogen peroxide
KLF9	• Expressed in cultured cardiomyocytes • Induced by ET-1
KLF10	• Expressed in cultured cardiomyocytes • Induced by ET-1 and hydrogen peroxide • Mice with systemic KLF10 deficiency develop spontaneous pathological hypertrophy by age 16 months in males but not females • Regulates expression of pituitary tumor transforming gene (*Pttg1*); however, functional significance of this observation is unknown
KLF11	• Expressed in cultured cardiomyocytes • Induced by ET-1
KLF13	• Expressed in developing cardiomyocytes • Can bind and activate the BNP promoter • Can transactivate multiple cardiac promoters in concert with GATA4 • Interacts with the zinc finger domain of GATA4 • Deficiency of KLF13 in *Xenopus* leads to cardiac developmental abnormalities (atrial septal defects and ventricular hypotrabeculation) • Cardiac phenotype of KLF13-deficient *Xenopus* embryos can be rescued by GATA4 overexpression
KLF15	• Expressed in cardiomyocytes and cardiac fibroblasts • Cardiac expression is low in developing heart and dramatically upregulated postnatally • Downregulated by ET-1 in cultured cardiomyocytes • Inhibitor of cardiac hypertrophy and fibrosis • Can inhibit GATA4 and MEF2 DNA binding and transcriptional activity • Mice with systemic KLF15 deficiency develop severe eccentric hypertrophy and exaggerated cardiac fibrosis with pressure overload • Inhibits TGF-β-induced CTGF expression in cardiac fibroblasts • Represses Smad3-mediated induction of the CTGF promoter in part via its ability to inhibit PCAF recruitment

BNP = brain natriuretic peptide; CTGF = connective tissue growth factor; ET-1 = endothelin-1; IL-1β = interleukin-1β; MEF2 = myocyte enhancing factor 2; PCAF = p300/CBP-associated factor; PDGF-A = platelet-derived growth factor A; TGF-β = transforming growth factor-β; TNF-α = tumor necrosis factor-α mice die near embryonic day 8.5 (E8.5), although the precise developmental defect in these mice has not been well characterized. KLF5$^{+/-}$ mice are viable into adulthood and demonstrate resistance to angiotensin-mediated cardiac remodeling. Mice with KLF5 haplo-insufficiency showed a blunted hypertrophic response to angiotensin II infusion with reduced cardiac mass, wall thickness, cardiac fibrosis, and PDGF-A expression (Fig. 1A) (Shindo et al. 2002). Furthermore, angiotensin II mediated induction of transforming growth factor-β (TGF-β) and collagen type IV were also blunted in KLF5 haplo-insufficient hearts. Interestingly, the investigators show that KLF5 is able to interact with the retinoic acid receptor-α (RARα), suggesting that RARα activation regulates KLF5 function. Taken together, these observations indicate that KLF5 plays an important role in cardiac remodeling and expands the repertoire of angiotensin-responsive transcription factors in the cardiovascular system (Shindo et al. 2002).

KLF15 in the Heart

KLF15 is expressed in multiple tissues, including liver, white and brown adipose, kidney, heart, and skeletal muscle (Gray et al. 2002). KLF15 has been implicated as a critical regulator of adipogenesis (Mori et al. 2005) and hepatic gluconeogenesis (Gray et al. 2007). Recently, studies from our group demonstrated that KLF15 is a novel negative regulator of cardiac hypertrophy (Fisch et al. 2007) and fibrosis (Fisch et al. 2007; Wang et al. 2008).

KLF15 expression in the developing heart is minimal and is detectable only at very low levels during the early postnatal period. However, cardiac KLF15 expression is robustly induced within the first several weeks postnatally. Interestingly, this period is a time when ANF, BNP, and cyclin-A are downregulated (Fisch et al. 2007). KLF15 levels are reduced dramatically by pressure overload in murine models and in human hearts with left ventricular hypertrophy (LVH) due to valvular aortic stenosis (Fisch et al. 2007). Consistent with this observation, various pro-hypertrophic neurohormonal agonists such as phenylephrine and ET-1 also reduce KLF15 expression in cultured cardiac myocytes (Fisch et al. 2007).

Fig. 1 (continued) show exaggerated pathological remodeling in response to left ventricular (LV) pressure overload. Hearts from KLF15 knockout mice show eccentric hypertrophy (*left panels*). M-mode echocardiography shows severe LV dilation and systolic dysfunction (*middle panels*). Isolated KLF15$^{-/-}$ cardiomyocytes are enlarged compared to those of the wild-type controls (*right panels*). (Adapted from Fisch et al. 2007, with permission. © National Academy of Sciences, 2007.) **c** KLF15 knockout mice show exaggerated collagen deposition after 1 week of ascending aortic constriction. LV sections are stained by Masson's trichrome. (From Wang et al. 2008, with permission from Elsevier.) **d** KLF13 knockdown in *Xenopus* embryo causes atrial septal abnormalities and defects in ventricular trabeculation. (From Lavallee et al 2006, with permission from Macmillan Publishers.)

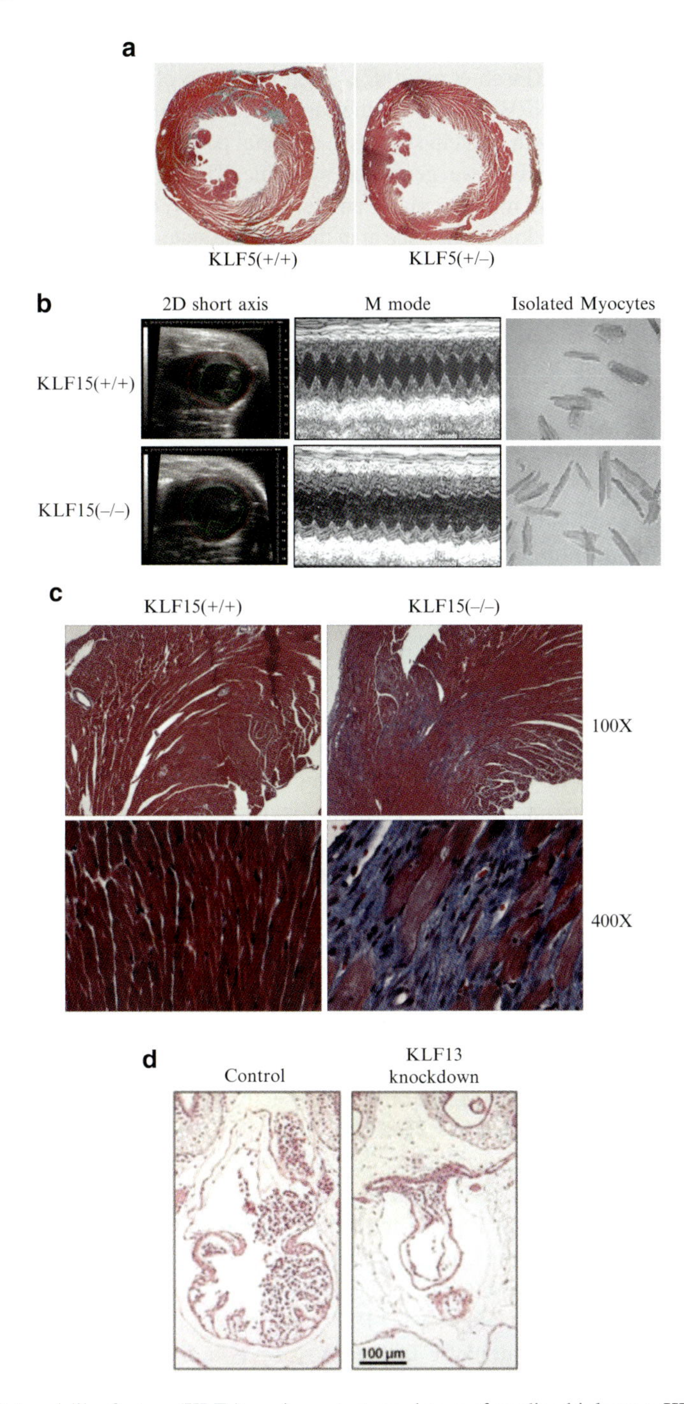

Fig. 1 Krüppel-like factors (KLFs) are important regulators of cardiac biology. **a** KLF5 haplo-insufficient mice show reduced perivascular fibrosis in response to angiotensin II infusion. (From Shindo et al. 2002, with permission from Macmillan Publishers, © 2002.) **b** KLF15 knock out mice

We performed gain- and loss-of-function studies to understand the detailed role of KLF15 in the heart (Fisch et al. 2007). Overexpression of KLF15 in neonatal rat ventricular myocytes (NRVMs) inhibits the cardinal features of cardiomyocyte hypertrophy, such as cell growth, protein synthesis, and fetal gene expression. To understand the role of KLF15 in vivo, we generated systemic KLF15 knockout mice. These KLF15 homozygous null mice are viable and fertile. No overt cardiac decompensation was observed in KLF15 knockout mice at in the baseline state (age 8–12 weeks), however, these animals show exaggerated hypertrophic remodeling in response to pressure overload. Hearts from KLF15 knockout mice after aortic constriction showed cavity enlargement, impaired systolic function, and exaggerated fetal gene expression (Fig. 1B). Furthermore, cardiac myocytes from these animals were large and elongated, which is suggestive of eccentric hypertrophic remodeling (Fig. 1B).

From a mechanistic standpoint, KLF15 can attenuate transcriptional activity of MEF2 and GATA4 (Fisch et al. 2007), transcription factors that are critical hypertrophic effectors (Czubryt and Olson 2004; Pikkarainen et al. 2004), in large part via their ability to bind and transactivate promoters of key pro-hypertrophic genes. KLF15 is able to inhibit the ability of these factors to bind target promoters; and further study is underway to elucidate the precise molecular mechanism underlying the inhibitory effect of KLF15 on MEF2 and GATA4.

Interestingly, KLF15 knockout mice have cardiac phenotypes similar to those of transgenic mice overexpressing both MEF2 or GATA4 in the heart. Transgenic overexpression of MEF2A and MEF2C in the heart causes dilated cardiomyopathy in a dose-dependent manner in response to pressure overload (Xu et al. 2006). Cardiomyocytes from these MEF2-transgenic hearts have large, elongated myocytes (Xu et al. 2006), similar to the cardiomyocytes derived from KLF15 knockout mice (Fisch et al. 2007). High-level overexpression of GATA4 in the heart results in severe cardiomyopathy and premature death (Liang et al. 2001). A spontaneous cardiomyopathy (increased heart mass and hypertrophic gene expression) is observed even with modest GATA4 overexpression (Liang et al. 2001). Taken together, it is likely that these two factors are activated in the absence of KLF15 and cause exaggerated cardiac remodeling in response to stress in KLF15-deficient mice. As described before, KLF15 has an intriguing expression pattern—notably its dramatic postnatal induction. Molkentin and colleagues (Molkentin and Markham 1993) showed an increase in MEF2 binding and activity during the postnatal period. As such, the inhibitory effect of KLF15 for MEF2 raises the possibility that KLF15 plays a regulatory role in postnatal cardiac maturation (Fisch et al. 2007).

Our group has also recently identified a role for KLF15 in the cardiac fibroblast as a negative regulator of connective tissue growth factor (CTGF) signaling. CTGF is expressed in both cardiomyocytes and cardiac fibroblasts (Chen et al. 2000) and plays an important role in the development of fibrosis in disease states such as atherosclerosis (Oemar et al. 1997) and heart failure (Chen et al. 2000). TGF-β1 is a major regulator of CTGF expression (Chen et al. 2000). The TGF-β receptor is a serine/threonine kinase transmembrane heteromeric type I and type II receptor complex that signals through Smad family transcription factors upon receptor activation. Smad proteins can be divided into three groups: receptor-activated type (Smads 1, 2, 3, 5, and 8), co-mediator type (Smads 4 and 10), and inhibitory type (Smads 6 and 7) (Khan and Sheppard, 2006). Among them, Smad3 has been shown to bind to a consensus element in the CTGF promoter by TGF-β1 stimulation (Chen et al. 2002; Grotendorst et al. 1996).

We demonstrated that KLF15 knockout mice show exaggerated collagen deposition and excess induction of CTGF (trichrome staining, Fig. 1C) in response to pressure overload (Wang et al. 2008). Furthermore, adenoviral overexpression of KLF15 inhibits CTGF induction by TGF-β1 in neonatal rat ventricular fibroblasts (NRVFs), and this repressive effect occurs at the promoter level. The electrophoretic mobility shift assay (EMSA) showed that this repressive effect was not due to inhibition of Smad3 binding to the CTGF promoter. As the protein P/CAF has been implicated as an important transcriptional co-activator of Smad3 target genes, we hypothesized that KLF15 may inhibit CTGF promoter activity via an inhibitory effect on P/CAF recruitment. Indeed, a co-immunoprecipitation assay demonstrated that KLF15 interacts with P/CAF, and a chromatin immunoprecipitation assay revealed that KLF15 overexpression inhibited recruitment of P/CAF to CTGF promoter. Moreover, repression of the CTGF promoter by KLF15 is rescued by P/CAF overexpression (Wang et al. 2008). These observations suggest that KLF15 is a negative regulator of CTGF expression in cardiac fibroblasts, in part via its ability to inhibit P/CAF–Smad3 signaling at the CTGF promoter.

KLF13 in the Heart

Expression of KLF13 (also known as FKLF-2/BTEB3) is restricted to erythroid cells, T lymphocytes, heart, and skeletal muscle (Asano et al. 2000; Song et al. 1999). KLF13 is detectable at low levels by reverse transcription-polymerase chain reaction (RT-PCR) in other adult mouse tissues (Scohy et al. 2000). Developmental expression of KLF13 is seen in the heart, cephalic mesenchyme, dermis, and epithelial layers of the gut and urinary bladder in the mouse embryo (Martin et al. 2001). Previous studies demonstrated that KLF13 plays an important role in regulation of erythroid gene expression (Feng and Kan 2005) and plays critical role in RANTES induction in activated T lymphocytes (Ahn et al. 2007; Song et al. 1999).

Nemer and colleagues demonstrated a role for KLF13 in the embryonic myocardium in studies of BNP gene regulation and *Xenopus* development (Lavallee et al. 2006). The investigators previously reported a proximal BNP promoter that can induce cardiac transcription maximally (Grepin et al. 1994). An essential KLF consensus site (CACCC) is located nearby GATA sites of this proximal BNP promoter and was shown to be essential for promoter activity. KLF13 was able to bind the CACCC element in the proximal BNP promoter, as demonstrated by electrophoretic mobility shift assay (EMSA). The authors further demostrated that KLF13 syngerizes with GATA4 to transactivate multiple cardiac promoters (BNP, ANF, β-MHC, cardiac α-actin). In addition, KLF13 was shown to be able to interact with the N-terminal zinc finger of GATA4 (Lavallee et al. 2006).

To gain a better understanding of the role of KLF13 in heart development, the expression pattern of KLF13 in the mouse embryonic heart was studied. Cardiac expression of KLF13 was first detected at E9.5. Subsequently, expression of KLF13 was seen in the developing atrial myocardium, ventricular trabeculae, atrioventricular (AV) cushions, and the truncus arteriosus. Postnatally, KLF13 expression was reduced in the heart and was restricted to the valves and interventricular septum. KLF13 knockdown in the *Xenopus* embryo was used to explore the role of KLF13 in heart development. KLF13-deficient embryos showed atrial septal defects and ventricular hypotrabeculation (Fig. 1D). This observation is consistent with the phenotype of humans with GATA4 mutation and mice with GATA4 deficiency (Epstein and Parmacek 2005). There was no correlation between this hypoplastic phenotype and increased apoptosis, suggesting that KLF13 may be involved in regulating cardioblast proliferation. Interestingly, GATA4 overexpression in these embryos could rescue these cardiac defects in a dose-dependent manner, suggesting that KLF13 and GATA4 are factors that can work synergistically in heart development. These findings using *Xenopus* as a model system indicate that KLF13 may be a novel candidate gene for human congenital heart disease.

However, we noted that the role of KLF13 in mammalian systems may be more complex. KLF13 knockout mice were recently developed in the Krensky Laboratory (Zhou et al. 2007) and were found to be viable. These investigators identified defects in T-lymphocyte survival. However, the role of KLF13 in cardiac biology has not been reported to date, .

KLF10 in the Heart

Role of KLF10/TIEG1 in the heart is not well understood. KLF10 was initially reported as a TGF-β inducible early gene 1 (TIEG-1) in osteoblasts (Subramaniam et al. 1995). Spelsberg and colleagues have shown that KLF10 plays an important role in the regulation of bone mineralization (Subramaniam et al. 2005), osteoclast differentiation (Subramaniam et al. 2005), and epithelial proliferation (Subramaniam et al. 1998; Tachibana et al. 1997). KLF10 regulates Smad signaling in osteoblasts, and KLF10 deficiency leads to osteopenia (Bensamoun et al. 2006) and impaired tendon healing (Tsubone et al. 2006).

KLF10 has been shown to be expressed in the adult heart at low levels, but its distribution within the myocardium is unknown (Subramaniam et al. 1995). In addition, its expression levels in the developing heart are not known. Spelsberg and colleagues reported the cardiac phenotype of KLF10 null mice (Rajamannan et al. 2007). These investigators observed spontaneous pathological cardiac hypertrophy in male (but not female) KLF10 knockout mice at 16 months of the age. Affected mice have increased heart mass, wall thickness, fibrosis, and myocyte disarray with preservation of LV systolic function. The exact timing of onset of this phenotype is not yet known. From a mechanistic standpoint, analysis of hypertrophic KLF10 knockout hearts revealed that KLF10 may regulate the pituitary tumor transforming gene (Pttg1), but the significance of this finding is unclear. There are several questions remaining questions regarding KLF10's role in the heart: (1) Is KLF10 expressed in cardiomyocytes, cardiac fibroblasts, or both? (2) Is KLF10 expression altered with mechanical or neurohormonal stress? (3) What are KLF10 target genes in the heart, and what mechanisms explain the pathology seen in KLF10-null hearts? Indeed, future studies will continue to elucidate the importance of KLF10 in cardiac remodeling.

Regulation of KLFs by Hypertrophic and Apoptotic Stimuli

Recent expression-profiling studies have reported differential regulation of KLFs in cultured cardiomyocytes in response to pharmacological stimuli and oxidative stress. Clerk and colleagues reported expression profiles of KLFs in response to endothelin (ET-1) stimulation in neonatal rat cardiomyocytes. Quantitative PCR analysis revealed that expressions of KLFs 2, 4, 5, 6, 9, and 10 are induced rapidly and transiently by ET-1, whereas expressions of KLFs 3, 11, and 15 are downregulated. As oxidative stress and cytokine stimulation are implicated in cardiac myocyte apoptosis, these investigators also examined the effects of hydrogen peroxide and inflammatory cytokines on the expression of KLFs. Hydrogen peroxide upregulated KLFs 2, 4, 5, 6, and 10 mRNA expression levels and reduced KLF15 expression in cultured cardiomyocytes (Clerk et al. 2006; Cullingford et al. 2008). In addition, KLF2 is downregulated by tumor necrosis factor-α (TNF-α) and interleukin-1β (IL-1β) (Cullingford et al. 2008). Although the physiological relevance of these findings is not yet known, these observations raise the possibility that KLFs other than KLFs 5, 10, 13, and 15 may be involved in regulating cardiac growth and the response to stress.

Future Directions

Although the accumulating evidence demonstrates that KLFs play an important role in cardiac development and remodeling, there are significant questions remaining that must be addressed. First, little is known about expression profiles of KLFs in the developing and postnatal heart. In addition, distribution of KLF expression among the multiple cellular subsets that comprise the myocardium (cardiomyocytes, fibroblasts, endothelial cells, vascular smooth muscle cells, immune cells) must be defined. The relative function of KLFs in these various cell types of the heart are of great interest and will undoubtedly be important in understanding the interplay between these tissues in heart disease. Tissue- and cell-type-specific gain- or loss-of-function approaches will be necessary to address these questions.

Recent studies have highlighted the importance of coupled cardiac angiogenesis as an adaptive feature of compensated cardiac hypertrophy (Sano et al. 2007).

A number of KLFs have already been implicated in the angiogenic response. For example, KLF2 is implicated as an antiangiogenic factor in endothelial cell biology (Bhattacharya et al. 2005). It is highly likely that this family of transcription factors has broad roles in regulating this process in the myocardium under hypertrophic and ischemic conditions.

Another area in which the KLFs are likely to play an important role is in the context of cardiac metabolism. There is certainly increasing appreciation that alterations in cardiac fatty acid and glucose utilization can affect the heart's response to stress, particularly in disease states such as diabetes or obesity. Recent studies have identified several members of the KLF family as important regulators of adipogenesis, glucose homeostasis, and energy metabolism. Among the KLFs implicated in cardiac biology, KLF5 and KLF15 have been shown to alter cellular metabolism. For example, KLF5 regulates genes involved in skeletal muscle lipid oxidation and energy coupling such as UCP and CPT1—genes that certainly affect cardiac energetics. Intriguingly, Oishi and colleagues showed that this regulation occurs in cooperation with PPARδ—a nuclear receptor that has been shown to regulate cardiac fatty acid and glucose utilization (Burkart et al. 2007; Oishi et al. 2008). KLF15 has been shown to critically regulate systemic glucose homeostasis through effects on hepatic amino acid catabolism, which certainly raises the possibility that this factor has an important role in cardiac metabolism (Gray et al. 2007). Indeed, it is exciting to postulate that cooperative interactions between KLFs and PPARs—two major transcription factor families—may critically regulate cardiac substrate utilization and consequently cardiac function.

Finally, it is of utmost importance to identify compounds that regulate KLFs or interact with KLFs in the heart. For example, KLF5's function in the heart can be modulated by RARα antagonists (Shindo et al. 2002). Furthermore, clear interplay between statins and KLF2 has been demonstrated in endothelial biology (Sen-Banerjee et al. 2005). Neurohormonal antagonists that are currently used in heart failure therapy may regulate KLFs (e.g., KLF5, KLF15) in the heart. As is the case with statins and KLF2 in the endothelium, it is possible that KLFs can mediate favorable myocardial effects of drugs such as β-blockers, angiotensin-converting enzyme (ACE) blockers, and angiotensin-II receptor (AT_1) blockers. These studies have important implications for the treatment of cardiomyopathic conditions.

Another critical issue is the delineation of overlapping and restricted roles of the multiple KLFs that are co-expressed in the heart. For example, KLF13 and KLF15 modulate GATA4 activity oppositely. KLF13 synergizes with GATA4 to activate multiple promoters (Lavallee et al. 2006), whereas KLF15 inhibits induction of these promoters by GATA4 (Fisch et al. 2007). These facts raise the possibility that KLF15 may regulate GATA4 activity in part through inhibition of KLF13's function. As has been shown in other tissues, it is likely that KLFs family members regulate the expression and function of each other in the same cell type (Funnell et al. 2007). Another example of potential interplay is between KLF5 and KLF15. Both are expressed in cardiac fibroblasts: KLF5 promotes fibrosis, and KLF15 inhibits it. It is possible that a tight balance of relative expression/activity of KLFs influences the heart's response to physiological and pathological stimuli.

Hence, it is important to identify common target genes or interacting proteins for KLFs that are co-expressed in the heart to better define their overlapping or divergent roles.

References

Adhikari N, Charles N, Lehmann U et al (2006) Transcription factor and kinase-mediated signaling in atherosclerosis and vascular injury. Curr Atheroscler Rep 8:252-260

Ahn YT, Huang B, McPherson L et al (2007) Dynamic interplay of transcriptional machinery and chromatin regulates "late" expression of the chemokine RANTES in T lymphocytes. Mol Cell Biol 27:253-266

Akazawa H, Komuro I (2003) Roles of cardiac transcription factors in cardiac hypertrophy. Circ Res 92:1079-1088

Asano H, Li XS, Stamatoyannopoulos G (2000) FKLF-2: a novel Krüppel-like transcriptional factor that activates globin and other erythroid lineage genes. Blood 95:3578-3584

Bensamoun SF, Hawse JR, Subramaniam M et al (2006) TGFbeta inducible early gene-1 knockout mice display defects in bone strength and microarchitecture. Bone 39: 1244-1251

Bhattacharya R, Senbanerjee S, Lin Z et al (2005) Inhibition of vascular permeability factor/vascular endothelial growth factor-mediated angiogenesis by the Kruppel-like factor KLF2. J Biol Chem 280:28848-28851

Braunwald E (2008) Biomarkers in heart failure. N Engl J Med 358:2148-2159.

Burkart EM, Sambandam N, Han X et al (2007) Nuclear receptors PPARbeta/delta and PPARalpha direct distinct metabolic regulatory programs in the mouse heart. J Clin Invest 117:3930-3939

Chen MM, Lam A, Abraham JA et al (2000) CTGF expression is induced by TGF- beta in cardiac fibroblasts and cardiac myocytes: a potential role in heart fibrosis. J Mol Cell Cardiol 32:1805-1819

Chen Y, Blom IE, Sa S et al (2002) CTGF expression in mesangial cells: involvement of SMADs, MAP kinase, and PKC. Kidney Int 62:1149-1159

Clerk A, Kemp TJ, Zoumpoulidou G et al (2006) Cardiac myocyte gene expression profiling during H2O2-induced apoptosis. Physiol Genomics 29:118-27

Cullingford TE, Butler MJ, Marshall AK et al (2008) Differential regulation of Kruppel-like factor family transcription factor expression in neonatal rat cardiac myocytes: Effects of endothelin-1, oxidative stress and cytokines. Biochim Biophys Acta 1783:1229-1236

Czubryt MP, Olson EN (2004) Balancing contractility and energy production: the role of myocyte enhancer factor 2 (MEF2) in cardiac hypertrophy. Recent Prog Horm Res 59:105-124

Epstein JA, Parmacek MS (2005) Recent advances in cardiac development with therapeutic implications for adult cardiovascular disease. Circulation 112:592-597

Feinberg MW, Lin Z, Fisch S et al (2004) An emerging role for Kruppel-like factors in vascular biology. Trends Cardiovasc Med 14:241-246

Feng D, Kan YW (2005) The binding of the ubiquitous transcription factor Sp1 at the locus control region represses the expression of beta-like globin genes. Proc Natl Acad Sci U S A 102:9896-9900

Finck BN, Kelly DP (2006) PGC-1 coactivators: inducible regulators of energy metabolism in health and disease. J Clin Invest 116:615-622

Fisch S, Gray S, Heymans S et al (2007) Kruppel-like factor 15 is a regulator of cardiomyocyte hypertrophy. Proc Natl Acad Sci U S A 104:7074-7079

Funnell AP, Maloney CA, Thompson LJ et al (2007) Erythroid kruppel-like factor directly activates the basic kruppel-like factor gene in erythroid cells. Mol Cell Biol 27: 2777-2790

Gray S, Feinberg MW, Hull S et al (2002) The Kruppel-like factor KLF15 regulates the insulin-sensitive glucose transporter GLUT4. J Biol Chem 277:34322-34328

Gray S, Wang B, Orihuela Y et al (2007) Regulation of Gluconeogenesis by Kruppel-like Factor 15. Cell Metab 5:305-312

Grepin C, Dagnino L, Robitaille L et al (1994) A hormone-encoding gene identifies a pathway for cardiac but not skeletal muscle gene transcription. Mol Cell Biol 14:3115-3129

Grotendorst GR, Okochi H, Hayashi N (1996) A novel transforming growth factor beta response element controls the expression of the connective tissue growth factor gene. Cell Growth Differ 7:469-480

Haldar SM, Ibrahim OA, Jain MK (2007) Kruppel-like Factors (KLFs) in muscle biology. J Mol Cell Cardiol 43:1-10

Heineke J, Molkentin JD (2006) Regulation of cardiac hypertrophy by intracellular signalling pathways. Nat Rev Mol Cell Biol 7:589-600

Jain MK, Ridker PM (2005) Anti-inflammatory effects of statins: clinical evidence and basic mechanisms. Nat Rev Drug Discov 4:977-987

Khan R, Sheppard R (2006) Fibrosis in heart disease: understanding the role of transforming growth factor-beta in cardiomyopathy, valvular disease and arrhythmia. Immunology 118:10-24

Lavallee G, Andelfinger G, Nadeau M et al (2006) The Kruppel-like transcription factor KLF13 is a novel regulator of heart development. Embo J 25:5201-5213

Liang Q, De Windt LJ, Witt SA et al (2001) The transcription factors GATA4 and GATA6 regulate cardiomyocyte hypertrophy in vitro and in vivo. J Biol Chem 276:30245-30253

Martin KM, Metcalfe JC, Kemp PR (2001) Expression of Klf9 and Klf13 in mouse development. Mech Dev 103:149-151

Molkentin JD, Markham BE (1993) Myocyte-specific enhancer-binding factor (MEF-2) regulates alpha-cardiac myosin heavy chain gene expression in vitro and in vivo. J Biol Chem 268:19512-19520

Mori T, Sakaue H, Iguchi H et al (2005) Role of Kruppel-like factor 15 (KLF15) in transcriptional regulation of adipogenesis. J Biol Chem 280:12867-12875

Oemar BS, Werner A, Garnier JM et al (1997) Human connective tissue growth factor is expressed in advanced atherosclerotic lesions. Circulation 95:831-839

Oettgen P (2006) Regulation of vascular inflammation and remodeling by ETS factors. Circ Res 99:1159-1166

Oishi Y, Manabe I, Tobe K et al (2008) SUMOylation of Kruppel-like transcription factor 5 acts as a molecular switch in transcriptional programs of lipid metabolism involving PPAR-delta. Nat Med 14:656-666

Perry C, Soreq H (2002) Transcriptional regulation of erythropoiesis. Fine tuning of combinatorial multi-domain elements. Eur J Biochem 269:3607-3618

Pikkarainen S, Tokola H, Kerkela R et al (2004) GATA transcription factors in the developing and adult heart. Cardiovasc Res 63:196-207

Puigserver P, Spiegelman BM (2003) Peroxisome proliferator-activated receptor-gamma coactivator 1 alpha (PGC-1 alpha): transcriptional coactivator and metabolic regulator. Endocr Rev 24:78-90

Rajamannan NM, Subramaniam M, Abraham TP et al (2007) TGFbeta inducible early gene-1 (TIEG1) and cardiac hypertrophy: Discovery and characterization of a novel signaling pathway. J Cell Biochem 100:315-325

Sano M, Minamino T, Toko H et al (2007) p53-induced inhibition of Hif-1 causes cardiac dysfunction during pressure overload. Nature 446:444-448

Scohy S, Gabant P, Van Reeth T et al (2000) Identification of KLF13 and KLF14 (SP6), novel members of the SP/XKLF transcription factor family. Genomics 70:93-101

Sen-Banerjee S, Mir S, Lin Z et al (2005) Kruppel-like factor 2 as a novel mediator of statin effects in endothelial cells. Circulation 112:720-726

Shindo T, Manabe I, Fukushima Y et al (2002) Kruppel-like zinc-finger transcription factor KLF5/BTEB2 is a target for angiotensin II signaling and an essential regulator of cardiovascular remodeling. Nat Med 8:856-863

Sogawa K, Imataka H, Yamasaki Y et al (1993) cDNA cloning and transcriptional properties of a novel GC box-binding protein, BTEB2. Nucleic Acids Res 21:1527-1532

Song A, Chen YF, Thamatrakoln K et al (1999) RFLAT-1: a new zinc finger transcription factor that activates RANTES gene expression in T lymphocytes. Immunity 10:93-103
Subramaniam M, Gorny G, Johnsen SA et al (2005) TIEG1 null mouse-derived osteoblasts are defective in mineralization and in support of osteoclast differentiation in vitro. Mol Cell Biol 25:1191-1199
Subramaniam M, Harris SA, Oursler MJ et al (1995) Identification of a novel TGF-beta-regulated gene encoding a putative zinc finger protein in human osteoblasts. Nucleic Acids Res 23:4907-4912
Subramaniam M, Hefferan TE, Tau K et al (1998) Tissue, cell type, and breast cancer stage-specific expression of a TGF-beta inducible early transcription factor gene. J Cell Biochem 68:226-236
Suzuki T, Aizawa K, Matsumura T et al (2005) Vascular implications of the Kruppel-like family of transcription factors. Arterioscler Thromb Vasc Biol 25:1135-1141
Tachibana I, Imoto M, Adjei PN et al (1997) Overexpression of the TGFbeta-regulated zinc finger encoding gene, TIEG, induces apoptosis in pancreatic epithelial cells. J Clin Invest 99:2365-2374
Tsubone T, Moran SL, Subramaniam M et al (2006) Effect of TGF-beta inducible early gene deficiency on flexor tendon healing. J Orthop Res 24:569-575
Wang B, Haldar SM, Lu Y et al (2008) The Kruppel-like factor KLF15 inhibits connective tissue growth factor (CTGF) expression in cardiac fibroblasts. J Mol Cell Cardiol (2008) 45:193-7
Watanabe N, Kurabayashi M, Shimomura Y et al (1999) BTEB2, a Kruppel-like transcription factor, regulates expression of the SMemb/Nonmuscle myosin heavy chain B (SMemb/NMHC-B) gene. Circ Res 85:182-191
Wei D, Kanai M, Huang S et al (2006) Emerging role of KLF4 in human gastrointestinal cancer. Carcinogenesis 27:23-31
Xu J, Gong NL, Bodi I et al (2006) Myocyte enhancer factors 2A and 2C induce dilated cardiomyopathy in transgenic mice. J Biol Chem 281:9152-9162
Zhou M, McPherson L, Feng D et al (2007) Kruppel-like transcription factor 13 regulates T lymphocyte survival in vivo. J Immunol 178:5496-5504

Chapter 14
Krüppel-like Factors in the Vascular Endothelium

Guillermo García-Cardeña and Guadalupe Villarreal, Jr.

Abstract Although Krüppel-like factors (KLFs) have been the subject of extensive biological investigation, the role of this family of transcription factors in the biology and pathophysiology of the vascular endothelium is just becoming apparent. Most investigative efforts thus far have focused on KLF2, its contribution to the endothelial vasoprotective phenotype, and its possible impact on atherogenesis. This chapter reflects on the current state of the field and highlights evolving areas where KLFs are emerging as important regulators of endothelial function.

Introduction

The Krüppel gene was first identified as a critical embryonic patterning gene by Nüsslein-Volhard and Wieschaus, who used a screen in *Drosophila* to identify mutants with segmentation defects (Nusslein-Volhard and Wieschaus 1980). While this locus defines a single gene responsible for the identity of abdominal and thoracic segments in *Drosophil*a, there are numerous Krüppel orthologues that appear in higher organisms and have acquired divergent expression patterns and functions. The isolation of a Krüppel-like factor (LKLF, KLF2) gene from mouse lung and its demonstrated similarity to the zinc finger region of the erythroid-specific Krüppel-like factor (EKLF, KLF1) led to the unveiling of the Krüppel-like family of transcription factors (Anderson et al. 1995). To date, 17 human members of this family have been identified. The Krüppel transcription factor family is characterized by a carboxy terminus containing three DNA-binding C_2H_2 zinc fingers (Suske et al. 2005). This region is highly conserved among the family members and with the

G. García-Cardeña (✉) and G. Villarreal Jr.
Laboratory for Systems Biology, Center for Excellence in Vascular Biology, Departments of Pathology, Harvard Medical School and Brigham and Women's Hospital, Boston, MA 02115, USA

G. Villarreal Jr.
Harvard-MIT Division of Health, Sciences and Technology, Boston, MA 02115, USA

R. Nagai et al. (eds.), *The Biology of Krüppel-like Factors*,
DOI 10.1007/978-4-431-87775-2_14, © Springer 2009

original *Drosophila* Krüppel protein. Outside of this domain, the amino-terminal transactivation domain exhibits high divergence among members of the Krüppel-like family, and therefore accounts for most of the functional specificity.

KLF2

Structure–Function Studies of KLF2

The cloning of human KLF2 demonstrated high sequence similarity to the murine homologue, with 85% nucleotide identity and 90% amino acid similarity, suggesting a conserved function. The human *KLF2* gene spans roughly 3 kb of genomic sequence and consists of three exons interrupted by two small introns (Wani et al. 1999). The overall genomic structure of the coding region, including splice donor/acceptor sites, is fully conserved between human and mouse genes. Furthermore, the proximal 5′ promoter sequence contains a canonical TATA box and an adjacent 75-bp sequence that exhibits 100% sequence conservation between species (Wani et al. 1999). Transcriptional reporter analysis via transient transfection of the human KLF2 promoter in a murine endothelial cell line demonstrated that deletion of this proximal region results in loss of KLF2 promoter activity, indicating a major function for this region in the regulation of KLF2 expression (Huddleson et al. 2004).

Human KLF2 encodes a 354 amino-acid protein containing both a conserved DNA-binding zinc finger carboxy terminus and a divergent amino-terminal transactivation domain, as described above. KLF2 has been shown to bind CACCC elements and Sp1/Sp3 motifs (Anderson et al. 1995; Wang et al. 2005), resulting in either activation or repression of target gene expression. Related to this dual function, the activation domain of KLF2 contains subdomains that provide activating or repressor effects, reminiscent of the duality of the *Drosophila* Krüppel (Sauer and Jackle 1993). Amino acids 1–110 of KLF2 are able to activate transcription more potently than the full-length protein, indicating an auto-inhibitory subdomain. Deletion analysis mapped the minimally inhibitory subdomain to amino acids 111–267, which appears to bind the ubiquitin ligase WW domain-containing protein 1 to mediate auto-inhibition (Conkright et al. 2001).

KLF2 in Vascular Development

The first evidence suggesting a critical role for KLF2 in vascular morphogenesis and function came from studies of its expression pattern and function during murine development (Kuo et al. 1997). KLF2 mRNA was first detected at E9.5 in the vascular endothelium throughout the embryo, and continued to be expressed in endothelial cells throughout development at apparently similar levels in arteries and veins. KLF2 expression was also documented in the vertebral column and the lung buds by E12.5. More detailed characterization determined that KLF2 expression in embryonic blood vessels was restricted to the CD34+ endothelial lineage and was absent in the tunica media, which largely consists of vascular smooth muscle cells. Importantly, examination of a KLF2 homologue in zebrafish, KLF2a, also demonstrated a vascular endothelium-specific expression pattern in the aorta and cardinal vein at 24 hours after fertilization (Oates et al. 2001).

Generation of KLF2-null mouse embryos resulted in embryonic lethality at approximately E12.5–E14.5, with the most pronounced gross phenotype being intraembryonic hemorrhaging near the outflow tract and into the abdomen at E12.5. Profuse bleeding into the amniotic cavity was observed in embryos surviving to E13.5 or E14.5 (Kuo et al. 1997). Using the endothelial marker PECAM-1, the authors found no evidence of abnormal vasculogenesis (capillary plexus formation) or angiogenesis (capillary sprouting) at E12.5, arguing against a role for these early vascular processes in mediating the KLF2-null phenotype. In contrast, later stages of vascular development, including vessel wall maturation of the umbilical vessels and dorsal aorta, were found to be severely defective before signs of hemorrhaging and were characterized histologically by impaired smooth muscle recruitment to the endothelium. The resulting impairment in the tunica media was associated with aneurysmal dilation of these larger vessels. Because KLF2 was found to be restricted to endothelium, this phenotype implied a nonautonomous role of KLF2 in signaling between endothelium and surrounding pericytes and/or vascular smooth muscle cells. KLF2 thus appeared to function as an endothelium-restricted critical regulator of vessel wall maturation and stability during development. Deletion of KLF2 in the endothelial, endocardial, and hematopoietic compartments in the mouse was later shown to result in embryonic heart failure due to elevated cardiac output (Lee et al. 2006). No evidence of anemia or arteriovenous malformation was found in these mice. Interestingly, administration of phenylephrine, a vasopressor, was sufficient to rescue the lethal embryonic phenotype, suggesting that endothelial expression of KLF2 is necessary for maintaining vascular tone and hemodynamic regulation (Lee et al. 2006).

Regulation of KLF2 Expression by Hemodynamic Forces

Expression of KLF2 was first shown to be upregulated in cultured endothelial cells exposed to laminar shear stress by Dekker and colleagues (Dekker et al. 2002). Using in situ hybridization of adult human arteries, these authors demonstrated that KLF2 expression was restricted to the endothelium in vivo and that the pattern of expression correlated with predicted types of shear stress. Specifically, endothelial expression of KLF2 was highest in regions that are predicted to be exposed to laminar shear stress and thus resistant to atherosclerosis. In contrast, KLF2 expression was absent in atherosclerosis-susceptible regions exposed to nonlaminar shear stress (e.g., bifurcations). Based on these observations, Dekker et al. hypothesized a potential role for KLF2 in atherogenesis (Dekker et al. 2002).

The response of KFL2 expression to laminar shear stress in vitro has since been confirmed by other groups (Huddleson et al. 2004; SenBanerjee et al. 2004). Furthermore, complex modeling of the arterial waveforms characteristic of the in vivo atherosclerosis-resistant (athero-protective flow) and atherosclerosis-susceptible (athero-prone flow) regions of the human carotid artery demonstrated selective upregulation of KLF2 by athero-protective flow as determined by transcriptional profiling (Dai et al. 2004). Dissection of the basic biomechanical parameters relevant to KLF2 induction revealed an increased effect of pulsatile flow compared to nonpulsatile laminar flow and no KLF2 response to cyclical stretch (Dekker et al. 2005). In vivo studies in silent heart mutant zebrafish, which lack blood flow yet remain viable and functional for several days, demonstrate a dramatic loss of expression of the vascular endothelium-specific isoform of KLF2 (KLF2a) in the aorta, cardinal vein, and intersomitic vessels (Parmar et al. 2006). These observations documented for the first time the strict flow-dependence of an endothelial gene in vivo. Other studies used a murine carotid artery collar model to demonstrate that KLF2 mRNA expression is sensitive to local changes in blood flow in the murine circulation (Dekker et al. 2005).

The signaling mechanisms underlying the shear stress-inducible expression of KLF2 are not fully defined, but some critical pathways have been elucidated. Huddleson et al. identified the proximal, highly conserved 160 bp of the KLF2 promoter, which contains the previously mentioned 75-bp region, as critical for flow-mediated upregulation of KLF2. This region contains a MEF2-binding site that is bound by MEF2A and MEF2C (Parmar et al. 2006). Importantly, early studies demonstrated that inhibition of MEF2 function blocks atheroprotective flow-mediated upregulation of KLF2. Mutations in MEF2A that reportedly confer a dominant negative effect have been implicated in the Mendelian inheritance of coronary artery disease in a family, although studies in larger populations have questioned the penetrance of these mutations as a modifier of vascular disease (Wang et al. 2003; Weng et al. 2005). Also, mouse embryos deficient in MEF2C exhibit vascular defects that resemble the KLF2 knockout (Bi et al. 1999). These data raise the possibility that the vascular phenotypes rendered by inactivation of MEF2 (e.g., susceptibility to atherosclerosis, developmental phenotypes) are due to reduced KLF2 expression. Upstream in the pathway, the mitogen-activated protein kinase, kinase 5/mitogen-activated protein kinase 7 (MEK5/ERK5) signaling cascade is both necessary and sufficient for athero-protective flow-mediated upregulation of KLF2, and it is possible that this pathway functions via MEF2, given that these transcription factors are well characterized substrates of ERK5 (Kinderlerer et al. 2008; Parmar et al. 2006). Athero-protective flow thus appears to induce KLF2 levels via a MEK5/ERK5/MEF2 signaling pathway (Fig. 1).

Regulation by Statins and Proinflammatory Stimuli

3-Hydroxy-3-methylglutaryl coenzyme A reductase inhibitors (statins) have been shown to confer clinical benefits beyond those explained by their reduction of plasma lipid levels (Liao and Laufs 2005). Many of these "pleiotropic" effects have been attributed to direct improvements in vascular function. Interestingly, endothelial KLF2 levels rapidly increase in response to statin treatment, with this class of drugs

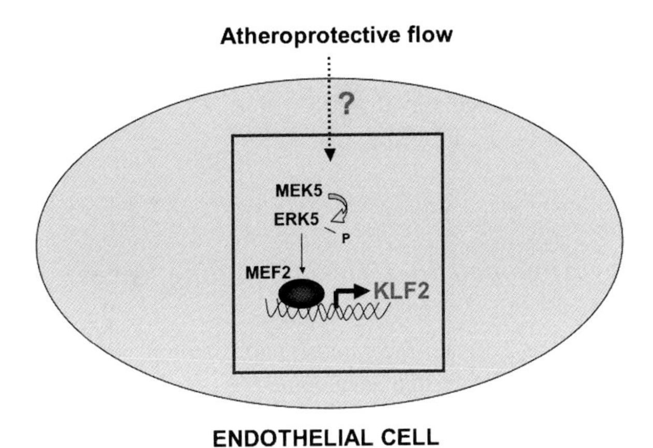

Fig. 1 Athero-protective flow-mediated induction of Krüppel-like factor 2 (KLF2). Experimental evidence has demonstrated involvement of the mitogen-activated protein kinase, kinase 5/mitogen-activated protein kinase 7/myocyte enhancer factor 2 (MEK5/ERK5/MEF2) pathway in the induction of KLF2 expression by biomechanical stimulation. Nevertheless, the proximal mechanosensor and its immediate downstream signaling mediators remain unknown

mediating an approximately eightfold increase in KLF2 at pharmacologically relevant doses in vitro (Parmar et al. 2005). Statins block production of mevalonate, which forms two major downstream products known as isoprenoids: farnesyl pyrophosphate (FPP) and geranylgeranyl pyrophosphate (GGPP). These isoprenoids can each prenylate distinct sets of proteins in the cell to enable their proper localization and signaling. Statin-mediated upregulation of KLF2 in human umbilical vein endothelial cells is dependent on depletion of GGPP (Parmar et al. 2005; Sen-Banerjee et al. 2005), which is well known to prenylate several members of the Rho superfamily. Thus, it is likely that one or more geranylgeranylated proteins mediate a tonic inhibition of KLF2 expression at the promoter level, the relief of which results in induction of this gene. Importantly, upregulation of KLF2 is critical for many statin-dependent transcriptional changes in endothelial cells, implicating KLF2 in the statin-mediated beneficial vascular effects.

In contrast to upregulation of KLF2 by shear stress and statins (two stimuli that confer athero-protective effects), KLF2 expression is strongly suppressed in response to proinflammatory stimuli such as interleukin-1β (IL-1β) or tumor necrosis factor-α (TNF-α), which are thought to be important for pathological inflammation and atherosclerosis (Kumar et al. 2005; SenBanerjee et al. 2004). The ability of flow to curtail the suppression of KLF2 expression under pathophysiologically relevant proinflammatory environments has not yet been explored in detail. This suppression has been shown to be mediated in part by MEF2-dependent recruitment of the histone deacetylases HDAC4 and HDAC5 to the KLF2 promoter to silence transcription of the gene, and this site is in fact likely the binding site relevant to the one described under flow conditions (Kumar et al. 2005).

Nonspecific HDAC inhibitors block the rapid downregulation of KLF2 by TNF-α. Interestingly, the involvement of MEF2 transcription factors in both upregulation by shear stress and down-regulation by TNF-α places the MEF2 family at a pivotal point of regulation of the KLF2 gene.

Regulation by Angiopoietin-1/Tie-2 Signaling

The angiopoietin-1/Tie-2 ligand-receptor signaling complex serves an important role in vascular formation during development as well as vascular quiescence (Brindle et al. 2006; Sako et al. 2008; Sato et al. 1995; Wong et al. 1997). Although Tie-2 signaling can act to maintain quiescence of mature blood vessels, it can also function to promote angiogenesis (Sako et al. 2008). The duality lies in its cell surface distribution, which is dependent on the absence or presence of cell–cell contacts. Under confluent conditions, angiopoietin-1 facilitates trans-association of Tie-2 across cell–cell junctions (Fukuhara et al. 2008). Under subconfluent conditions however, angiopoietin-1 anchors Tie-2 to the extracellular matrix (Fukuhara et al. 2008). The localization of the Tie-2 receptor results in activation of distinct pathways whose unique functional outputs may explain the seemingly opposing roles of Tie-2. For example, angiopoietin-1/Tie-2 signaling was found to induce the expression of KLF2 specifically under confluent conditions. Similar to angiopoietin-1, KLF2 promotes an antiinflammatory, antipermeability state (Brindle et al. 2006; Parmar et al. 2006; Sako et al. 2008). Importantly, Sako et al. demonstrated that KLF2 is necessary for the blockade of vascular endothelial growth factor (VEGF)-induced inflammation via angiopoietin-1, suggesting that the presence or absence of KLF2 may be critical in controlling the functional output of angiopoietin-1/Tie-2 signaling. The upregulation of KLF2 expression by angiopoietin-1 was found to ocurr via a PI3K/AKT/MEF2-dependent pathway. Interestingly, induction of KLF2 by angiopoietin-1 was ERK5-independent, in contrast to the studies on flow-mediated KLF2 expression (Kinderlerer et al. 2008; Parmar et al. 2006). It remains unknown, though, how the PI3K/AKT pathway activates MEF2 transcriptional activity. Nevertheless, to date, MEF2 remains central for all the stimuli characterized to lead to the upregulation of endothelial KLF2.

Downstream Transcriptional Targets and Functions of KLF2 in Endothelium

KLF2 regulates a vast array of genes of major functional importance in the endothelium, and it is becoming apparent that the sum of its complex actions may confer a phenotype that promotes endothelial homeostasis (Dekker et al. 2005; Parmar et al. 2006). KLF2 overexpression in cultured human endothelial cells inhibits

IL-1β-dependent induction of the proinflammatory adhesion molecules, vascular cell adhesion molecule (VCAM)-1 and E-selectin (SenBanerjee et al. 2004). As a consequence, T-cell adhesion and rolling on endothelial monolayers overexpressing KLF2 are markedly attenuated. These observations suggested a potential role for KLF2 as an antiinflammatory regulator in blood vessels. Subsequent studies revealed an important role for KLF2 in regulating certain key genes involved in thrombosis. For example, KLF2 overexpression upregulates thrombomodulin (TM), an endothelial cell surface molecule that produces the potent anticoagulant activated protein C. In addition, plasminogen activator inhibitor 1 was downregulated by KLF2 overexpression. Importantly, KLF2 overexpression in endothelial cells blocked TNF-α-mediated induction of tissue factor (Lin et al. 2005).

KLF2 also plays a role in VEGF-dependent signaling and angiogenesis. KLF2 overexpression blocked VEGF-mediated angiogenesis and tissue edema in a nude mouse ear model, with the proposed mechanism involving transcriptional downregulation of VEGF recepter 2 (VEGFR2) (Bhattacharya et al. 2005). Other studies have pointed to a role for KLF2 in mediating vessel tone. Dekker et al. demonstrated that pulsatile shear stress induces expression of the vasodilatory gene endothelial nitric oxide synthase (eNOS) in cultured human umbilical vein endothelial cells while suppressing expression of the vasoconstrictive genes endothelin-1 and adrenomedullin in a KLF2-dependent manner (Dekker et al. 2005).

Experiments using a systems biology approach unveiled KLF2 as a critical integrator of the global transcriptional responses of endothelial cells to athero-protective shear stresses (Parmar et al. 2006). KLF2 overexpression experiments revealed that this transcription factor orchestrates transcriptional programs involved in blood vessel development, inflammation, thrombosis, and vascular tone. For example, KLF2 is able to repress the induction of 32 cytokines/chemokines in response to proinflammatory stimuli, in addition to promoting a robust antiinflammatory program involving the protective cytokines ELAFIN and IL-11. KLF2 also regulates a coordinated antithrombotic phenotype that parallels earlier studies involving thrombomodulin (TM) and tissue factor. Importantly, aside from eNOS and TM, none of the direct transcriptional targets of KLF2 in endothelium is known.

As a consequence of the coordinated transcriptional changes described above, KLF2 expression is critical for the endothelial functional phenotype. For example, endothelial–leukocyte interactions are suppressed when KLF2 expression is forced in IL-1β-treated endothelial cells (Parmar et al. 2006). This is important because endothelial–leukocyte interactions are a critical step for the development of inflammatory sites, and inhibition of this process by KLF2 may be a key mechanism underlying flow-mediated athero-protection. KLF2 overexpression in endothelial cells also blocks oxidative stress-mediated cell injury (Parmar et al. 2006). In addition, KLF2 expression functions as an important modulator of vascular remodeling and maturation in vitro and in vivo (Mack et al. 2008; Wu et al. 2008). Regulation of endothelial KLF2 expression by flow controls the release of endothelium-derived signals that affect smooth muscle cell migration and vessel wall stabilization, establishing an important mechanistic link between flow, endothelial expression, and smooth muscle cell function (Mack et al. 2008).

Pathophysiological Studies of KLF2 in Murine Models of Atherosclerosis

The importance of KLF2 for athero-protection in vivo was recently documented (Atkins et al. 2008). In this study, ApoE-null mice bearing a hemizygous deficiency in KLF2 exhibited a 31%–37% increase in atherosclerotic lesion area relative to littermate controls. Aortic segments displayed no significant differences in endothelial expression of thrombomodulin, VCAM-1, or eNOS. However, macrophages from hemizygous KLF2 mice showed enhanced lipid uptake as well as increased expression of the lipid chaperone aP2/FABP4, a protein thought to contribute to the development of atherosclerosis (Makowski et al. 2001). Interestingly, a 39% increase in KLF4 mRNA levels was found in lung tissue harvested from KLF2 hemizygous mice. As described in the following section, KLF4 has been shown to bestow antiinflammatory properties to the endothelium, and this compensatory increase may explain the lack of difference in the endothelial parameters measured (Hamik et al. 2007). Although the study by Atkins et al. was critical in establishing the global importance of KLF2 for athero-protection in vivo, the focal importance of KLF2 in each cell type (endothelial cells, macrophages, T cells) remains to be defined (Homeister and Patterson 2008).

KLF4

Expression of KLF4 in the vascular endothelium was first described more than a decade ago (Yet et al. 1998), but early studies of KLF4 knockout mice, which die shortly after birth owing to a loss of skin barrier function, failed to reveal any vascular abnormalities (Segre et al. 1999). In vitro, a DNA microarray screen identified the flow-responsive property of endothelial KLF4 (McCormick et al. 2001), but its function was not explored. Recently however, KLF4 was found to confer an antiinflammatory (Methe et al. 2007) and antithrombotic phenotype to the endothelium (Hamik et al. 2007). Silencing endogenous KLF4 expression resulted in decreased levels of eNOS and thrombomodulin as well as increased levels of procoagulant and proinflammatory factors in response to inflammatory stimuli (Hamik et al. 2007). Furthermore, overexpression of KLF4 was sufficient to induce antithrombotic factors and inhibit secretion of various inflammatory mediators (Hamik et al. 2007). These data suggested that KLF4 functions as a regulator for endothelial activation (Hamik et al. 2007; Methe et al. 2007).

One of the interesting properties of endothelial KLF4 is its induction by inflammatory mediators, in contrast to KLF2, whose expression was previously found to be suppressed by these mediators (Hamik et al. 2007; SenBanerjee et al. 2004). Such a disparity may be reconciled by KLF4 acting as a molecular rheostat to regulate the degree of endothelial activation. Interestingly, the signaling mechanisms mediating the induction of KLF4 by flow are MEK5-dependent (G.V. and

G.G.C., unpublished observations, 2008). The specific requirement of ERK5 and MEF2 remains unknown, but it will be interesting to determine if this critical pathway for flow-mediated induction of KLF2 is also conserved for KLF4. Indeed, phylogenetic analysis reveals that among the Krüppel-like family, KLF2 and KLF4 are the most related (Bieker 2001). Although the functional redundancy of KLF2 and KLF4 has been described in other cellular contexts (Jiang et al. 2008), it is of particular interest to determine their interactions in endothelial cells. Recent studies suggest that KLF4 may play some compensatory role for KLF2 in the context of atherogenesis, but specifically how this interaction is mediated, and its importance, remain unclear (Atkins et al. 2008).

Other KLFs

Although most of the focus has been placed on KLF2 and KLF4, other Krüppel-like family members are expressed in endothelial cells as well, including KLF5, KLF6, KLF7, and KLF11. Our understanding of the roles of these KLFs, in most cases, is limited; however, several groups have gained some insigths into their function. KLF11, for instance, is thought to function as an antagonist for the sterol-responsive element-binding proteins (SREBPs) and Sp1 proteins, which mediate transcriptional activation of cholesterol-dependent genes (Cao et al. 2005). KLF6, meanwhile, is upregulated during vascular injury and induces expression of a broad range of genes important in the response to injury, including endoglin, urokinase plasminogen activator, and transforming growth factor-β1 (TGF-β1) (Botella et al. 2002; Kojima et al. 2000). Additionally, KLF6, in conjunction with Sp2, has been shown to regulate negatively the expression of matrix metalloproteinase 9 (MMP-9) (Das et al. 2006). Activation of the farnesoid X receptor, however, upregulates small heterodimeric partner (SHP), which disrupts binding of the Sp2–KLF6 complex to the metalloproteinase promoter (Das et al. 2006). This results in activation of MMP-9 and enhanced endothelial cell motility, an important process in vascular remodeling (Das et al. 2006). Although we have gained some insights into the role of these other members of the KLF family, our understanding of their importance in endothelial cell biology is still in the early stages.

Conclusions

KLF2 and KLF4 functions in the vascular endothelium are critical for two main reasons: (1) their rapid, robust regulation by distinct types of shear stress present in the circulation; and (2) the pleiotropy of their downstream transcriptional and cellular effects resulting in an athero-protective endothelial phenotype. Understanding the mechanisms underlying the athero-protective shear stress and the pharmacologically-induced expression of KLF2 is critical for designing endothelium-targeted vasoprotective therapies. Furthermore, establishing whether statins and a proinflammatory milieu (e.g., sepsis) can result in robust changes in endothelial KLF2 expression in vivo is of great interest. Regarding the latter context, it is unknown whether downregulation of KLF2 plays an important role in allowing the endothelial cell to become fully "activated." Because this appears to be its role in T lymphocytes and macrophages, it is tempting to speculate about such a scenario in the vascular compartment. Most significantly, identifying in vivo a functional consequence of the two remarkable features of KLF2 and KLF4 —flow-dependent regulation and coordinated regulation of athero-protective transcriptional programs—may yield insights into the long-observed correlation between local hemodynamics and atherogenesis.

References

Anderson KP, Kern CB, Crable SC, and Lingrel JB (1995) Isolation of a gene encoding a functional zinc finger protein homologous to erythroid Kruppel-like factor: identification of a new multigene family. Mol Cell Biol *15*, 5957-5965.

Atkins GB, Wang Y, Mahabeleshwar GH, Shi H, Gao H, Kawanami D, Natesan V, Lin Z, Simon DI, and Jain MK (2008) Hemizygous deficiency of Kruppel-like factor 2 augments experimental atherosclerosis. Circ Res *103*, 690-693.

Bhattacharya R, Senbanerjee S, Lin Z, Mir S, Hamik A, Wang P, Mukherjee P, Mukhopadhyay D, and Jain MK (2005) Inhibition of vascular permeability factor/vascular endothelial growth factor-mediated angiogenesis by the Kruppel-like factor KLF2. J Biol Chem *280*, 28848-28851.

Bi W, Drake CJ, and Schwarz JJ (1999) The transcription factor MEF2C-null mouse exhibits complex vascular malformations and reduced cardiac expression of angiopoietin 1 and VEGF. Dev Biol *211*, 255-267.

Bieker JJ (2001) Kruppel-like factors: three fingers in many pies. J Biol Chem *276*, 34355-34358.

Botella LM, Sanchez-Elsner T, Sanz-Rodriguez F, Kojima S, Shimada J, Guerrero-Esteo M, Cooreman MP, Ratziu V, Langa C, Vary CP et al (2002) Transcriptional activation of endoglin and transforming growth factor-beta signaling components by cooperative interaction between Sp1 and KLF6: their potential role in the response to vascular injury. Blood *100*, 4001-4010.

Brindle NP, Saharinen P, and Alitalo K (2006) Signaling and functions of angiopoietin-1 in vascular protection. Circ Res *98*, 1014-1023.

Cao S, Fernandez-Zapico ME, Jin D, Puri V, Cook TA, Lerman LO, Zhu XY, Urrutia R, and Shah V (2005) KLF11-mediated repression antagonizes Sp1/sterol-responsive element-binding protein-induced transcriptional activation of caveolin-1 in response to cholesterol signaling. J Biol Chem *280*, 1901-1910.

Conkright MD, Wani MA, and Lingrel JB (2001) Lung Kruppel-like factor contains an autoinhibitory domain that regulates its transcriptional activation by binding WWP1, an E3 ubiquitin ligase. J Biol Chem *276*, 29299-29306.

Dai G, Kaazempur-Mofrad MR, Natarajan S, Zhang Y, Vaughn S, Blackman BR, Kamm RD, Garcia-Cardena G, and Gimbrone MA Jr (2004) Distinct endothelial phenotypes evoked by arterial waveforms derived from atherosclerosis-susceptible and -resistant regions of human vasculature. Proc Natl Acad Sci U S A *101*, 14871-14876.

Das A, Fernandez-Zapico ME, Cao S, Yao J, Fiorucci S, Hebbel RP, Urrutia R, and Shah VH (2006) Disruption of an SP2/KLF6 repression complex by SHP is required for farnesoid X receptor-induced endothelial cell migration. J Biol Chem *281*, 39105-39113.

Dekker RJ, van Soest S, Fontijn RD, Salamanca S, de Groot PG, VanBavel E, Pannekoek H, and Horrevoets AJ (2002) Prolonged fluid shear stress induces a distinct set of endothelial cell genes, most specifically lung Kruppel-like factor (KLF2). Blood *100*, 1689-1698.

Dekker RJ, van Thienen JV, Rohlena J, de Jager SC, Elderkamp YW, Seppen J, de Vries CJ, Biessen EA, van Berkel TJ, Pannekoek H, and Horrevoets AJ (2005) Endothelial KLF2 links local arterial shear stress levels to the expression of vascular tone-regulating genes. Am J Pathol *167*, 609-618.

Fukuhara S, Sako K, Minami T, Noda K, Kim HZ, Kodama T, Shibuya M, Takakura N, Koh GY, and Mochizuki N (2008) Differential function of Tie2 at cell-cell contacts and cell-substratum contacts regulated by angiopoietin-1. Nat Cell Biol *10*, 513-526.

Hamik A, Lin Z, Kumar A, Balcells M, Sinha S, Katz J, Feinberg MW, Gerzsten RE, Edelman ER, and Jain MK (2007) Kruppel-like factor 4 regulates endothelial inflammation. J Biol Chem *282*, 13769-13779.

Homeister JW, and Patterson C (2008) Zinc fingers in the pizza pie aorta. Circ Res *103*, 687-689.

Huddleson JP, Srinivasan S, Ahmad N, and Lingrel JB (2004) Fluid shear stress induces endothelial KLF2 gene expression through a defined promoter region. Biol Chem *385*, 723-729.

Jiang J, Chan YS, Loh YH, Cai J, Tong GQ, Lim CA, Robson P, Zhong S, and Ng HH (2008) A core Klf circuitry regulates self-renewal of embryonic stem cells. Nat Cell Biol *10*, 353-360.

Kinderlerer AR, Ali F, Johns M, Lidington EA, Leung V, Boyle JJ, Hamdulay SS, Evans PC, Haskard DO, and Mason JC (2008) KLF2-dependent, shear stress-induced expression of CD59: a novel cytoprotective mechanism against complement-

Kojima S, Hayashi S, Shimokado K, Suzuki Y, Shimada J, Crippa MP, and Friedman SL (2000) Transcriptional activation of urokinase by the Kruppel-like factor Zf9/COPEB activates latent TGF-beta1 in vascular endothelial cells. Blood *95*, 1309-1316.

Kumar A, Lin Z, SenBanerjee S, and Jain MK (2005) Tumor necrosis factor alpha-mediated reduction of KLF2 is due to inhibition of MEF2 by NF-kappaB and histone deacetylases. Mol Cell Biol *25*, 5893-5903.

Kuo CT, Veselits ML, Barton KP, Lu MM, Clendenin C, and Leiden JM (1997) The LKLF transcription factor is required for normal tunica media formation and blood vessel stabilization during murine embryogenesis. Genes Dev *11*, 2996-3006.

Lee JS, Yu Q, Shin JT, Sebzda E, Bertozzi C, Chen M, Mericko P, Stadtfeld M, Zhou D, Cheng L et al (2006) Klf2 is an essential regulator of vascular hemodynamic forces in vivo. Dev Cell *11*, 845-857.

Liao JK, and Laufs U (2005) Pleiotropic effects of statins. Annu Rev Pharmacol Toxicol *45*, 89-118.

Lin Z, Kumar A, SenBanerjee S, Staniszewski K, Parmar K, Vaughan DE, Gimbrone MA Jr, Balasubramanian V, Garcia-Cardena G, and Jain MK (2005) Kruppel-like factor 2 (KLF2) regulates endothelial thrombotic function. Circ Res *96*, e48-57.

Mack PJ, Zhang Y, Chung S, Vickerman V, Kamm RD, and Garcia-Cardena G (2008) Biomechanical regulation of endothelial-dependent events critical for adaptive remodeling. J Biol Chem.

Makowski L, Boord JB, Maeda K, Babaev VR, Uysal KT, Morgan MA, Parker RA, Suttles J, Fazio S, Hotamisligil GS, and Linton MF (2001) Lack of macrophage fatty-acid-binding protein aP2 protects mice deficient in apolipoprotein E against atherosclerosis. Nat Med *7*, 699-705.

McCormick SM, Eskin SG, McIntire LV, Teng CL, Lu CM, Russell CG, and Chittur KK (2001) DNA microarray reveals changes in gene expression of shear stressed human umbilical vein endothelial cells. Proc Natl Acad Sci U S A *98*, 8955-8960.

Methe H, Balcells M, Alegret Mdel C, Santacana M, Molins B, Hamik A, Jain MK, and Edelman ER (2007) Vascular bed origin dictates flow pattern regulation of endothelial adhesion molecule expression. Am J Physiol Heart Circ Physiol *292*, H2167-2175.

Nusslein-Volhard C, and Wieschaus E (1980) Mutations affecting segment number and polarity in Drosophila. Nature *287*, 795-801.

Oates AC, Pratt SJ, Vail B, Yan Y, Ho RK, Johnson SL, Postlethwait JH, and Zon LI (2001) The zebrafish klf gene family. Blood *98*, 792-801.

Parmar KM, Larman HB, Dai G, Zhang Y, Wang ET, Moorthy SN, Kratz JR, Lin Z, Jain MK, Gimbrone MA Jr, and Garcia-Cardena G (2006) Integration of flow-dependent endothelial phenotypes by Kruppel-like factor 2. J Clin Invest *116*, 49-58.

Parmar KM, Nambudiri V, Dai G, Larman HB, Gimbrone MA Jr, and Garcia-Cardena G (2005) Statins exert endothelial atheroprotective effects via the KLF2 transcription factor. J Biol Chem *280*, 26714-26719.

Sako K, Fukuhara S, Minami T, Hamakubo T, Song H, Kodama T, Fukamizu A, Gutkind JS, Koh GY, and Mochizuki N (2008) Angiopoietin-1 induces Kruppel-like factor 2 expression through a phosphoinositide 3-kinase/AKT-dependent activation of myocyte enhancer factor 2. J Biol Chem.

Sato TN, Tozawa Y, Deutsch U, Wolburg-Buchholz K, Fujiwara Y, Gendron-Maguire M, Gridley T, Wolburg H, Risau W, and Qin Y (1995) Distinct roles of the receptor tyrosine kinases Tie-1 and Tie-2 in blood vessel formation. Nature *376*, 70-74.

Sauer F, and Jackle H (1993) Dimerization and the control of transcription by Kruppel. Nature *364*, 454-457.

Segre JA, Bauer C, and Fuchs E (1999) Klf4 is a transcription factor required for establishing the barrier function of the skin. Nat Genet *22*, 356-360.

SenBanerjee S, Lin Z, Atkins GB, Greif DM, Rao RM, Kumar A, Feinberg MW, Chen Z, Simon DI, Luscinskas FW et al (2004) KLF2 Is a novel transcriptional regulator of endothelial proinflammatory activation. J Exp Med *199*, 1305-1315.

Sen-Banerjee S, Mir S, Lin Z, Hamik A, Atkins GB, Das H, Banerjee P, Kumar A, and Jain MK (2005) Kruppel-like factor 2 as a novel mediator of statin effects in endothelial cells. Circulation *112*, 720-726.

Suske G, Bruford E, and Philipsen S (2005) Mammalian SP/KLF transcription factors: bring in the family. Genomics *85*, 551-556.

Wang F, Zhu Y, Huang Y, McAvoy S, Johnson WB, Cheung TH, Chung TK, Lo KW, Yim SF, Yu MM et al (2005) Transcriptional repression of WEE1 by Kruppel-like factor 2 is involved in DNA damage-induced apoptosis. Oncogene *24*, 3875-3885.

Wang L, Fan C, Topol SE, Topol EJ, and Wang Q (2003) Mutation of MEF2A in an inherited disorder with features of coronary artery disease. Science *302*, 1578-1581.

Wani MA, Conkright MD, Jeffries S, Hughes MJ, and Lingrel JB (1999) cDNA isolation, genomic structure, regulation, and chromosomal localization of human lung Kruppel-like factor. Genomics *60*, 78-86.

Weng L, Kavaslar N, Ustaszewska A, Doelle H, Schackwitz W, Hebert S, Cohen JC, McPherson R, and Pennacchio LA (2005) Lack of MEF2A mutations in coronary artery disease. J Clin Invest *115*, 1016-1020.

Wong AL, Haroon ZA, Werner S, Dewhirst MW, Greenberg CS, and Peters KG (1997) Tie2 expression and phosphorylation in angiogenic and quiescent adult tissues. Circ Res *81*, 567-574.

Wu J, Bohanan CS, Neumann JC, and Lingrel JB (2008) KLF2 transcription factor modulates blood vessel maturation through smooth muscle cell migration. J Biol Chem *283*, 3942-3950.

Yet SF, McA'Nulty MM, Folta SC, Yen HW, Yoshizumi M, Hsieh CM, Layne MD, Chin MT, Wang H, Perrella MA et al (1998) Human EZF, a Kruppel-like zinc finger protein, is expressed in vascular endothelial cells and contains transcriptional activation and repression domains. J Biol Chem *273*, 1026-1031.

Chapter 15
Krüppel-like Factors KLF2, KLF4, and KLF5: Central Regulators of Smooth Muscle Function

Christopher W. Moehle and Gary K. Owens

Abstract The vascular smooth muscle cell (SMC) plays a vital role in mammalian physiology through its regulation of blood pressure via contraction and relaxation. In response to vascular injury, it is capable of rapidly and reversibly modulating its phenotype to a cell type capable of performing a number of functions key to wound healing and vascular inflammation including migration, proliferation, matrix synthesis, chemokine production, and protein synthesis. Recent work has identified three Krüppel-like factors—KLF2, KLF4, KLF5—as intricately involved in all of these processes. This review provides a brief overview of the role these factors play in regulating these and other key SMC functions.

Introduction

The smooth muscle cell (SMC) develops from embryonic precursors in a process characterized by the expression of a number of SMC-selective genes including smooth muscle α-actin (SMαA), smooth muscle myosin heavy chain (SM-MHC), SM22α, calponin, and caldesmon. These genetic markers are part of the contractile apparatus, contributing to the primary function of differentiated SMCs—the regulation of blood pressure through contraction and relaxation. The SMC is not terminally differentiated, however, and can rapidly and reversibly change its phenotype in response to changes in its extracellular environment. These phenotypically modulated SMCs are associated with a marked increase in proliferation, migration, and protein synthesis compared to cells in a fully differentiated state that occur in blood vessels of mature animals.

C.W. Moehle and G.K. Owens (✉)
Department of Molecular Physiology and Biological Physics, University of Virginia, 415 Lane Road, Charlottesville, VA 22908, USA
Robert M. Berne Cardiovascular Research Center, University of Virginia, Charlottesville, VA 22908, USA

R. Nagai et al. (eds.), *The Biology of Krüppel-like Factors*,
DOI 10.1007/978-4-431-87775-2_15,

Phenotypic modulation of SMCs plays a key role in repair of vascular injury and is thought also to underlie the pathogenesis of many vascular diseases including atherosclerosis and restenosis following balloon angioplasty or stent placement. Migration and proliferation of phenotypically modulated medial SMCs play a critical role in neointima formation during the early stages of these processes. In addition, synthesis of matrix components, chemokines, and other proteins by SMCs is believed to be a main determinant in the composition of atherosclerotic lesions. Later in the progression of plaques, the migration, matrix deposition, and reinvestment of phenotypically modulated SMCs is thought to play a significant role in the formation of stable fibrous caps. Formation of such caps is believed to help prevent plaque rupture, thrombosis, and subsequent clinical outcomes. Understanding the molecular mechanisms that control SMC phenotypic modulation is therefore a vital topic in modern biomedical research.

Recent work has identified three closely related transcription factors—KLF2, KLF4, KLF5—as central to many key SMC functions. Their influence starts during early development, where both KLF2 and KLF5 loss leads to impaired SMC investment around new blood vessels. The utility of these factors continue into adulthood, as KLF5 is a vital regulator of proliferation, in part by controlling a number of genes strongly associated with vascular remodeling. Although KLF2 has not been directly studied in adult SMCs to our knowledge, mouse embryonic fibroblasts (MEFs) lacking KLF2 showed impaired migration and proliferation in response to platelet derived growth factor-BB (PDGF)-BB. PDGF-BB-dependent migration is important to many types of vascular remodeling, suggesting a potential role for KLF2 in this process. KLF4, in contrast, has been studied primarily in adult functions, where it has been implicated in vitro and in vivo in SMC phenotypic modulation and progression through the cell cycle. Moreover, as discussed in Chapter 10 in this book, there is tremendous excitement regarding KLF4 because it is one of four factors that, when overexpressed, can induce dermal fibroblasts and other mesenchymal cells into a embryonic stem cell-like state (Takahashi and Yamanaka 2006). The overall goal of this chapter is to review these and other roles played by these three closely related KLF transcription factors in controlling SMC development, function, and/or phenotypic modulation. Particular emphasis is placed on results that are supported by functional evidence in vascular cells in vivo.

KLF2: Required for Normal Formation and Function of Arteries During Development

KLF2, originally known as LKLF, was one of the first KLFs shown to be required for proper investment of differentiated SMCs in vivo. Expression is detected as early as E9.5 in the endothelial cells (ECs) of the developing vasculature. In both embryonic and adult vessels, KLF2 induction is reportedly a consequence of laminar shear stress (Dekker et al. 2002; Lee et al. 2006). Conventional knockout of this factor is embryonic lethal between E12.5 and E14.5, with severe intraembryonic and intraamniotic hemorrhaging as the most overt phenotype. Angiogenesis and vasculogenesis appear normal, suggesting that KLF2 loss strongly hinders proper vascular maturation but not de novo formation of nascent blood vessels (i.e., capillary tubes). Consistent with this, ECs and some SMCs form around developing arteries in KLF2 knockout (KO) mice. However, the tunica media in KLF2 KO animals is irregular and poorly formed, with a drastic reduction in the total number of invested SMCs and pericytes. The few SMαA-positive cells surrounding the arteries of knockout animals are also much less organized than their wild-type counterparts, failing to exhibit the tight, elongated morphology that is a hallmark of properly invested SMCs. Importantly, KLF2 loss did not affect erythroid or myelomonocytic development, even though it may affect function of those cell lineages later in development (Kuo et al. 1997). Taken together, these results support the hypothesis that KLF2 loss impairs vascular maturation through defective cell signaling in the vessel wall. Unfortunately, it does not distinguish whether this defect lies in SMCs, ECs, or both.

There is evidence that KLF2 signaling plays its earliest role in SMC development via its effects in the endothelium. For example, careful histologic analysis of both conventional KLF2 KO and EC-targeted Tie-2 Cre conditional KLF2 KO mice by Lee et al. showed a slow, then absent, heartbeat prior to the appearance of SMC defects (Lee et al. 2006). This evidence is suggestive of heart failure as the primary cause of death. Importantly, Doppler flow studies between embryonic days 11.5 and 13.5 in EC-targeted Tie-2 Cre KLF2 KO mice showed an initial high cardiac output state followed by a precipitous drop in ejection fraction and heart rate. Maternal administration of the SMC contractile agonist phenylephrine rescued the embryonic lethality in some Tie-2 Cre embryos, suggesting that mortality is due to high output heart failure caused by decreased peripheral vascular resistance. Thus, KLF2 signaling in the endothelium appears necessary for establishment of vascular tone. The exact mechanisms whereby loss of KLF2 signaling in ECs may result in defective SMC contraction are not known. However, several putative mediators have been implicated based on adenoviral and lentiviral KLF2 overexpression studies in cultured ECs including endothelial nitric oxide synthase (eNOS) (Dekker et al. 2006; SenBanerjee et al. 2004). However, Lee et al. did not detect changes in expression of eNOS or other vasoactive genes in Tie-2 Cre KLF2 KO mice based on in situ hybridization or quantitative polymerase chain reaction (qPCR) analyses. As these authors astutely pointed out, the effects of KLF2 loss on vascular tone may therefore be due to either an unknown gene target or simply the summative effect of changes in multiple genes. It is also possible that factors, including other members of the KLF family, may compensate for KLF2 loss, thereby obscuring the factor initially required for maintenance of SMC tone. Despite the uncertainty regarding the precise mechanism of KLF2's hemodynamic effects, there is compelling evidence that endothelial KLF2 plays an essential developmental role via altering SMC tone.

Recent evidence suggests that KLF2 may also signal directly within SMCs. Further studies of the conventional KLF2 KO mouse by Wu et al. showed defective SMC investment that was localized around the dorsal aorta (Wu et al. 2008). This led the authors to hypothesize that KLF2 may play a role in migration as SMC investment of the dorsal aorta is dependent on migration of nascent SMCs from the surrounding mesenchyme. Consistent with this hypothesis, in vitro analysis of MEFs extracted from these mice showed them to be deficient in PDGF-BB-stimulated proliferation and migration (Wu et al. 2008), a process important for SMC/pericyte investment during angiogenesis (Cao et al. 2003; Kano et al. 2005), arteriogenesis (Cao et al. 2003), the response to vascular injury (Buetow et al. 2003), and atherosclerosis (Kozaki et al. 2002; Sano et al. 2001). This observation not only provides a potential SMC-intrinsic mechanism for the hemorrhagic defects seen in KLF2 $^{-/-}$ mice, it suggests that KLF2 has the potential to mediate a number of key functions of phenotypically modulated SMCs during normal development, repair of vascular injury, and progression of vascular disease.

Lee et al. did attempt to address the direct role of KLF2 signaling in SMCs using SMC-selective KLF2 KO mice generated by crossing KLF2 floxed mice with SM22 Cre transgenic mice. These mice developed normally and survived into adulthood with no overt phenotype (Lee et al. 2006). Although these results were interpreted as evidence that KLF2 does not play an essential developmental role in SMCs, a major limitation in these studies is that SMC must first advance through the early stages of differentiation for SM22 to be activated (Li et al. 1996). Hence, the studies do not address the possibility that KLF2 might play a key role in initial recruitment of SMCs/pericytes to nascent blood vessels. Indeed, the observations of Wu et al. that KLF2 may impair the early process of SMC migration make this a distinct possibility. Moreover, in vitro studies have shown that MEFs lacking KLF2 exhibit impaired migration in response to PDGF-BB, a factor whose knockout also produces leaky, hemorrhagic vessels (Leveen et al. 1994; Lindahl et al. 1997; Wu et al. 2008). Further studies are needed using a Cre recombinase strategy that selectively targets early SMC/pericyte progenitor cells, such as PDGF-β receptor Cre mice (Foo et al. 2006) or Wnt-1 Cre mice (Chai et al. 2000; Huang et al. 2008) for mesenchymal and neural crest-derived cells, respectively. In addition, although Kuo et al. and Lee et al. did report vascular in situ hybridization analysis for KLF2 at embryonic days 9.5, 11.5, and 12.5 (Kuo et al. 1997; Lee et al. 2006), higher-resolution imaging of serial sections of the dorsal section of the developing aorta between days 9.5 and 14.5 is needed to ascertain if KLF2-positive mesenchymal cells appear to be migrating toward the developing aorta. Similarly, there are no data that address whether impaired migration in KLF2$^{-/-}$ MEFs translates into a cell autonomous defect in migration in either SMCs or cells capable of becoming SMCs. Thus, expanding the experiments of Wu et al. to adult SMCs or to known SMC precursor populations (e.g., proepicardial cells) would be valuable for further defining the cell autonomous functions of KLF2 in SMC and SMC precursors. In addition, it would be of interest to determine if KLF2 KO cells show defective investment in nascent blood vessels in the context of chimeric KO mice generated by injecting lineage-tagged KLF2 KO embryonic stem cells (ESCs) into wild-type blastocysts. Taken together, the preceding results provide clear evidence that KLF2 plays a critical role in normal SMC development, but further studies are needed to elucidate the precise underlying mechanisms and/or if KLF2 has direct functions within SMCs.

KLF2: Influences SMC Phenotype in Adult Animals Through Effects on Vascular Inflammation

There is evidence based on studies in cultured ECs and macrophages (see Chapter 14 for more details) that KLF2 may play an important role in regulation of vascular inflammation. SenBanerjee et al., for example, demonstrated that KLF2 promoted expression of antiinflammatory genes such as eNOS while simultaneously repressing expression of proinflammatory genes such as vascular cell adhesion molecule-1 (VCAM-1) in cultured ECs through effects on nuclear factor κB (NF-κB)-mediated gene expression (SenBanerjee et al. 2004). Similarly, KLF2 has been shown to inhibit NF-κB and AP-1 responsive promoter expression in cultured macrophages. The resulting inhibition of macrophage activation and phagocytosis did not interfere with monocyte recruitment (Das et al. 2006). NF-κB has been shown to be important for the production of matrix metalloproteinase 9 (MMP9) and mononcyte chemotactic protein 1 (MCP-1) by SMCs (de Martin et al. 2000; Knipp et al. 2004; Marumo et al. 1997), but the role of KLF2 in regulating NF-κB-mediated gene expression in SMCs has not been reported to our knowledge. Nevertheless, activation of proinflammatory properties of ECs and macrophages is strongly associated with the phenotypic modulation of SMCs, making this regulatory effect of KLF2 of major interest to the study of SMC phenotype.

Although KLF2 may affect some processes such as PDGF-BB-induced migration that are initially atherogenic, one would hypothesize that its antiinflammatory actions in ECs and macrophages makes it atheroprotective. Results using heterozygous KLF2 KO mice bred to an ApoE $^{-/-}$ background are consistent with this hypothesis. In contrast to their homozygous KLF2$^{-/-}$ counterparts, hemizygous KLF2 mice survive to adulthood and are apparently normal. When fed a high fat, high cholesterol diet for 20 weeks, ApoE$^{-/-}$ mice hemizygous for KLF2 showed a 31% increase in atherosclerotic lesion area as measured by Sudan IV staining of whole aortas. Interestingly, KLF4 expression was shown to be significantly increased in KLF2$^{+/-}$ mice. As the authors pointed out, the inability to detect changes in known KLF2 regulated genes in KLF2$^{+/-}$ mice may be due to KLF4 compensating for KLF2 loss. Some lasting changes were detected, however, as macrophages isolated from these mice showed enhanced uptake of low-density lipoprotein (LDL) in culture (Atkins et al. 2008). Although these data provides a possible mechanism for KLF2's observed athero-protective effect, it by no means defines it as the only, or the major, protective mechanism KLF2 mediates. Indeed, further in vivo study of KLF2's potential to repress vascular inflammation and modulate SMC migration is necessary to characterize fully KLF2's role in complex, multicellular processes such as atherosclerosis.

In summary, KLF2 is clearly an important factor in SMC maturation and vascular disease, although the bulk of the current evidence indicates that these effects may be mediated indirectly through KLF2-dependent effects in ECs and macrophages. There is also evidence based on studies in cultured MEFs that KLF2 may additionally regulate PDGF-BB-induced migration and proliferation. However, there have been no detailed studies of KLF2's role in regulating these processes in SMCs, nor is there any evidence it does so in vivo in the context of development or disease.

KLF5: Upregulated Expression in Cardiovascular Disease and Positive Regulation of Neointima Formation in Experimental Disease Models

KLF5, formerly known as BTEB2 and IKLF, is strongly expressed in the medial layer of fetal arteries in humans and rabbits. This expression seems to wane by the time SMCs reach maturity as KLF5 expression is virtually absent in adult vessels. Expression is potently reinduced, however, in a number of vascular disorders including coronary atherosclerosis, the response to vascular injury, and vein graft hyperplasia (Bafford et al. 2006; Hoshino et al. 2000). Furthermore, patients expressing KLF5 at the time of coronary atherectomy have significantly higher rates of stenosis 4 months later as well as a dramatic increase in both the rate and recurrence of restenosis (Hoshino et al. 2000). Taken together, these important initial studies of KLF5's expression patterns implicate it as a potential regulator of both pathologic and normal SMC growth and proliferation. Consistent with this, KLF5 has been shown to be involved in the signaling of a number of growth factors. An increase in KLF5 expression occurs following treatment of SMCs or NIH3T3 fibroblasts with angiotensin II, tumor necrosis factor-α (TNF-α), and fibroblast growth factor-2 (FGF-2). Although each of these factors could be the subject of its own review, all have been shown to stimulate growth of SMCs in vitro and are implicated in vascular remodeling in vivo. Furthermore, KLF5 is induced 8 hours after adenoviral overexpression of the anti-apoptotic protein survivin (SVV) (Bafford et al. 2006). In addition, induction of a KLF5 promoter-reporter construct was upregulated by adenovirus-mediated overexpression of Egr-1, which could be blocked by mutation of an Egr-1-binding site in the KLF5 promoter. Similarly, adenovirus-mediated overexpression of MAP kinase (MAPK) cascade factor mitogen-activated protein kinase kinase-1 (MEK1) leads to activation of the KLF5 reporter. Complementing these data, a small molecule MEK1 inhibitor blocks phorbol 12-myristate 13-acetate (PMA)-induced expression of KLF5 mRNA (Kawai-Kowase et al. 1999). Taken together, these results show that KLF5 is involved in the signaling of a variety of intracellular and extracellular growth and remodeling cues.

In light of evidence implicating KLF5 in the regulation of SMC growth in vitro, there has been considerable interest in its potential role in mediating SMC growth and/or phenotypic modulation in vivo. Unfortunately, these efforts have been hampered by the lethality of KLF5$^{-/-}$ mice prior to embryonic day 8.5. Heterozygous knockouts, however, are viable but show moderate thinning of the medial and adventitial layers compared to their wild-type counterparts. In addition, the villi of the intestines do not form properly, and there is a noticeable reduction in cellularity of the intestinal mesenchyme. Importantly, the heterozygous mice show a marked reduction in neointima formation and granulation tissue formation following placement of a cuff around the femoral artery. Heterozygous KLF5 KO mice also show a dramatic reduction in angiogenesis following implantantion of S180 tumor cells (Fig. 1). The effects of KLF5 loss are by no means limited to the vessel wall, as angiotensin II-induced cardiac hypertrophy and fibrosis was also significantly reduced (Shindo et al. 2002). Taken together, results provide

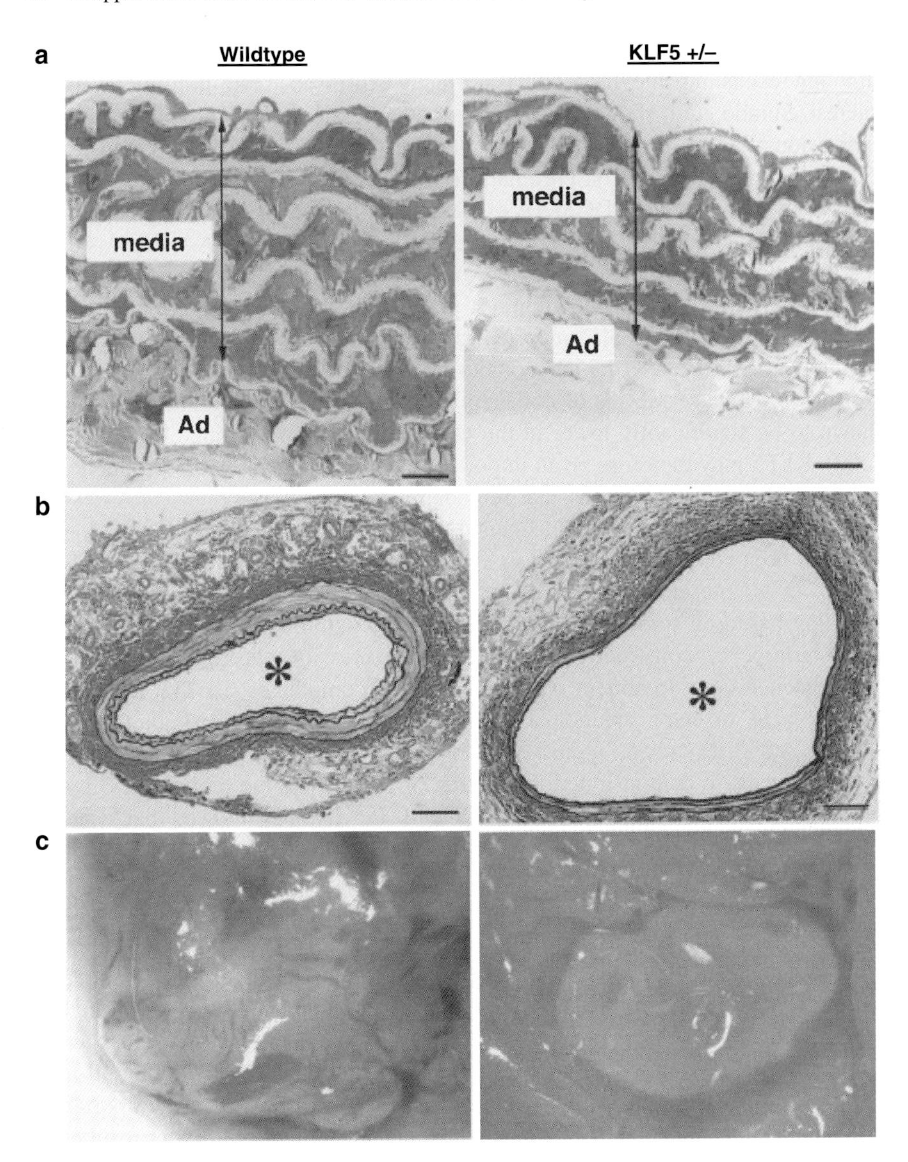

Fig. 1 KLF5 hemizygous mice show medial thinning, impaired neointima formation after cuff placement, and reduced angiogenesis following tumor implantation. Compared to their wild-type counterparts, $KLF5^{+/-}$ mice show marked reduction in medial thickness by transmission electron microscopy. **a** Neointima formation following cuff placement also appears drastically reduced in hemizygous mice. Ad = Adventitia; * = lumen **b** Furthermore, there is a striking reduction in angiogeneis following implantation of tumors into $KLF^{+/-}$ mice. **c** Taken together, this indicates KLF5 is functionally implicated in vessel formation and proliferation (Compiled from Shindo et al. 2002.)

compelling evidence that KLF5 is important in the development of a number of organ systems and may contribute to the pathogenesis of several cardiovascular diseases.

The mechanisms by which loss of one KLF5 allele causes these effects is unclear, although results indicate that KLF5 is normally expressed at rate-limiting levels. Of interest, Shindo et al. showed that KLF5 overexpression can significantly increase expression of a PDGF-A promoter-reporter construct in HeLa cells and binds to the PDGF-A promoter based on chromatin IP (ChIP) analyses in cardiac fibroblasts. KLF5 may also regulate PDGF-AA expression in vivo, as its expression is reduced in the intestinal villi of KLF5 $^{+/-}$ mice. Furthermore, PDGF-AA expression is nearly absent in the hearts of angiotensin-II-infused KLF5$^{+/-}$ mice and the femoral arteries of KLF5$^{+/-}$ mice following cuff placement. Transforming growth factor-β (TGF-β) expression was also reported to be significantly reduced in the hearts of angiotensin-II-infused KLF5 hemizygous mice, although these important data were not shown. However, levels of collagen IV, a known TGF-β regulated gene, were shown to be significantly lower in the vessel wall of KLF5$^{+/-}$ mice (Shindo et al. 2002). KLF5 may therefore be an important factor in controlling the production of one or more factors key to vascular remodeling.

Due to KLF5's apparently detrimental role in a number of vascular pathologies, there has been considerable interest in identifying the factors that regulate its activity. Shindo et al. presented evidence that retinoids could potently modulate KLF5 signaling based on screening cells co-transfected with both a PDGF-A luciferase promoter/reporter construct and a KLF5 overexpression plasmid for changes in luminescence upon treatment with a number of compounds. LE135, a synthetic retinoic acid receptor antagonist, increased activity of the promoter/reporter; whereas Am80, a synthetic agonist of the retinoic acid receptor, repressed activity. Of interest, they also demonstrated interaction of KLF5 with the retinoic acid receptor based on immunoprecipitation experiments. Finally, they showed that systemic administration of LE135 reversed the intestinal defects seen in KLF$^{+/-}$ mice and promoted neointimal formation after cuff injury. Conversely, systemic administration of Am80 suppressed neointima formation, granulation tissue formation, and cardiac hypertrophy in wild-type mice (Shindo et al. 2002).

Subsequent studies from the Nagai group found that KLF5 exists in complexes with a RAR/RXR heterodimer on the PDGF-A promoter in cultured SMCs. This transcriptionally active complex was disrupted by the addition of Am80, thereby abrogating expression of PDGF-A. In addition, treatment with either all-trans retinoic acid (ATRA) or Am80 decreased levels of KLF5 mRNA. Importantly, oral administration of this factor in a rabbit model of stent placement resulted in a significant decrease in neointima formation and resulting stenosis. Although the in vivo results detailed in this and the Shindo et al. study can by no means be restricted to a SMC-specific effect, they do have clear ramifications regarding overall effects on SMC phenotype. Investigation of cultured SMCs shows an increase in transcription of both SMaA and SM-MHC following Am80 treatment—a result complemented by their observation that KLF5 siRNA blocks serum-induced downregulation of the same SMC markers. Of interest, they also found that treatment of rabbits with Am80 was associated with higher levels of SMC markers in in-stent lesions consistent with Am80 lessening SMC phenotypic modulation in these lesions. Expression of the potent SRF co-activator myocardin was unchanged in these studies, suggesting that Am80 is removing an inhibitor of SMC differentiation rather than promoting expression of a known activator. This phenotypic modulation correlates with functional consequences, at least in vitro, as Am80 inhibited proliferation of rat aortic SMCs and human coronary SMC cultures in an apparently dose-dependent manner. Furthermore, fetal bovine serum-induced SMC migration was also potently inhibited by Am80. Importantly, the same doses of Am80 had no apparent effect on the proliferation of human umbilical vein ECs (Fujiu et al. 2005).

Although the preceding observations show evidence of some SMC selectivity in Am80's effects, KLF5 has been reported also to activate a number of non-SMC genes (e.g., MCP-1 in ECs) (Kumekawa et al. 2008; Shinoda et al. 2008) that are also important in vascular inflammation and remodeling. Furthermore, Am80's effects on other KLFs are yet to be fully studied. Because it has not been reported whether siRNA to KLF5 abrogates the effects of Am80, one cannot definitively conclude that Am80's effects are due entirely to loss of KLF5 function despite similarities between the $KLF5^{+/-}$- and Am80-treated phenotypes. Although they are of tremendous importance in evaluating Am80 as a preclinical candidate, these studies do not fully elucidate KLF5's functions in vascular biology. Conditional, targeted knockout of a floxed KLF5 allele using an inducible Cre recombinase system would be an excellent way to address this deficiency initially. Furthermore, fetal liver transplants from $KLF5^{-/-}$ animals and cell type-specific KLF5 knockout would be excellent ways to distinguish the effects of KLF5 knockout that are due to bone marrow cells versus non-bone-marrow-derived cells.

Given the implications of KLF5 signaling in experimental models of cardiovascular disease, there is much interest in the intracellular mechanisms whereby it is regulated. Recent work has identified a number of factors capable of directly interacting with KLF5 and modulating its function. Importantly, KLF5's DNA-binding domain is acetylated by transcriptional co-activator p300 in vitro. Point mutation of this zinc finger reduces p300's ability to positively regulate KLF5's function. This enhancement of KLF5 signaling is potently opposed by SET, which binds to the DNA-binding domain and inhibits acetylation (Miyamoto et al. 2003). In addition, HDAC1 co-immunoprecipitates from SMCs with KLF5 and is shown to repress KLF5-driven activation of the PDGF-A promoter. Interestingly, HDAC1 and p300 competitively interact with KLF5 at the same zinc finger (Matsumura et al. 2005). Taken together, this indicates that KLF5's DNA-binding domain is a key point of regulation in vitro as function is positively enhanced by p300-mediated acetylation and opposed by HDAC1 and SET. These observations are of further interest as there is extensive evidence that epigenetic controls, including histone modifications, play a key role in SMC phenotypic modulation (reviewed in McDonald and Owens 2007). Although the ramifications of these fiindings in vivo are still speculative, it nevertheless identifies an important mechanism for regulating SMC proliferation that may limit pathologic SMC growth.

In summary, KLF5 seems to be a potent pro-growth and remodeling factor. KLF5 is strongly expressed in the developing vasculature. Although expression fades by the time arteries mature, KLF5 is potently induced in a number of pathologies including coronary atherosclerosis, vascular injury, and vein graft hyperplasia. Results of studies in heterozygous KLF5 KO mice indicate that KLF5 plays an important role in positively regulating neointima formation and secretion of PDGF-AA and other pro-growth factors. Much has been done to understand not only the implications of KLF5 expression but also the factors that are capable of regulating or modulating KLF5 function; however, there are many unresolved questions. They include the effects of targeted KLF5 loss in adult function and the dominant cell type(s) of KLF5's actions. Cell type-specific and conditional knockout of a floxed KLF5 allele is an excellent way to rectify these issues. Such experiments are important for both fully understanding KLF5's biological functions and definitively analyzing the usefulness of KLF5 antagonism as a potential therapeutic strategy for in-stent restenosis and vein graft hyperplasia.

KLF4: Repression of SMC Markers and Inhibition of Proliferation in the Injured Vasculature

KLF4 signaling within SMCs has been most heavily investigated in the context of phenotypic modulation of differentiated cells. Expression of KLF4, formerly known as GKLF, is virtually absent in the healthy, quiescent vessel wall but is rapidly induced in the medial and intimal layers following vascular injury (Yoshida et al. 2008b). In culture, this induction of KLF4 expression has been linked to a number of cytokines and growth factors including PDGF-BB (Liu et al. 2005) and PDGF-DD (Thomas et al. 2008), potent SMC mitogens strongly associated with wound healing and atherosclerosis; and 1-palmitoyl-2-(5-oxovaleroyl)-*sn*-glycero-3-phosphocholine (POVPC) (Pidkovka et al. 2007), a lipid oxidation product known to accumulate in atherosclerotic lesions. Thus, KLF4 is linked directly to vascular injury and induced by a number of secreted factors strongly associated with multiple vascular diseases.

The direct examination of KLF4's role in vascular injury, as was the case with KLF2 and KLF5, was initially hampered by the lethality of the conventional KLF4 knockout mouse. While these mice are born at expected Mendelian ratios, they die early after birth due to improper skin barrier formation (Segre et al. 1999). This is by no means the only phenotype of these mice, as KLF4 (-/-) mice also display a 90% reduction in goblet cells in the colon (Katz et al. 2002) illustrating that KLF4 loss may have very broad developmental affects. While no vascular phenotype has been reported in these mice, it is unclear whether this tissue has been carefully scrutinized in the conventional knockout. Furthermore, the apparent broad developmental defects caused by KLF4 loss may obscure any vascular abnormalities.

In vitro analysis has provided a mechanistic link between KLF4 and the repression of SMC markers. Importantly, this process involves specific *cis* elements within marker gene promoters. Initial studies identified KLF4 as capable of binding to the TGF-β Control Element (TCE) within the SM22 and SMaA promoters (Adam et al. 2000). KLF4 binds the TCE based on electrophoretic mobility shift assays (Liu et al. 2003), and also binds the TCE region of SMC promoters within intact chromatin following treatment of cultured SMC with PDGF-BB (McDonald et al. 2006), PDGF-DD (Thomas et al. 2008), or POVPC (Pidkovka et al. 2007), and in vivo following vessel injury (Yoshida et al. 2008b) based on ChIP assays. Unfortunately, the resolution of a ChIP assay is not sufficient to ascertain if KLF4 is binding directly to the TCE element. Of interest, transfection of 10T1/2 cells with KLF4 antagonizes TGF-β induced expression of both SM22 and SMaA promoter/reporter constructs. Interestingly, KLF5 positively regulates activity of this SM22 reporter (Adam et al. 2000). Complementarily, RNA interference to KLF4 with antisense oligos increases expression of both SMaA and SM-MHC. Taken together, these studies suggest KLF4 represses expression of SMC markers, potentially through DNA binding at the TCE. This may directly antagonize KLF5's actions at the same *cis* element. Unfortunately, mutation of the TCE within both the SMaA (Liu et al. 2003) and SM22 (Adam et al. 2000) promoter abolishes expression of a transgenic reporter in vivo. While this suggests in vivo relevance for the interaction between the TCE and positive regulators of SMC differentiation such as KLF5, it unfortunately limits the usefulness of this system in studying the repressive *cis/trans* interactions of KLF4 at the TCE.

KLF4 can also remove key activating cues for marker gene expression, most notably myocardin, which homodimerizes at a leucine zipper motif, allowing it to act as a potent co-activator for serum response factor (SRF) bound to paired CArG boxes in the promoter of marker genes (Du et al. 2003; Wang et al. 2003; Yoshida et al. 2003). KLF4 overexpression has been shown to potently repress levels of myocardin mRNA by qPCR. Furthermore, immunoprecipitation analysis indicates a direct, physical interaction between KLF4 and SRF. Taken together, KLF4 can reduce activation of marker gene promoters via both direct interaction with SRF and repression of its co-activator myocardin. Indeed, adenoviral overexpression of KLF4 potently represses SRF enrichment of the 5′-CArG region within the SMaA promoter of SMCs (Liu et al. 2005). Thus, KLF4's repressive affects on SMC markers seem to be affected in part via the removal of activating cues for marker gene expression.

Part of KLF4's antagonism of myocardin/SRF function is mediated via altered chromatin dynamics. Based on ChIP evidence both in vitro and in vivo, the activation and repression of SMC marker genes is characterized by a number of chromatin modifications that appear specific to CArG box chromatin within SMCs. Activation is strongly associated with Histone acetylation – a mark associated with the increased interaction of SRF with H3K4dMe at CArG boxes of SMC promoters. KLF4, in turn, can encourage histone deacetylation to prevent association of SRF with methylated histones in a process postulated to be due to KLF4's recruitment of HDAC2. Similar repressive chromatin dynamics are seen following PDGF-BB treatment and vascular injury, indicating that these mechanisms may have broad importance to SMC function. Interestingly, neither KLF4, myocardin,

PDGF-BB, nor vascular injury affect the presence of this H3K4 di-methylation – despite their ability to dynamically change histone acetylation - indicating this as a potential "lineage mark" that helps SMCs retain their identity during phenotypic switching (McDonald and Owens 2007; McDonald et al. 2006). Although further characterization of this potential memory mark is needed, it thus appears changes in chromatin dynamics play a central role in SMC phenotypic changes. KLF4 appears to be a central modulator of these changes.

Recent work has illustrated interaction with Elk-1 as an additional mechanism of KLF4-induced repression of SMC markers. KLF4 and Elk-1 are bound to the same region of the SMaA promoter in response to POVPC treatment, as evidenced by ChIP analysis. This indicates the possibility of cooperative interaction between KLF4 and Elk-1. Indeed, combining genetic knockout of KLF4 and pharmacologic inhibition of MEK or ERK, kinases upstream of Elk-1, completely abrogates POVPC induced suppression of marker genes. Furthermore, POVPC treatment was associated with histone H4 hypoacetylation and recruitment of HDAC 2 and 5 (Yoshida et al. 2008a). This is similar to results seen with PDGF-BB, where siRNA studies showed its repressive effects to be dependent on ERK/Elk-1 and HDACs 2, 4, and 5 (Yoshida et al. 2007). Thus, multiple factors can repress markers through processes shown to separately involve KLF4/Elk-1 and histone H4 hypoacetylation. However, the interdependence of these processes has yet to be fully tested. Knocking down KLF4 with siRNA or mutating histones would be interesting first steps to address whether KLF4 is required for histone modifications or whether histone modifications are required for the full repressive affects of KLF4.

Significantly, Yoshida et al. have recently generated tamoxifen conditional KLF4 KO mice to investigate the possible role of KLF4 in regulation of SMC phenotypic switching and growth in a carotid ligation model of vascular injury (Yoshida et al. 2008b). Consistent with the hypothesis that KLF4 is important in regulating these processes in vivo, tamoxifen Cre conditional KLF4 KO mice exhibited a transient delay in repression of SMaA and SM22 following vascular injury (Fig. 2). Furthermore, KLF4 binding to the SMaA and SM22 promoters was heavily enriched 3 days post injury indicating this effect may be, at least in part, transcriptionally mediated (Yoshida et al. 2008b). Taken together, these results indicate that KLF4 is rate limiting for phenotypic switching early in this injury model but apparently is either not required late, or there is activation of alternative compensatory pathways that mediate SMC phenotypic modulation in its absence. Whatever the case, these findings are highly significant as they are the first and only studies to date that define a specific transcriptional regulatory pathway required for SMC phenotypic modulation in vivo, even though the results cannot be narrowed to KLF4 within SMCs. A number of other mediators of phenotypic switching have been identified based on studies in cultured SMC including HERP and the ERK-mediated phosphorylation of Elk-1 [see review by (Kawai-Kowase and Owens 2007)]. Many of the factors shown in vitro to induce KLF4, including PDGF-BB and PDGF-DD, can also activate ERK/Elk-1 (Kawai-Kowase and Owens 2007). Thus, this is a prime candidate for the eventual downregulation of SMC markers in the KLF4 knockouts. Given the similarity among KLFs, it is also quite possible that another member of the KLF family can compensate for the loss of KLF4. Indeed, likely candidates are KLF2 and KLF5, as these have been shown to compensate for KLF4 loss in maintaining stem cell pluripotency (Jiang et al. 2008). Nevertheless, KLF4 seems to be a key effector of SMC phenotypic switching in vivo regardless of the factor that eventually compensates for it.

In addition to the delay in repression of SMC markers, the conditional KLF4 KO mice showed marked hyperproliferation and enhanced neointima formation compared to WT mice (Fig. 2). Of interest, authors showed that this was likely through loss of KLF4 dependent activation of the cyclin dependent kinase inhibitor p21 in that they observed KLF4, along with p53, binding to the promoter of p21 at 3 days post injury based on in vivo ChIP assays. Importantly, the p21 expression seen following infection of SMCs with a KLF4 expressing Adenovirus was abrogated by a p53 siRNA. Taken together, these implicate a synergistic activation of p21 by KLF4 and p53 as a possible mechanistic explanation for KLF4's effects on neointima formation. Furthermore, injured conditional KO arteries showed increased BrdU uptake and no significant changes in TUNEL staining, indicating enhanced proliferation, with no changes in cell death (Yoshida et al. 2008b). Thus, the induction of KLF4 expression following vascular injury may be part of a protective mechanism that limits excess proliferation via enhancement of p21 expression.

It is becoming increasingly apparent that KLF4's affects extend beyond regulation of the cell cycle and repression of marker gene expression. Important recent data has indicated the SMC produces inflammatory mediators including MCP-1 and TGF-β following POVPC treatment (Pidkovka et al. 2007). This supplements earlier reports of inflammatory mediator production by SMC following exposure to cytokines such as IL-1β and TNFα (Loppnow et al. 2008). While KLF4 has not been directly linked to inflammatory gene expression in SMCs, it is both pro- and anti-inflammatory in other cell types [see Chapter 14]. In macrophages, for example, KLF4 enhance NF-κB p65-mediated inflammatory gene expression in vitro, at least in the example of the iNOS promoter, while simultaneously abrogating the interaction between Smad3 and p300. This prevents anti-inflammatory signaling initiated by factors such as TGF-β, creating a cumulative effect that appears shifts macrophages towards an inflammatory phenotype in vitro (Feinberg et al. 2005). KLF4 in the endothelium, in contrast, has been shown to repress inflammation via activation of anti-inflammatory genes such as endothelial nitrous oxide synthase (eNOS) and apparent inhibition of NF-κB p65 regulated inflammatory genes such as VCAM-1 (Hamik et al. 2007). Thus, there is reason to think KLF4 may play a role in regulating inflammation within the SMC, although it is entirely unclear whether it would be pro- or anti-inflammatory.

In summary, KLF4 is an important regulatory factor in a number of processes important to both healthy and pathologic SMC function. It appears to be a central regulator in SMC phenotypic switching and proliferation – roles that likely contribute to its apparent limiting of neointima formation in vivo. Despite the strength of current evidence for KLF4's importance in SMC biology, there is still much about KLF4 yet to be fully understood. KLF4 can regulate vascular inflammation in other cell types, although its importance within the SMC has not been fully established.

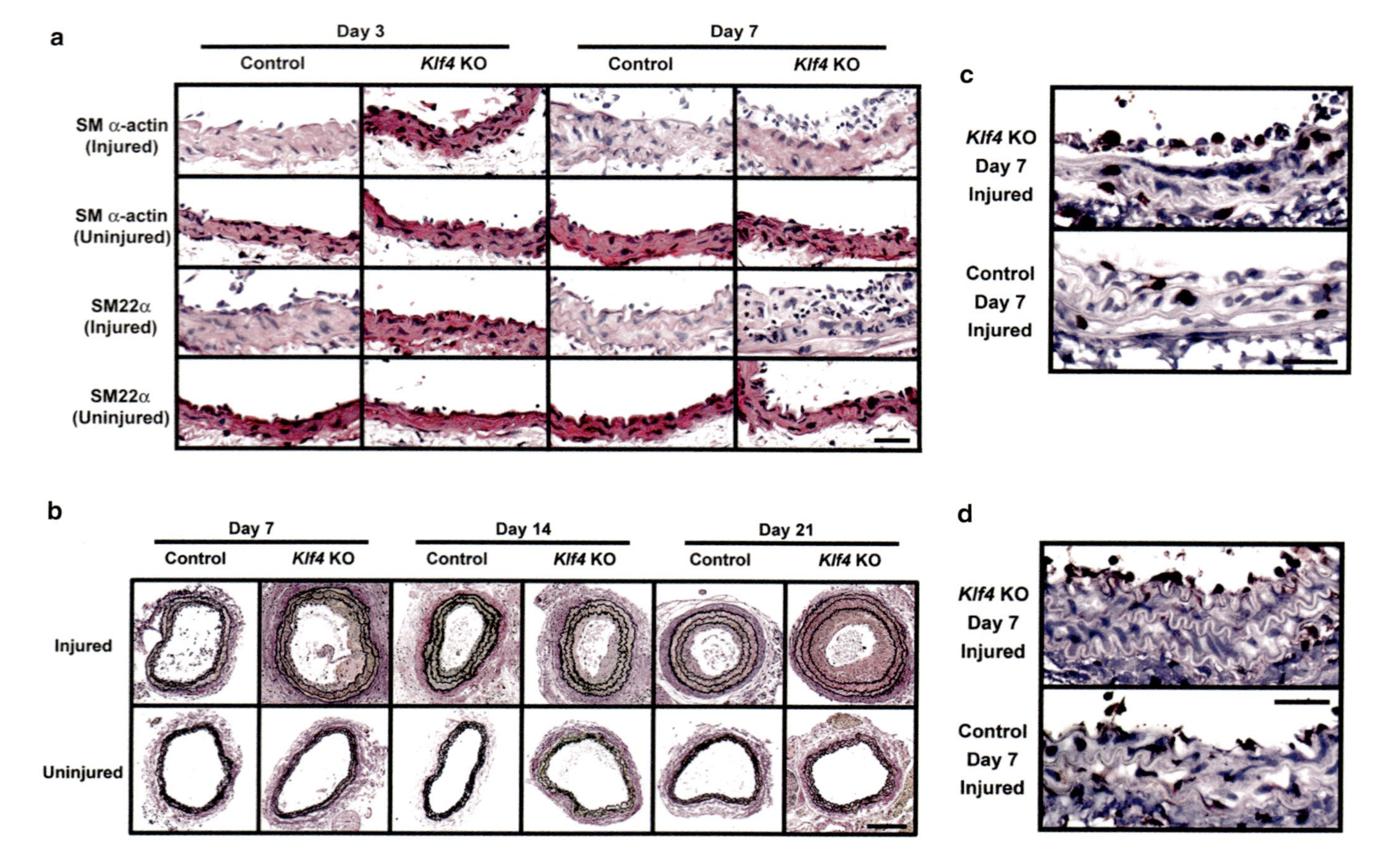

Fig.2 Conditional KLF4 knockout (*KO*) leads to delayed repression of smooth muscle markers and increased neointima formation following ligation injury in vivo. Conditional KO of KLF4 leads to a delay in repression of SMαA and SM22 following ligation injury. **a** This is accompanied by an increase in neointima formation. **b** BrdU staining shows increased proliferation in the vessel of wall of KLF4 KO mice. **c** Immunohistochemistry also indicates a decrease in p21 in the KLF4 KO mice. **d** These last two observations may partially explain the increased neointima formation in **b**. (Compiled from Yoshida et al. 2008.)

Furthermore, the precise cell type(s) that contribute to KLF4's effects on vascular injury need to be elucidated with cell type specific-knockout. This needs to be complemented with further study into the mechanisms whereby KLF4 itself is regulated - particularly the epigenetic modifications that control access to KLF4's own promoter. The coming years should therefore yield much further insight into both the effects of KLF4 on SMC phenotype and the molecular mechanisms whereby KLF4 function is regulated (Fig. 3).

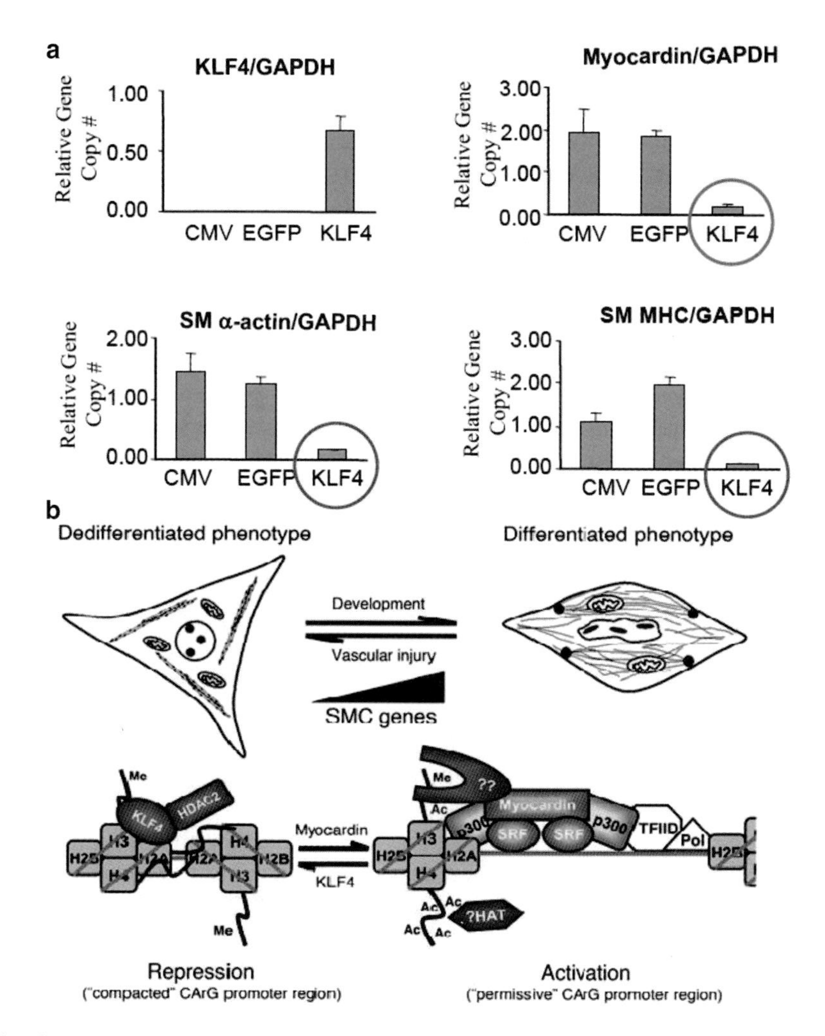

Fig. 3 Summary of KLF4's repression of myocardin-induced activation of SMC marker genes. Adenoviral overexpression of KLF4 potently represses expression of myocardin, SMαA, and SM-MHC mRNA in cultured SMCs. All results normalized to GAPDH. **a** This plastic suppression includes prevention of the interaction of SRF with CArG box elements at SMC promoters in part through repressive histone modifications. Modeled in **b**. (Compiled from Liu et al. 2005 and McDonald et al. 2006)

Conclusion

At least three KLFs are integral to SMC biology. KLF2, at least within MEFs, regulates migration and proliferation in response to PDGF-BB. While this relationship has not been explored in SMC, KLF2 has been directly implicated in adult vascular biology as KLF2 hemizygous mice develop increased aortic atherosclerosis. KLF5, on the other hand, works directly within cultured SMC to promote proliferation. It is strongly implicated in normal and pathologic vascular growth in vivo, including the neointimal hyperplasia seen following experimental models of vascular injury, stent placement, and vein grafting. The pro-growth affects of KLF5 are apparently in opposition to KLF4, which appears to limit pathologic vascular growth following ligation injury. This may be explained by KLF5's ability to promote p21 expression via cooperative interaction with p53 in cultured SMC. Taken together, individual KLFs seems to play unique activating roles in regulating key SMC processes. While KLFs are understudied in some important vascular pathologies, most notably atherosclerosis, the urrent evidence nevertheless implicates KLF2, 4, and 5 as central regulatory factors in SMC and vascular biology both in vitro and in vivo.

These apparent niches for different KLFs in adult SMC are particularly interesting given recent developments regarding the role of these three KLFs in stem cells [see Chapter 10]. Interestingly, whereas overexpression of KLF4, Sox2, c-Myc and Oct3/4 can induce adult cells into a pluripotent state (Takahashi and Yamanaka 2006), triple knockdown of KLF2, KLF4, and KLF5 is required to induce embryonic stem cells to differentiate. SiRNA to one or two of these KLFs led to occupancy of target promoters by the other and no discernable effect. Overexpression of KLF10 could not rescue the triple knockdown whereas siRNA-resistant KLF2, 4, or 5 can, arguing this phenomena is specific to KLF2, 4, and 5 (Jiang et al. 2008). Thus, these factors appear to work together to maintain pluripotency.

This is particularly interesting given the observations in both conventional KLF2 knockout mice and KLF5 hemizygous mice of impaired SMC investment or growth around blood vessels (Shindo et al. 2002; Wu et al. 2008). While no developmental vascular phenotype has been reported with KLF4, there nevertheless appears to be deficiencies in defined cell subsets such as skin cells and colonic goblet cells (Katz et al. 2002; Segre et al. 1999). Thus, it appears that these KLFs switch from overlapping repressors of differentiation to factors essential to the formation of specific cellular subpopulations. Furthermore, KLFs appear to acquire cell type specific effects by adulthood. This is most strikingly apparent in the case of KLF4-mediated regulation of inflammatory gene expression, where is reportedly both pro- and anti-inflammatory in different cell types (Autieri 2008). Indeed, all three KLFs discussed in this chapter exert transcriptional regulatory affects in multiple cell types. Taken together, it appears the consequences of KLF2, 4, and 5 are altered as cells progress down their differentiation pathway.

The precise cues that lead to these altered affects and the intracellular effector mechanisms that bring them about are still speculative at this point. One likely hypothesis is that this is a function of altered chromatin dynamics. As stem cells respond to cues from their extracellular milieu and begin to differentiate, KLF-responsive promoters required for pluripotency and the earlier stages of development are silenced into heterochromatin and new KLF-responsive elements are opened into euchromatin. In the case of plastic cell types such as SMC, lineage marks may be acquired that help cells retain their identity. The H3K4dMe mark – thus far invariant during SMC phenotypic switching – is a potential example. These marks may help in promoting cell specific functions, as with the example of H3K4dMe aiding in the CArG/SRF interaction.

Despite this lineage commitment and subsequent silencing of genes required for pluripotency, KLF4 nevertheless can confer upon SMCs some properties of pluripotent cells. Indeed, analysis of chromatin modifications at marker gene loci show a marked similarity between embryonic stem cells and phenotypically modulated SMCs (McDonald and Owens 2007). While KLF5 has also been linked to phenotypic switching, its ability to induce such epigenetic changes does not appear to have been investigated to date. It is therefore interesting to speculate that KLF4, and potentially KLF2 and 5, may induce the SMC into a "more embryonic" state that may be multipotential in the appropriate extracellular environment. Indeed, phenotypically modified mouse SMCs can give rise to macrophage-like cells in vitro (Rong et al. 2003). Whether the H3K4dMe SMC "memory mark" is still present in these SMC derived macrophage-like cells, and the role of KLFs in their formation, are unanswered questions.

While investigating the nature of the putative SMC memory mark, and its presence in SMC-derived macrophages in vitro, would help explain how KLFs obtain functions specific to individual cell types and lineages – and just how plastic this commitment is – it still leaves the matter of what relative consequence these differing functions play in a complex multicellular process such as atherosclerosis. Thus while further global, conditional knockout will be important initial studies to elucidate the summative role of each KLF, further define the chromatin modifications associated with KLFs in vivo, and evaluate KLF inhibition's potential as a pre-clinical therapy, they will be inadequate to fully understand the physiology of each factor. Cell type specific knockout will therefore be an important follow-up to any conditional knockout study. These important experiments will serve as crucial augmentations to existing evidence of the significant role KLF2, 4, and 5 plays in both SMC and vascular biology.

References

Adam PJ, Regan CP, Hautmann MB, and Owens GK (2000) Positive- and negative-acting Kruppel-like transcription factors bind a transforming growth factor β control element required for expression of the smooth muscle cell differentiation marker SM22α in vivo. J. Biol Chem. 275:37798–37806

Atkins GB, Wang Y, Mahabeleshwar GH, Shi H, Gao H, Kawanami D, Natesan V, Lin Z, Simon DI, and Jain MK (2008) Hemizygous Deficiency of Kruppel- Like Factor 2 Augments Experimental Atherosclerosis. Circ. Res. 103;690–693

Autieri MV (2008) Kruppel-Like Factor 4: Transcriptional Regulator of Proliferation, or Inflammation, or Differentiation, or All Three? Circ Res 102:1455–1457

Bafford R, Sui XX, Wang G, and Conte M (2006) Angiotensin II and tumor necrosis factor-α upregulate survivin and Kruppel-like factor 5 in smooth muscle cells: Potential relevance to vein graft hyperplasia. Surgery 140:289–296

Buetow BS, Tappan KA, Crosby JR, Seifert RA, and Bowen-Pope DF (2003) Chimera Analysis Supports a Predominant Role of PDGFRβ in Promoting Smooth-Muscle Cell Chemotaxis after Arterial Injury. Am J Pathol 163:979–984

Cao R, Brakenhielm E, Pawliuk R, Wariaro D, Post MJ, Wahlberg E, Leboulch P, and Cao Y (2003) Angiogenic synergism, vascular stability and improvement of hind-limb ischemia by a combination of PDGF-BB and FGF-2. Nat Med 9:604–613

Chai Y, Jiang X, Ito Y, Bringas P, Jr., Han J, Rowitch DH, Soriano P, McMahon AP, and Sucov HM (2000) Fate of the mammalian cranial neural crest during tooth and mandibular morphogenesis. Development 127:1671–1679

Das H, Kumar A, Lin Z, Patino WD, Hwang PM, Feinberg MW, Majumder PK, and Jain MK (2006) Kruppel-like factor 2 (KLF2) regulates proinflammatory activation of monocytes. Proc. Natl. Acad. Sci. U.S.A 103:6653–6658

de Martin R, Hoeth M, Hofer-Warbinek R, and Schmid JA (2000) The transcription factor NF-kappa B and the regulation of vascular cell function. Arterioscler Thromb Vasc Biol 20:E83-E88

Dekker RJ, van SS, Fontijn RD, Salamanca S, de Groot PG, VanBavel E, Pannekoek H, and Horrevoets AJ (2002) Prolonged fluid shear stress induces a distinct set of endothelial cell genes, most specifically lung Kruppel-like factor (KLF2). Blood 100:1689–1698

Dekker RJ, Boon RA, Rondaij MG, Kragt A, Volger OL, Elderkamp YW, Meijers JCM, Voorberg J, Pannekoek H, and Horrevoets AJG (2006) KLF2 provokes a gene expression pattern that establishes functional quiescent differentiation of the endothelium. Blood 107:4354–4363

Du KL, Ip HS, Li J, Chen M, Dandre F, Yu W, Lu MM, Owens GK, and Parmacek MS (2003) Myocardin Is a Critical Serum Response Factor Cofactor in the Transcriptional Program Regulating Smooth Muscle Cell Differentiation. Mol. Cell. Biol. 23:2425–2437

Feinberg MW, Cao Z, Wara AK, Lebedeva MA, SenBanerjee S, and Jain MK (2005) Kruppel-like Factor 4 Is a Mediator of Proinflammatory Signaling in Macrophages. J. Biol. Chem. 280:38247–38258

Foo SS, Turner CJ, Adams S, Compagni A, Aubyn D, Kogata N, Lindblom P, Shani M, Zicha D, and Adams RH (2006) Ephrin-B2 controls cell motility and adhesion during blood-vessel-wall assembly. Cell 124:161–173

Fujiu K, Manabe I, Ishihara A, Oishi Y, Iwata H, Nishimura G, Shindo T, Maemura K, Kagechika H, Shudo K, and Nagai R (2005) Synthetic Retinoid Am80 Suppresses Smooth Muscle Phenotypic Modulation and In-Stent Neointima Formation by Inhibiting KLF5. Circ. Res. 97:1132–1141

Hamik A, Lin Z, Kumar A, Balcells M, Sinha S, Katz J, Feinberg MW, Gerszten RE, Edelman ER, and Jain MK (2007) Kruppel-like Factor 4 Regulates Endothelial Inflammation. J.Biol. Chem. 282:13769–13779

Hoshino Y, Kurabayashi M, Kanda T, Hasegawa A, Sakamoto H, Okamoto Ei, Kowase K, Watanabe N, Manabe I, Suzuki T, Nakano A, Takase Si, Wilcox JN, and Nagai R (2000) Regulated expression of the BTEB2 transcription factor in vascular smooth muscle cells : Analysis of developmental and pathological expression profiles shows implications as a predictive factor for restenosis. Circulation 102:2528–2534

Huang J, Cheng L, Li J, Chen M, Zhou D, Lu MM, Proweller A, Epstein JA, and Parmacek MS (2008) Myocardin regulates expression of contractile genes in smooth muscle cells and is required for closure of the ductus arteriosus in mice. J. Clin. Invest 118:515–525

Jiang J, Chan YS, Loh YH, Cai J, Tong GQ, Lim CA, Robson P, Zhong S, and Ng HH (2008) A core Klf circuitry regulates self-renewal of embryonic stem cells. Nat Cell Biol 10:353–360

Kano MR, Morishita Y, Iwata C, Iwasaka S, Watabe T, Ouchi Y, Miyazono K, and Miyazawa K (2005) VEGF-A and FGF-2 synergistically promote neoangiogenesis through enhancement of endogenous PDGF-B-PDGFRβ signaling. J Cell Sci 118:3759–3768

Katz JP, Perreault N, Goldstein BG, Lee CS, Labosky PA, Yang VW, and Kaestner KH (2002) The zinc-finger transcription factor Klf4 is required for terminal differentiation of goblet cells in the colon. Development 129:2619–2628

Kawai-Kowase K, Kurabayashi M, Hoshino Y, Ohyama Y, and Nagai R (1999) Transcriptional Activation of the Zinc Finger Transcription Factor BTEB2 Gene by Egr-1 Through Mitogen-Activated Protein Kinase Pathways in Vascular Smooth Muscle Cells. Circ Res 85:787–795

Kawai-Kowase K and Owens GK (2007) Multiple repressor pathways contribute to phenotypic switching of vascular smooth muscle cells. Am J Physiol Cell Physiol 292:C59–C69

Knipp BS, Ailawadi G, Ford JW, Peterson DA, Eagleton MJ, Roelofs KJ, Hannawa KK, Deogracias MP, Ji B, Logsdon C, Graziano KD, Simeone DM, Thompson RW, Henke PK, Stanley JC, and Upchurch GR, Jr. (2004) Increased MMP-9 expression and activity by aortic smooth muscle cells after nitric oxide synthase inhibition is associated with increased nuclear factor-kappaB and activator protein-1 activity. J. Surg. Res. 116:70–80

Kozaki K, Kaminski WE, Tang J, Hollenbach S, Lindahl P, Sullivan C, Yu JC, Abe K, Martin PJ, Ross R, Betsholtz C, Giese NA, and Raines EW (2002) Blockade of Platelet-Derived Growth Factor or Its Receptors Transiently Delays but Does Not Prevent Fibrous Cap Formation in ApoE Null Mice. Am J Pathol 161:1395–1407

Kumekawa M, Fukuda G, Shimizu S, Konno K, and Odawara M (2008) Inhibition of monocyte chemoattractant protein-1 by Kruppel-like factor 5 small interfering RNA in the tumor necrosis factor-α-activated human umbilical vein endothelial cells. Biol Pharm. Bull. 31:1609–1613

Kuo CT, Veselits ML, Barton KP, Lu MM, Clendenin C, and Leiden JM (1997) The LKLF transcription factor is required for normal tunica media formation and blood vessel stabilization during murine embryogenesis. Genes Dev. 11:2996–3006

Lee JS, Yu Q, Shin JT, Sebzda E, Bertozzi C, Chen M, Mericko P, Stadtfeld M, Zhou D, Cheng L, Graf T, MacRae CA, Lepore JJ, Lo CW, and Kahn ML (2006) Klf2 is an essential regulator of vascular hemodynamic forces in vivo. Developmental Cell 11:845–857

Leveen P, Pekny M, Gebre-Medhin S, Swolin B, Larsson E, and Betsholtz C (1994) Mice deficient for PDGF B show renal, cardiovascular, and hematological abnormalities. Genes Dev. 8:1875–1887

Li L, Miano JM, Cserjesi P, and Olson EN (1996) SM22α, a marker of adult smooth muscle, is expressed in multiple myogenic lineages during embryogenesis. Circ. Res. 78:188–195

Lindahl P, Johansson BR, Leveen P, and Betsholtz C (1997) Pericyte loss and microaneurysm formation in PDGF-B-deficient mice. Science 277:242–245

Liu Y, Sinha S, McDonald OG, Shang Y, Hoofnagle MH, and Owens GK (2005) Kruppel-like factor 4 abrogates myocardin-induced activation of smooth muscle gene expression. J. Biol. Chem. 280:9719–9727

Liu Y, Sinha S, and Owens G (2003) A Transforming Growth Factor-β Control element required for SM α-actin expression in vivo also partially mediates GKLF-dependent transcriptional repression. J.Biol.Chem. 278:48004–48011

Loppnow H, Werdan K, and Buerke M (2008) Vascular cells contribute to atherosclerosis by cytokine- and innate-immunity-related inflammatory mechanisms. Innate. Immun. 14:63–87

Marumo T, Schini-Kerth VB, Fisslthaler B, and Busse R (1997) Platelet-derived growth factor-stimulated superoxide anion production modulates activation of transcription factor NF-κB and expression of monocyte chemoattractant protein 1 in human aortic smooth muscle cells. Circulation 96:2361–2367

Matsumura T, Suzuki T, Aizawa K, Munemasa Y, Muto S, Horikoshi M, and Nagai R (2005) The Deacetylase HDAC1 Negatively Regulates the Cardiovascular Transcription Factor Kruppel-like Factor 5 through Direct Interaction. J. Biol. Chem. 280:12123–12129

McDonald OG and Owens GK (2007) Programming smooth muscle plasticity with chromatin dynamics. Circ. Res. 100:1428–1441

McDonald OG, Wamhoff BR, Hoofnagle MH, and Owens GK (2006) Control of SRF binding to CArG box chromatin regulates smooth muscle gene expression in vivo. Journal of Clinical Investigation. 116(1):36–48

Miyamoto S, Suzuki T, Muto S, Aizawa K, Kimura A, Mizuno Y, Nagino T, Imai Y, Adachi N, Horikoshi M, and Nagai R (2003) Positive and Negative Regulation of the Cardiovascular Transcription Factor KLF5 by p300 and the Oncogenic Regulator SET through Interaction and Acetylation on the DNA-Binding Domain. Mol. Cell. Biol. 23:8528–8541

Pidkovka NA, Cherepanova OA, Yoshida T, Alexander MR, Deaton RA, Thomas JA, Leitinger N, and Owens GK (2007) Oxidized phospholipids induce phenotypic switching of vascular smooth muscle cells in vivo and in vitro. Circ. Res. 101:792–801

Rong JX, Shapiro M, Trogan E, and Fisher EA (2003) Transdifferentiation of mouse aortic smooth muscle cells to a macrophage-like state after cholesterol loading. Proc. Natl. Acad. Sci. U.S.A 100:13531–13536

Sano H, Sudo T, Yokode M, Murayama T, Kataoka H, Takakura N, Nishikawa S, Nishikawa SI, and Kita T (2001) Functional blockade of platelet-derived growth factor receptor-β but not of receptor-α prevents vascular smooth muscle cell accumulation in fibrous cap lesions in Apolipoprotein E-deficient mice. Circulation 103:2955–2960

Segre JA, Bauer C, and Fuchs E (1999) Klf4 is a transcription factor required for establishing the barrier function of the skin. Nat. Genet. 22:356–360

SenBanerjee S, Lin Z, Atkins GB, Greif DM, Rao RM, Kumar A, Feinberg MW, Chen Z, Simon DI, Luscinskas FW, Michel TM, Gimbrone MA, Jr., Garcia-Cardena G, and Jain MK (2004) KLF2 is a novel transcriptional regulator of endothelial proinflammatory activation. J. Exp. Med. 199:1305–1315

Shindo T, Manabe I, Fukushima Y, Tobe K, Aizawa K, Miyamoto S, Kawai-Kowase K, Moriyama N, Imai Y, Kawakami H, Nishimatsu H, Ishikawa T, Suzuki T, Morita H, Maemura K, Sata M, Hirata Y, Komukai M, Kagechika H, Kadowaki T, Kurabayashi M, and Nagai R (2002) Kruppel-like zinc-finger transcription factor KLF5/BTEB2 is a target for angiotensin II signaling and an essential regulator of cardiovascular remodeling. Nat Med 8:856–863

Shinoda Y, Ogata N, Higashikawa A, Manabe I, Shindo T, Yamada T, Kugimiya F, Ikeda T, Kawamura N, Kawasaki Y, Tsushima K, Takeda N, Nagai R, Hoshi K, Nakamura K, Chung Ui, and Kawaguchi H (2008) Kruppel-like factor 5 causes cartilage degradation through transactivation of matrix metalloproteinase 9. J. Biol. Chem. 283:24682–24689

Takahashi K and Yamanaka S (2006) Induction of pluripotent stem cells from mouse embryonic and adult fibroblast cultures by defined factors. Cell 126:663–676

Thomas JA, Deaton RA, Hastings NE, Shang Y, Moehle CW, Eriksson UJ, Topouzis S, Wamhoff BR, Blackman BR, and Owens GK (2008) PDGF-DD, a novel mediator of smooth muscle cell phenotypic modulation, is upregulated in endothelial cells exposed to atherosclerotic-prone flow patterns. Am. J. Physiol Heart Circ. Physiol 296: H442–H452

Wang Z, Wang DZ, Pipes GC, and Olson EN (2003) Myocardin is a master regulator of smooth muscle gene expression. Proc. Natl. Acad. Sci. U.S.A 100:7129–7134

Wu J, Bohanan CS, Neumann JC, and Lingrel JB (2008) KLF2 transcription factor modulates blood vessel maturation through smooth muscle cell migration. J. Biol. Chem. 283: 3942–3950

Yoshida T, Gan Q, and Owens GK (2008a) Kruppel-like factor 4, Elk-1, and histone deacetylases cooperatively suppress smooth muscle cell differentiation markers in response to oxidized phospholipids. Am. J. Physiol. Cell Physiol. 295:C1175–C1182

Yoshida T, Gan Q, Shang Y, and Owens GK (2007) Platelet-derived growth factor-BB represses smooth muscle cell marker genes via changes in binding of MKL factors and histone deacetylases to their promoters. Am. J. Physiol. Cell Physiol. 292:C886–C895

Yoshida T, Kaestner KH, and Owens GK (2008b) Conditional deletion of kruppel-like factor 4 delays downregulation of smooth muscle cell differentiation markers but accelerates neointimal formation following vascular injury. Circ. Res. 102:1548–1557

Yoshida T, Sinha S, Dandre F, Wamhoff BR, Hoofnagle MH, Kremer BE, Wang DZ, Olson EN, and Owens GK (2003) Myocardin is a key regulator of CArG-dependent transcription of multiple smooth muscle marker genes. Circ. Res. 92:856–864

Chapter 16
Krüppel-like Factors in Cancers

Vincent W. Yang

Abstract Krüppel-like factors (KLFs) are zinc finger-containing transcription factors that play important roles in diverse physiological and pathophysiological processes. A major function of many KLFs is to regulate cell growth, proliferation, and differentiation. It is therefore not surprising that some of the KLFs are involved in tumorigenesis of various organs and tissues. This chapter reviews the pathobiological roles of KLFs in several cancers, including those of the gastrointestinal tract, breast, skin, and pancreas. Understanding the functions of KLFs in cancers may help gain insight into the pathogenesis of cancers and provide novel therapeutic approaches to their treatment.

Introduction

Krüppel-like factors (KLFs) belong to the family of zinc finger-containing transcription factors that share homology to the *Drosophila melanogaster* gap gene product, Krüppel (Bieker 2001; Black et al 2001; Dang et al 2000b; Kaczynski et al 2003; Lomberk and Urrutia 2005; Philipsen and Suske 1999). Since identification and isolation of the prototypic mammalian KLF—KLF1 or erythroid Krüppel-like factor (EKLF)—a decade and half ago (Bieker 1996; Miller and Bieker 1993), there has been an explosion of research devoted to the identification, isolation, and characterization of many additional KLF family members. To date, there are approximately 17 identified mammalian KLFs (excluding the Sp1 and Sp1-related proteins) (Kaczynski et al. 2003). Together, these KLFs have been shown to exert important regulatory functions in numerous biological and physiological processes.

V.W. Yang (✉)
Division of Digestive Diseases, Department of Medicine and Department of Hematology and Medical Oncology, Emory University School of Medicine, 210 Whitehead Biomedical Research Building, 615 Michael Street, Atlanta, GA 30322, USA

R. Nagai et al. (eds.), *The Biology of Krüppel-like Factors*,
DOI 10.1007/978-4-431-87775-2_16, © Springer 2009

Expression or activities of the KLFs are also frequently perturbed in pathological events. One of the main roles of many of the KLFs is their involvement in the regulation of cell growth, proliferation, differentiation, and development. As such, KLF expression and activities are often abnormal in neoplastic processes including cancers. Here we review the roles played by representative KLFs in several tumors including those of the gastrointestinal tract, breast, skin, and pancreas. The functions of KLFs in cancers of the liver, prostate, and ovaries are described elsewhere in this book (see Chapters 11 and 17).

Colorectal Cancer

Colorectal cancer is a common form of cancer and one of the leading causes of cancer mortality, with more than 655,000 deaths per year worldwide (Cancer. World Health Organization, February 2006). Clinical and epidemiological evidence indicates that colorectal cancer is preceded by a benign precursor lesion, an adenoma (Levin et al. 2008). Much progress has been made in understanding the genetics and pathogenesis of colorectal cancer at a molecular level (de la Chapelle 2004; Rustgi 2007). However, recent studies point to the complex, heterogeneous nature of colorectal cancer, which involves close to 200 genes that are mutated at a significant frequency (Sjoblom et al. 2006; Wood et al. 2007).

KLF4

Several KLFs have been implicated in the pathogenesis of colorectal cancer (Ghaleb and Yang 2008; Wei et al. 2006). Among these, KLF4 is the most extensively studied. KLF4 (also called gut-enriched Krüppel-like factor or GKLF) was initially identified as a gene whose expression is enriched in epithelial tissues, including the intestine and epidermis (Garrett-Sinha et al. 1996; Shields et al. 1996). In vivo studies in transgenic mice that are null for the *Klf4* alleles indicate that KLF4 is required for the terminal differentiation of goblet cells in the colon and for the barrier function of the skin in neonates (Katz et al. 2002; Segre et al. 1999). Studies also indicate that expression of KLF4 is primarily located in the postmitotic, differentiated cells of epithelial tissues (Garrett-Sinha et al. 1996; McConnell et al. 2007; Shie et al. 2000b; Shields et al. 1996). This growth arrest-specific pattern of expression is also observed in cultured cells in vitro (Shields et al. 1996). Consequently, ectopic expression of KLF4 in cultured cells results in growth arrest (Chen et al. 2001; Shields et al. 1996). Additional conditions that are known to cause growth arrest in cultured colonic epithelial cells—such as DNA damage and treatment with interferon-γ, sodium butyrate, or 15-deoxy-Δ(12,14) prostaglandin J2 (15d-PGJ2)—all lead to the induction of KLF4 expression (Chen et al. 2000, 2004; Chen and Tseng 2005; Yoon et al. 2003; Yoon and Yang 2004; Zhang et al. 2000).

The growth-suppressive activity of KLF4 and its activation upon conditions that elicit growth arrest suggest that KLF4 may have a tumor-suppressive function. Indeed, overexpression of KLF4 in the human colon cancer cell line RKO reduces its tumorigenicity in vivo (Dang et al. 2003). The levels of *KLF4* mRNA have also been shown to be reduced in intestinal adenomas of $Apc^{Min/+}$ mice, a model of intestinal tumorigenesis (Moser et al. 1990), in colonic adenomas from patients with familial adenomatous polyposis, and in colorectal cancer when compared to the respectively matched normal tissues (Dang et al. 2000a; Ton-That et al. 1997; Zhao et al. 2004). Moreover, loss of KLF4 protein is relatively common in colorectal cancer as assessed by immunohistochemistry (Choi et al. 2006). The causes for the loss of or reduced expression of KLF4 in colorectal cancer have been shown to be mediated at different levels including loss of heterozygosity (LOH), promoter hypermethylation, and point mutations that reduce protein activity (Zhao et al. 2004), all of which are representative features of tumor suppressors. Finally, results of recent genetic studies demonstrating that haploinsufficiency of the *Klf4* alleles in mice promotes intestinal tumorigenesis in $Apc^{Min/+}$ mice are highly indicative of the tumor-suppressive nature of KLF4 in vivo (Ghaleb et al. 2007b).

A number of studies have demonstrated the mechanism by which KLF4 exerts a growth-suppressive effect (Ghaleb et al 2005, 2007a; McConnell et al. 2007). Upon its identification, KLF4 was shown to inhibit DNA synthesis when overexpressed in transfected cells (Shields et al. 1996). When examined in the context of an inducible system, induction of KLF4 inhibits cell proliferation by blocking the G_1/S progression of the cell cycle (Chen et al. 2001). This effect is correlated with the induction of the gene encoding the cell cycle inhibitor $p21^{WAF1/CIP1}$ (Chen et al. 2001). Similarly, when growth arrest is caused by serum starvation or DNA damage, expression of both KLF4 and $p21^{WAF1/CIP1}$ are concurrently induced although the increase in *KLF4* mRNA levels precedes that of $p21^{WAF1/CIP1}$ (Zhang et al. 2000). In addition, the induction of both *KLF4* and $p21^{WAF1/CIP1}$ are dependent on p53 (Zhang et al. 2000). Significantly, KLF4 activates the $p21^{WAF1/CIP1}$ promoter through a specific Sp1-like *cis*-element in the $p21^{WAF1/CIP1}$ proximal promoter (Zhang et al. 2000). This element is necessary for p53 to activate the $p21^{WAF1/CIP1}$ promoter, even though p53 does not directly bind to it (Zhang et al. 2000). Instead, KLF4 and p53 physically interact with each other and synergistically induce activity of the $p21^{WAF1/CIP1}$ proximal promoter (Zhang et al. 2000). The physiological significance of KLF4 in mediating p53-dependent activation of $p21^{WAF1/CIP1}$ is further demonstrated by the ability of antisense KLF4 oligonucleotides to block the induction of $p21^{WAF1/CIP1}$ in response to p53 activation (Zhang et al. 2000). Subsequently, the p53-dependent induction of KLF4 was shown to be essential for DNA damage-induced arrest at both the G_1/S and G_2/M checkpoints of the cell cycle (Yoon et al. 2003; Yoon and Yang 2004). These results indicate that KLF4 is an essential mediator of p53 in controlling cell cycle progression following DNA damage.

In addition to activating $p21^{WAF1/CIP1}$, KLF4 has been shown to repress a number of genes that are involved in cell cycle progression or DNA synthesis, including cyclin D1 (Shie et al. 2000a), cyclin B1 (Evans et al. 2007; Yoon and Yang 2004), cyclin E (Yoon et al. 2005), Cdc2 (Yoon and Yang 2004), and ornithine decarboxylase

(Chen et al. 2002), all of which contributing to the inhibitory effect of KLF4 on cell proliferation. The transcriptional targets of KLF4 have further been elaborated by gene profiling experiments using an inducible system for KLF4 expression (Chen et al. 2003; Whitney et al 2006). A major cluster of genes whose expression is significantly affected by KLF4 induction are those involved in cell cycle regulation. Within this cluster, many genes activated by KLF4 are inhibitors of the cell cycle. Conversely, many downregulated genes are promoters of the cell cycle. These results indicate that KLF4 controls cell proliferation by eliciting changes in expression of numerous cell cycle-regulatory genes in a coordinated manner (Chen et al. 2003; Whitney et al. 2006). Unexpectedly, several other groups of genes that are repressed by KLF4 are involved in the synthesis of macromolecules such as protein, RNA, and cholesterol (Whitney et al. 2006). These results suggest that KLF4 exerts a global inhibitory effect on macromolecular biosynthesis that is beyond its role as a cell cycle inhibitor.

The adenomatous polyposis coli (APC) tumor suppressor is the gatekeeper for colorectal carcinogenesis (Kinzler and Vogelstein 1996). APC, a crucial component of the Wnt signaling pathway that regulates cell proliferation, prevents nuclear localization of β-catenin, thus preventing its pro-proliferative activity (Dang et al. 2001). The finding that haplo insufficiency of *Klf4* in $Apc^{Min/+}$ mice promotes intestinal tumorigenesis (Ghaleb et al. 2007b) suggests that KLF4 is involved in the pathway of APC tumor suppression. Indeed, expression of KLF4 has been shown to be activated by APC (Dang et al. 2001). Conversely, overexpression of KLF4 reduces β-catenin levels (Stone et al. 2002). Moreover, KLF4 physically interacts with β-catenin and represses β-catenin-mediated gene expression (Zhang et al. 2006). These results strongly support a role for KLF4 in mediating the Wnt/β-catenin pathway that is involved in normal intestinal epithelial homeostasis and tumor suppression.

KLF5

Similar to KLF4, expression of KLF5 is developmentally regulated and is enriched in epithelial tissues of adults including the intestine and epidermis (Conkright et al. 1999; Ohnishi et al. 2000). In contrast to KLF4, KLF5 is primarily expressed in the proliferating crypt epithelial cells and basal cells of the intestine and epidermis, respectively (Conkright et al. 1999; McConnell et al. 2007; Ohnishi et al. 2000). These findings suggest that KLF5 may positively regulate cell proliferation, contrary to the antiproliferative activity of KLF4 (Ghaleb et al. 2005; McConnell et al. 2007). Results from several experimental systems support this notion. For example, expression of KLF5 is strongly upregulated in activated smooth muscle cells and myofibroblasts in the aorta following balloon injury or in vascular lesions (Hoshino et al. 2000; Watanabe et al. 1999). In response to external stress, mice with haplo insufficiency for *Klf5* exhibit diminished levels of arterial wall thickening, angiogenesis, cardiac hypertrophy, and interstitial hypertrophy, indicating that KLF5 is a key element to linking external stress and cardiovascular modeling (Shindo et al. 2002).

In the mouse, infection by the mouse pathogen *Citrobactor rodentium* results in hyperproliferation on colonic crypt epithelial cells (Luperchio and Schauer 2001), which is accompanied by induction of KLF5 expression in colonic crypt epithelial cells (McConnell et al. 2008). Infection of mice heterozygous for the *Klf5* alleles with *C. rodentium* shows attenuated induction of KLF5 that is accompanied by a reduced hyperproliferative response in the colonic crypts, suggesting that KLF5 is a key mediator of crypt cell proliferation in response to pathogenic bacterial infection (McConnell et al. 2008).

Direct evidence in support of a pro-proliferative role of KLF5 is primarily derived from in vitro studies using cultured cells. Expression of KLF5 in serum-deprived fibroblasts is rapidly activated when the cells are stimulated with serum, epidermal growth factor (EGF), or the phorbol ester phorbol 12-myristate 13-acetate (PMA) (Sun et al. 2001). Similarly, KLF5 is upregulated in vascular smooth muscle cells through the mitogen-activated protein kinase (MAPK) pathway when treated with PMA or basic fibroblast growth factor (bFGF) (Kawai-Kowase et al. 1999). Indeed, constitutive expression of KLF5 in transfected fibroblasts increases their rate of proliferation and leads to anchorage-independent growth (Sun et al. 2001). Moreover, KLF5 level is significantly increased in oncogenic HRAS-transformed fibroblasts due to elevated MAPK activity (Nandan et al. 2004). Importantly, the increased KLF5 level in HRAS-transformed cells is responsible for increased transcription of the genes encoding cyclin D1, cyclin B1, and Cdc2 and thus directly mediates the pro-proliferative and transforming activity of oncogenic HRAS (Nandan et al. 2004, 2005). Pertinent to colorectal cancer, KLF5 expression is also increased in intestinal epithelial cells containing an inducible oncogenic KRAS, which is present in approximately half of all colorectal cancers (Nandan et al. 2008). KLF5 levels are similarly elevated in intestinal tumors from mice transgenic for an intestine-specific oncogenic KRAS, human colorectal cancer cell lines containing oncogenic KRAS, and primary human colorectal cancer containing oncogenic KRAS (Nandan et al. 2008). In human colorectal cancer cell lines, inhibition of KLF5 leads to reduced proliferation and transformation, suggesting that KLF5 is a key mediator of colorectal carcinogenesis, at least in tumors containing mutated, oncogenic KRAS (Nandan et al. 2008).

In addition to the stimulatory effects on KLF5 expression by the various agonists stated above, such as serum, PMA, and bFGF, KLF5 is regulated by several other stimuli, which may also explain its role in regulating cell proliferation in intestinal epithelial cells. Among them is lysophosphatidic acid (LPA), a phospholipid that stimulates proliferation of colon cancer cells (Zhang et al. 2007a). Activation of KLF5 expression by LPA in colon cancer cells is mediated by LPA2 and LPA3 receptors (Zhang et al. 2007a). Silencing of KLF5 significantly attenuates LPA-stimulated proliferation of colon cancer cells (Zhang et al. 2007a). Conversely, expression of KLF5 in intestinal epithelial cells is inhibited by all-*trans* retinoic acid (ATRA), which inhibits cell proliferation (Chanchevalap et al. 2004). Constitutive ectopic expression of KLF5 intestinal epithelial cells abrogates the inhibitory effect of ATRA (Chanchevalap et al. 2004). Adding further relevant to tumorigenesis, KLF5 has been shown to be a target of the Wnt signaling pathway (Taneyhill and Pennica 2004; Ziemer et al 2001).

The biological functions of KLF5 can also be inferred from the target genes that it regulates and from how KLF5 is posttranscriptionally regulated. Studies cited above in HRAS-transformed fibroblasts indicate that several cell cycle-regulatory genes, including cyclin D1, cyclin B1, and Cdc2, are transcriptional targets of KLF5 (Nandan et al. 2004, 2005).

Examples of other target genes, many of which are epithelial in origin, stimulated by KLF5 include platelet-derived growth factor (PDGF)-A chain (Aizawa et al 2004; Shindo et al. 2002), lactoferrin (Teng et al. 1998), laminin α-1 (Piccinni et al. 2004), and decay-accelerating factor (DAF) (Shao et al. 2008). Interestingly, the antiproliferative KLF4 is negatively regulated by KLF5 (Dang et al. 2002). KLF5 has also been shown to regulate the proinflammatory and antiapoptotic gene NF-κB (Chanchevalap et al. 2006) and to cooperate with NF-κB in regulating target gene expression (Aizawa et al. 2004; Sur et al. 2002). Moreover, KLF5 physically interacts with a number of other regulators such as p300/CBP (Miyamoto et al 2003), retinoic acid receptor (Shindo et al. 2002), and the protein inhibitor of activated STAT1 (PIAS1) (Du et al. 2007) to modulate physiologically relevant processes. Finally, KLF5 has been shown to be posttranslationally modified by phosphorylation (Zhang and Teng 2003), SUMOylation (Oishi et al. 2008), and ubiquitination (Chen et al. 2005a,b). Given that many of the molecules or processes cited here are involved in cell proliferation, these results further support an important function of KLF5 in regulating proliferation.

KLF6

Similar to KLF4, KLF6 has been shown to be an inhibitor of cell proliferation because of its ability to activate expression of the *p21*$^{WAF1/CIP1}$ gene and to disrupt the cyclin D1/cyclin-dependent kinase 4 (Cdk4) complexes (Benzeno et al. 2004; Li et al. 2005; Narla et al. 2001). The first evidence that KLF6 is a tumor suppressor is derived from the study of prostate cancer (Narla et al. 2001). Here, LOH of the KLF6 alleles is frequently detected in a cohort of primary prostate cancer, and mutation of KLF6 in the remaining allele is also common. Of the mutations detected, many result in reduced ability to activate *p21*$^{WAF1/CIP1}$ expression (Narla et al. 2001). Since this discovery, KLF6 has been shown to function as a tumor suppressor in myriads of cancers, including colorectal cancer (this chapter), hepatocellular cancer (HCC) (see Chapter 11), brain cancer (Kimmelman et al. 2004), and lung cancer (Ito et al. 2004), to name a few. In addition to the traditional mechanisms of loss of tumor suppressor functions such as LOH, mutations, and promoter methylation, KLF6 exhibits a unique mechanism by which a gain of function is achieved due to alternative splicing (Narla et al. 2005). The clinical relevance and therapeutic implications of these findings are discussed in Chapter 17.

Reeves et al. were the first to demonstrate that inactivation of KLF6 by LOH and mutations is a common event in both sporadic and inflammatory bowel disease-associated colorectal cancer (Reeves et al. 2004). Subsequent studies in independent cohorts of colorectal cancer samples confirmed these findings (Cho et al. 2006a,b). Another study compared the difference in the role of KLF6 LOH between sporadic and hereditary colorectal cancers and found that although KLF6 is frequently lost in sporadic cancers, especially during the late stage, its loss is relatively uncommon in hereditary cancer including familial adenomatous polyposis (FAP) and hereditary nonpolyposis colon cancer (HNPCC) (Yamaguchi et al. 2006). Finally, mutations of KLF6, along with p53, are commonly found in nonpolypoid colorectal cancer but mutations of KRAS and BRAF are not (Mukai et al. 2007).

Gastric Cancer

The first in vivo evidence that KLF4 regulates proliferation of gastric epithelial cells came from studies involving tissue-specific ablation of *Klf4* from the gastric epithelium of transgenic mice (Katz et al. 2005). *Klf4* mutant mice survive to adulthood and show increased proliferation and altered differentiation of their gastric epithelia (Katz et al. 2005). In addition, KLF4 expression is drastically decreased in both intestinal and diffuse-type human gastric cancer (Katz et al. 2005). The loss of expression of KLF4 in gastric cancer was subsequently validated by additional independent studies (Cho et al. 2007; Wei et al. 2005). It is of interest to note that loss of KLF4 expression contributes to Sp1 overexpression, which is directly correlated with the angiogenic potential of and poor prognosis for human gastric cancer (Kanai et al. 2006).

A single study examined the expression of KLF5 in human gastric cancer (Kwak et al. 2008). It showed that the expression rate of KLF5 is significantly higher in early-stage gastric cancer, in gastric cancer without lymph node metastasis, and in tumors < 5 cm in size (Kwak et al. 2008). Interestingly, the 5-year survival rate of patients with KLF5-positive tumors is higher than those of patients with KLF5-negative tumors, although the difference is not statistically significant (Kwak et al. 2008).

Esophageal Cancer

Much information on the relation between KLFs and esophageal cancer is derived from studying the role of KLF5 in squamous cell carcinoma of the esophagus. In the adult, KLF5 is expressed in basal cells of the squamous epithelium of the esophagus (Conkright et al. 1999; Ohnishi et al. 2000). A transgenic mouse model for KLF5 overexpression throughout the esophageal epithelium, developed using the ED-L2 promoter of the Epstein-Barr virus, exhibits evidence of increased proliferation in the basal layer but not the suprabasal layer of the esophageal epithelium (Goldstein et al. 2007). Subsequent studies in mouse primary esophageal keratinocytes indicated that KLF5 activates the mitogen-activated protein kinase kinase (MEK)/extracellular signal-regulated kinase (ERK) signaling pathways via the epidermal growth factor receptor (EGFR) to stimulate proliferation

(Yang et al. 2007). KLF5 also controls keratinocyte migration by activating integrin-linked kinase (Yang et al. 2008). In contrast, overexpression of KLF5 in a poorly differentiated esophageal squamous cancer cell line, TE2, inhibits proliferation and invasion and increases apoptosis following DNA damage (Yang et al. 2005). In this regard, KLF5 may function as a context-dependent regulator of cell proliferation in a manner similar to that reported for intestinal tumor progression (Bateman et al. 2004).

Breast Cancer

KLF4 has been shown to be involved in the pathogenesis of breast cancer. However, in contrast to the tumor-suppressive role of KLF4 in colorectal cancer, studies suggest that KLF4 may be oncogenic in breast cancer. The levels of KLF4 mRNA and protein are often elevated in neoplastic breast tissues, including both ductal carcinoma in situ and invasive carcinoma, when compared to adjacent normal mammary tissues (Foster et al. 2000). In addition, nuclear localization of KLF4 in breast cancer cells is associated with an aggressive phenotype in early-stage infiltrating ductal carcinoma (Pandya et al. 2004). The mucin 1 (MUC1) protein is often overexpressed in human breast cancer and induces transformation (Li et al. 2003). MUC1 binds to KLF4 to repress transcription of p53, which may explain the oncogenic effect of MUC1 and possibly KLF4 in breast cancer (Wei et al. 2007). However, it should be noted that results contrary to the above have also been reported. KLF4 levels are low or absent in many breast cancer cell lines, which may explain the absence of laminin-5, a major extracellular matrix protein produced by mammary epithelial cells, from many breast cancer cells (Miller et al. 2001). Additionally, growth of breast cancer cells is suppressed by okadaic acid, which induces apoptosis by activating transcription of c-Myc, which is mediated in part by KLF4 (Zhang et al. 2007b). The definitive growth effect (activating or suppressing) of KLF4 on breast cancer cells therefore requires additional clarification.

The context-dependent nature by which KLF4 acts as an oncogene has been explored. KLF4 was identified in a functional screen for genes that bypass oncogenic RAS^{V12}-induced senescence (Rowland et al. 2005). Although KLF4 is a potent inhibitor of proliferation in untransformed cells, KLF4-induced senescence is bypassed by RAS^{V12} or by the RAS^{V12} target, cyclin D1. Inactivation of the cyclin D1 target, $p21^{WAF1/CIP1}$, not only neutralizes the cytostatic action of KLF4 but collaborates with KLF4 in oncogenic transformation. KLF4 suppresses expression of p53 by directly acting on its promoter, allowing RAS^{V12}-mediated transformation. Depletion of KLF4 from breast cancer cells restores p53 levels and causes p53-dependent apoptosis (Rowland et al. 2005). These studies underscore the importance of $p21^{WAF1/CIP1}$ acting as a switch that determines the outcome of KLF4 signaling (Rowland and Peeper 2006). These results are consistent with a recent report showing that KLF4 exhibits antiapoptotic activity following γ-radiation-induced DNA damage by inhibiting the ability of p53 to trans-activate the proapoptotic gene BAX (Ghaleb et al. 2007a).

A limited number of studies have been published that examined the role of KLF5 in breast cancer. Like KLF4, the results of these studies are conflicting. KLF5 is thought to be a tumor suppressor in breast cancer because its expression levels are low in breast cancer cell lines due to LOH (Chen et al. 2002). In addition, the reduced KLF5 levels are attributed to ubiquitin-mediated proteasome degradation in breast cancer cell lines (Chen et al. 2005a,b). On the other hand, as a prognostic factor in breast cancer, patients with high KLF5 expression levels have shorter disease-free survival and overall survival than patients with low KLF5 expression, suggesting that KLF5 may not be a classic tumor suppressor in breast cancer (Tong et al. 2006).

Finally, several studies has been implicated KLF6 (Guo et al. 2007), KLF8 (Wang et al. 2007), and TIEG1 (KLF10) (Subramaniam et al. 1998) in the pathogenesis of breast cancer.

Skin Cancer

In a manner similar to the ability of KLF4 to bypass oncogenic RAS-mediated senescence (Rowland et al. 2005), KLF4 is found to cooperate with adenovirus E1A oncoprotein to transform epithelial cells (Foster et al. 1999). In oral squamous epithelium KLF4 is detected in the upper, differentiated cell layers (Foster et al. 1999). In contrast, the KLF4 level is increased in dysplastic epithelium and is diffusely expressed throughout the entire epithelium, indicating that KLF4 is misexpressed in the basal compartment early during tumor progression (Foster et al. 1999). In a mouse model, conditional expression of KLF4 in the basal keratinocytes in the skin leads to squamous epithelial dysplasia (Foster et al. 2005; Huang et al. 2005). A similar pattern of maturation-independent expression of KLF4 has been observed in human squamous cell carcinoma (Huang et al. 2005). As such, nuclear KLF4 expression has been correlated with progression and metastasis of human squamous cell carcinoma (Chen et al. 2008).

In contrast to the potential oncogenic role of KLF4 in skin cancer, KLF6 continues to behave as a tumor suppressor in human head and neck squamous cell carcinoma (Chen et al. 2008). Here, allelic loss of KLF6 is frequent and strongly correlated with tumor recurrence and decreased patient survival (Chen et al. 2008).

Pancreatic Cancer

The role of KLFs in pancreatic cancer has recently been reviewed (Buttar et al. 2006; Cook and Urrutia 2000). Among the best studied KLF members involved in the pathogenesis of pancreatic cancer is KLF11, also called transforming growth factor-β (TGF-β)-inducible early response gene 2 (TIEG2). KLF11, the expression of which is enriched in the pancreas, is an inhibitor of cell proliferation including pancreatic cells both in vivo and in vitro (Cook et al. 1998; Fernandez-Zapico et al. 2003).

Conversely, KLF11 expression is reduced in several human tumors including pancreatic cancer (Fernandez-Zapico et al. 2003). The mechanism by which KLF11 exerts its effect is to serve as a critical component of the TGF-β growth-inhibitory signaling in normal epithelial cells, an effect that is inactivated by oncogenic extracellular signal-regulated kinase/mitogen-activated protein kinase (ERK/MAPK) in pancreatic cancer cells (Ellenrieder et al. 2004).

Conclusion

Substantial progress has been made since the first mammalian prototype Krüppel-like factor, KLF1, was identified some 15 years ago. The combined current family of 17 KLF members is shown to exhibit important functions in diverse physiological processes, including proliferation, differentiation, development, angiogenesis, and embryonic stem cell renewal. Many of the KLFs are also featured prominently in pathobiological processes including inflammation and carcinogenesis. This chapter reviewed the current knowledge on the role of KLFs in certain human cancers. They function either as tumor suppressors or oncoproteins. Some have functions that are dependent on the context. Further investigation on the role of KLFs in cancer will provide additional novel insight into the mechanism of tumorigenesis and may provide potential therapeutic approaches to the treatment of cancer.

Acknowledgment This work was supported in part by grants from the U.S. National Institutes of Health (DK52230, DK64399, CA84197).

References

Aizawa K, Suzuki T, Kada N et al (2004) Regulation of platelet-derived growth factor-A chain by Krüppel-like factor 5: new pathway of cooperative activation with nuclear factor-kappaB. J Biol Chem 279:70-76.

Bateman NW, Tan D, Pestell RG et al (2004) Intestinal tumor progression is associated with altered function of KLF5. J Biol Chem 279:12093-12101.

Benzeno S, Narla G, Allina J et al (2004) Cyclin-dependent kinase inhibition by the KLF6 tumor suppressor protein through interaction with cyclin D1. Cancer Res 64:3885-3891.

Bieker JJ (1996) Isolation, genomic structure, and expression of human erythroid Krüppel-like factor (EKLF). DNA Cell Biol 15:347-352.

Bieker JJ (2001) Krüppel-like factors: three fingers in many pies. J Biol Chem 276:34355-34358.

Black AR, Black JD, Azizkhan-Clifford J (2001) Sp1 and Krüppel-like factor family of transcription factors in cell growth regulation and cancer. J Cell Physiol 188:143-160.

Buttar NS, Fernandez-Zapico ME, Urrutia R (2006) Key role of Krüppel-like factor proteins in pancreatic cancer and other gastrointestinal neoplasias. Curr Opin Gastroenterol 22:505-511.

Chanchevalap S, Nandan MO, McConnell BB et al (2006) Krüppel-like factor 5 is an important mediator for lipopolysaccharide-induced proinflammatory response in intestinal epithelial cells. Nucleic Acids Res 34:1216-1223.

Chanchevalap S, Nandan MO, Merlin D et al (2004) All-trans retinoic acid inhibits proliferation of intestinal epithelial cells by inhibiting expression of the gene encoding Krüppel-like factor 5. FEBS Lett 578:99-105.

Chen C, Bhalala HV, Qiao H et al (2002) A possible tumor suppressor role of the KLF5 transcription factor in human breast cancer. Oncogene 21:6567-6572.

Chen C, Sun X, Guo P et al (2005a) Human Krüppel-like factor 5 is a target of the E3 ubiquitin ligase WWP1 for proteolysis in epithelial cells. J Biol Chem 280:41553-41561.

Chen C, Sun X, Ran Q et al (2005b) Ubiquitin-proteasome degradation of KLF5 transcription factor in cancer and untransformed epithelial cells. Oncogene 24:3319-3327.

Chen X, Johns DC, Geiman DE et al (2001) Krüppel-like factor 4 (gut-enriched Krüppel-like factor) inhibits cell proliferation by blocking G1/S progression of the cell cycle. J Biol Chem 276:30423-30428.

Chen X, Whitney EM, Gao SY et al (2003) Transcriptional profiling of Krüppel-like factor 4 reveals a function in cell cycle regulation and epithelial differentiation. J Mol Biol 326:665-677.

Chen YJ, Wu CY, Chang CC et al (2008) Nuclear Krüppel-like factor 4 expression is associated with human skin squamous cell carcinoma progression and metastasis. Cancer Biol Ther 7.

Chen ZY, Rex STseng CC (2004) Krüppel-like factor 4 is transactivated by butyrate in colon cancer cells. J Nutr 134:792-798.

Chen ZY, Shie J, Tseng C (2000) Up-regulation of gut-enriched Krüppel-like factor by interferon-gamma in human colon carcinoma cells. FEBS Lett 477:67-72.

Chen ZY, Shie JL, Tseng CC (2002) Gut-enriched Krüppel-like factor represses ornithine decarboxylase gene expression and functions as checkpoint regulator in colonic cancer cells. J Biol Chem 277:46831-46839.

Chen ZY, Tseng CC (2005) 15-deoxy-Delta12,14 prostaglandin J2 up-regulates Krüppel-like factor 4 expression independently of peroxisome proliferator-activated receptor gamma by activating the mitogen-activated protein kinase kinase/extracellular signal-regulated kinase signal transduction pathway in HT-29 colon cancer cells. Mol Pharmacol 68:1203-1213.

Cho YG, Choi BJ, Kim CJ et al (2006a) Genetic alterations of the KLF6 gene in colorectal cancers. Apmis 114:458-464.

Cho YG, Choi BJ, Song JW et al (2006b) Aberrant expression of Krüppel-like factor 6 protein in colorectal cancers. World J Gastroenterol 12:2250-2253.

Cho YG, Song JH, Kim CJ et al (2007) Genetic and epigenetic analysis of the KLF4 gene in gastric cancer. Apmis 115:802-808.

Choi BJ, Cho YG, Song JW et al (2006) Altered expression of the KLF4 in colorectal cancers. Pathol Res Pract 202:585-589.

Conkright MD, Wani MA, Anderson KP et al (1999) A gene encoding an intestinal-enriched member of the Krüppel-like factor family expressed in intestinal epithelial cells. Nucleic Acids Res 27:1263-1270.

Cook T, Gebelein B, Mesa K et al (1998) Molecular cloning and characterization of TIEG2 reveals a new subfamily of transforming growth factor-beta-inducible Sp1-like zinc finger-encoding genes involved in the regulation of cell growth. J Biol Chem 273:25929-25936.

Cook T, Urrutia R (2000) TIEG proteins join the Smads as TGF-beta-regulated transcription factors that control pancreatic cell growth. Am J Physiol Gastrointest Liver Physiol 278:G513-521.

Dang DT, Bachman KE, Mahatan CS et al (2000a) Decreased expression of the gut-enriched Krüppel-like factor gene in intestinal adenomas of multiple intestinal neoplasia mice and in colonic adenomas of familial adenomatous polyposis patients. FEBS Lett 476:203-207.

Dang DT, Chen X, Feng J et al (2003) Overexpression of Krüppel-like factor 4 in the human colon cancer cell line RKO leads to reduced tumorigenecity. Oncogene 22:3424-3430.

Dang DT, Mahatan CS, Dang LH et al (2001) Expression of the gut-enriched Krüppel-like factor (Krüppel-like factor 4) gene in the human colon cancer cell line RKO is dependent on CDX2. Oncogene 20:4884-4890.

Dang DT, Pevsner J, Yang VW (2000b) The biology of the mammalian Krüppel-like family of transcription factors. Int J Biochem Cell Biol 32:1103-1121.

Dang DT, Zhao W, Mahatan CS et al (2002) Opposing effects of Krüppel-like factor 4 (gut-enriched Krüppel-like factor) and Krüppel-like factor 5 (intestinal-enriched Krüppel-like factor) on the promoter of the Krüppel-like factor 4 gene. Nucleic Acids Res 30:2736-2741.

de la Chapelle A (2004) Genetic predisposition to colorectal cancer. Nat Rev Cancer 4:769-780.

Du JX, Yun CC, Bialkowska A et al (2007) Protein inhibitor of activated STAT1 interacts with and up-regulates activities of the pro-proliferative transcription factor Krüppel-like factor 5. J Biol Chem 282:4782-4793.

Ellenrieder V, Buck A, Harth A et al (2004) KLF11 mediates a critical mechanism in TGF-beta signaling that is inactivated by Erk-MAPK in pancreatic cancer cells. Gastroenterology 127:607-620.

Evans PM, Zhang W, Chen X et al (2007) Krüppel-like factor 4 is acetylated by p300 and regulates gene transcription via modulation of histone acetylation. J Biol Chem 282:33994-34002.

Fernandez-Zapico ME, Mladek A, Ellenrieder V et al (2003) An mSin3A interaction domain links the transcriptional activity of KLF11 with its role in growth regulation. Embo J 22:4748-4758.

Foster KW, Frost AR, McKie-Bell P et al (2000) Increase of GKLF messenger RNA and protein expression during progression of breast cancer. Cancer Res 60:6488-6495.

Foster KW, Liu Z, Nail CD et al (2005) Induction of KLF4 in basal keratinocytes blocks the proliferation-differentiation switch and initiates squamous epithelial dysplasia. Oncogene 24:1491-1500.

Foster KW, Ren S, Louro ID et al (1999) Oncogene expression cloning by retroviral transduction of adenovirus E1A-immortalized rat kidney RK3E cells: transformation of a host with epithelial features by c-MYC and the zinc finger protein GKLF. Cell Growth Differ 10:423-434.

Garrett-Sinha LA, Eberspaecher H, Seldin MF et al (1996) A gene for a novel zinc-finger protein expressed in differentiated epithelial cells and transiently in certain mesenchymal cells. J Biol Chem 271:31384-31390.

Ghaleb AM, Katz JP, Kaestner KH et al (2007a) Krüppel-like factor 4 exhibits antiapoptotic activity following gamma-radiation-induced DNA damage. Oncogene 26:2365-2373.

Ghaleb AM, McConnell BB, Nandan MO et al (2007b) Haploinsufficiency of Krüppel-like factor 4 promotes adenomatous polyposis coli dependent intestinal tumorigenesis. Cancer Res 67:7147-7154.

Ghaleb AM, Nandan MO, Chanchevalap S et al (2005) Krüppel-like factors 4 and 5: the yin and yang regulators of cellular proliferation. Cell Res 15:92-96.

Ghaleb AM, Yang VW (2008) The Pathobiology of Krüppel-like Factors in Colorectal Cancer. Curr Colorectal Cancer Rep 4:59-64.

Goldstein BG, Chao HH, Yang Y et al (2007) Overexpression of Krüppel-like factor 5 in esophageal epithelia in vivo leads to increased proliferation in basal but not suprabasal cells. Am J Physiol Gastrointest Liver Physiol 292:G1784-1792.

Guo H, Lin Y, Zhang H et al (2007) Tissue factor pathway inhibitor-2 was repressed by CpG hypermethylation through inhibition of KLF6 binding in highly invasive breast cancer cells. BMC Mol Biol 8:110.

Hoshino Y, Kurabayashi M, Kanda T et al (2000) Regulated expression of the BTEB2 transcription factor in vascular smooth muscle cells: analysis of developmental and pathological expression profiles shows implications as a predictive factor for restenosis. Circulation 102:2528-2534.

Huang CC, Liu Z, Li X et al (2005) KLF4 and PCNA identify stages of tumor initiation in a conditional model of cutaneous squamous epithelial neoplasia. Cancer Biol Ther 4:1401-1408.

Ito G, Uchiyama M, Kondo M et al (2004) Krüppel-like factor 6 is frequently down-regulated and induces apoptosis in non-small cell lung cancer cells. Cancer Res 64:3838-3843.

Kaczynski J, Cook T, Urrutia R (2003) Sp1- and Krüppel-like transcription factors. Genome Biol 4:206.

Kanai M, Wei D, Li Q et al (2006) Loss of Krüppel-like factor 4 expression contributes to Sp1 overexpression and human gastric cancer development and progression. Clin Cancer Res 12:6395-6402.

Katz JP, Perreault N, Goldstein BG et al (2005) Loss of Klf4 in mice causes altered proliferation and differentiation and precancerous changes in the adult stomach. Gastroenterology 128:935-945.

Katz JP, Perreault N, Goldstein BG et al (2002) The zinc-finger transcription factor Klf4 is required for terminal differentiation of goblet cells in the colon. Development 129:2619-2628.
Kawai-Kowase K, Kurabayashi M, Hoshino Y et al (1999) Transcriptional activation of the zinc finger transcription factor BTEB2 gene by Egr-1 through mitogen-activated protein kinase pathways in vascular smooth muscle cells. Circ Res 85:787-795.
Kimmelman AC, Qiao RF, Narla G et al (2004) Suppression of glioblastoma tumorigenicity by the Krüppel-like transcription factor KLF6. Oncogene 23:5077-5083.
Kinzler KW, Vogelstein B (1996) Lessons from hereditary colorectal cancer. Cell 87:159-170.
Kwak MK, Lee HJ, Hur K et al (2008) Expression of Krüppel-like factor 5 in human gastric carcinomas. J Cancer Res Clin Oncol 134:163-167.
Levin B, Lieberman DA, McFarland B et al (2008) Screening and Surveillance for the Early Detection of Colorectal Cancer and Adenomatous Polyps, 2008: A Joint Guideline from the American Cancer Society, the US Multi-Society Task Force on Colorectal Cancer, and the American College of Radiology. CA Cancer J Clin 58:130-160.
Li D, Yea S, Dolios G et al (2005) Regulation of Krüppel-like factor 6 tumor suppressor activity by acetylation. Cancer Res 65:9216-9225.
Li Y, Liu D, Chen D et al (2003) Human DF3/MUC1 carcinoma-associated protein functions as an oncogene. Oncogene 22:6107-6110.
Lomberk G, Urrutia R (2005) The family feud: turning off Sp1 by Sp1-like KLF proteins. Biochem J 392:1-11.
Luperchio SA, Schauer DB (2001) Molecular pathogenesis of Citrobacter rodentium and transmissible murine colonic hyperplasia. Microbes Infect 3:333-340.
McConnell BB, Ghaleb AM, Nandan MO et al (2007) The diverse functions of Krüppel-like factors 4 and 5 in epithelial biology and pathobiology. Bioessays 29:549-557.
McConnell BB, Klapproth JM, Sasaki M et al (2008) Krüppel-like factor 5 mediates transmissible murine colonic hyperplasia caused by Citrobacter rodentium infection. Gastroenterology 134:1007-1016.
Miller IJ, Bieker JJ (1993) A novel, erythroid cell-specific murine transcription factor that binds to the CACCC element and is related to the Krüppel family of nuclear proteins. Mol Cell Biol 13:2776-2786.
Miller KA, Eklund EA, Peddinghaus ML et al (2001) Krüppel-like factor 4 regulates laminin alpha 3A expression in mammary epithelial cells. J Biol Chem 276:42863-42868.
Miyamoto S, Suzuki T, Muto S et al (2003) Positive and negative regulation of the cardiovascular transcription factor KLF5 by p300 and the oncogenic regulator SET through interaction and acetylation on the DNA-binding domain. Mol Cell Biol 23:8528-8541.
Moser AR, Pitot HC, Dove WF (1990) A dominant mutation that predisposes to multiple intestinal neoplasia in the mouse. Science 247:322-324.
Mukai S, Hiyama T, Tanaka S et al (2007) Involvement of Krüppel-like factor 6 (KLF6) mutation in the development of nonpolypoid colorectal carcinoma. World J Gastroenterol 13:3932-3938.
Nandan MO, Chanchevalap S, Dalton WB et al (2005) Krüppel-like factor 5 promotes mitosis by activating the cyclin B1/Cdc2 complex during oncogenic Ras-mediated transformation. FEBS Lett 579:4757-4762.
Nandan MO, McConnell BB, Ghaleb AM et al (2008) Krüppel-like factor 5 mediates cellular transformation during oncogenic KRAS-induced intestinal tumorigenesis. Gastroenterology 134:120-130.
Nandan MO, Yoon HS, Zhao W et al (2004) Krüppel-like factor 5 mediates the transforming activity of oncogenic H-Ras. Oncogene 23:3404-3413.
Narla G, Difeo A, Reeves HL et al (2005) A germline DNA polymorphism enhances alternative splicing of the KLF6 tumor suppressor gene and is associated with increased prostate cancer risk. Cancer Res 65:1213-1222.
Narla G, Heath KE, Reeves HL et al (2001) KLF6, a candidate tumor suppressor gene mutated in prostate cancer. Science 294:2563-2566.
Ohnishi S, Laub F, Matsumoto N et al (2000) Developmental expression of the mouse gene coding for the Krüppel-like transcription factor KLF5. Dev Dyn 217:421-429.

Oishi Y, Manabe I, Tobe K et al (2008) SUMOylation of Krüppel-like transcription factor 5 acts as a molecular switch in transcriptional programs of lipid metabolism involving PPAR-delta. Nat Med 14:656-666.

Pandya AY, Talley LI, Frost AR et al (2004) Nuclear localization of KLF4 is associated with an aggressive phenotype in early-stage breast cancer. Clin Cancer Res 10:2709-2719.

Philipsen S, Suske G (1999) A tale of three fingers: the family of mammalian Sp/XKLF transcription factors. Nucleic Acids Res 27:2991-3000.

Piccinni SA, Bolcato-Bellemin AL, Klein A et al (2004) Krüppel-like factors regulate the Lama1 gene encoding the laminin alpha1 chain. J Biol Chem 279:9103-9114.

Reeves HL, Narla G, Ogunbiyi O et al (2004) Krüppel-like factor 6 (KLF6) is a tumor-suppressor gene frequently inactivated in colorectal cancer. Gastroenterology 126:1090-1103.

Rowland BD, Bernards R, Peeper DS (2005) The KLF4 tumour suppressor is a transcriptional repressor of p53 that acts as a context-dependent oncogene. Nat Cell Biol 7:1074-1082.

Rowland BD, Peeper DS (2006) KLF4, p21 and context-dependent opposing forces in cancer. Nat Rev Cancer 6:11-23.

Rustgi AK (2007) The genetics of hereditary colon cancer. Genes Dev 21:2525-2538.

Segre JA, Bauer C, Fuchs E (1999) Klf4 is a transcription factor required for establishing the barrier function of the skin. Nat Genet 22:356-360.

Shao J, Yang VW, Sheng H (2008) Prostaglandin E(2) and Krüppel-like transcription factors synergistically induce the expression of decay-accelerating factor in intestinal epithelial cells. Immunology.

Shie JL, Chen ZY, Fu M et al (2000a) Gut-enriched Krüppel-like factor represses cyclin D1 promoter activity through Sp1 motif. Nucleic Acids Res 28:2969-2976.

Shie JL, Chen ZY, O'Brien MJ et al (2000b) Role of gut-enriched Krüppel-like factor in colonic cell growth and differentiation. Am J Physiol Gastrointest Liver Physiol 279:G806-814.

Shields JM, Christy RJ, Yang VW (1996) Identification and characterization of a gene encoding a gut-enriched Krüppel-like factor expressed during growth arrest. J Biol Chem 271:20009-20017.

Shindo T, Manabe I, Fukushima Y et al (2002) Krüppel-like zinc-finger transcription factor KLF5/BTEB2 is a target for angiotensin II signaling and an essential regulator of cardiovascular remodeling. Nat Med 8:856-863.

Sjoblom T, Jones S, Wood LD et al (2006) The consensus coding sequences of human breast and colorectal cancers. Science 314:268-274.

Stone CD, Chen ZY, Tseng CC (2002) Gut-enriched Krüppel-like factor regulates colonic cell growth through APC/beta-catenin pathway. FEBS Lett 530:147-152.

Subramaniam M, Hefferan TE, Tau K et al (1998) Tissue, cell type, and breast cancer stage-specific expression of a TGF-beta inducible early transcription factor gene. J Cell Biochem 68:226-236.

Sun R, Chen X, Yang VW (2001) Intestinal-enriched Krüppel-like factor (Krüppel-like factor 5) is a positive regulator of cellular proliferation. J Biol Chem 276:6897-6900.

Sur I, Unden AB, Toftgard R (2002) Human Krüppel-like factor5/KLF5: synergy with NF-kappaB/Rel factors and expression in human skin and hair follicles. Eur J Cell Biol 81:323-334.

Taneyhill L, Pennica D (2004) Identification of Wnt responsive genes using a murine mammary epithelial cell line model system. BMC Dev Biol 4:6.

Teng C, Shi H, Yang N et al (1998) Mouse lactoferrin gene. Promoter-specific regulation by EGF and cDNA cloning of the EGF-response-element binding protein. Adv Exp Med Biol 443:65-78.

Ton-That H, Kaestner KH, Shields JM et al (1997) Expression of the gut-enriched Krüppel-like factor gene during development and intestinal tumorigenesis. FEBS Lett 419:239-243.

Tong D, Czerwenka K, Heinze G et al (2006) Expression of KLF5 is a prognostic factor for disease-free survival and overall survival in patients with breast cancer. Clin Cancer Res 12:2442-2448.

Wang X, Zheng M, Liu G et al (2007) Krüppel-like factor 8 induces epithelial to mesenchymal transition and epithelial cell invasion. Cancer Res 67:7184-7193.

Watanabe N, Kurabayashi M, Shimomura Y et al (1999) BTEB2, a Krüppel-like transcription factor, regulates expression of the SMemb/Nonmuscle myosin heavy chain B (SMemb/NMHC-B) gene. Circ Res 85:182-191.

Wei D, Gong W, Kanai M et al (2005) Drastic down-regulation of Krüppel-like factor 4 expression is critical in human gastric cancer development and progression. Cancer Res 65:2746-2754.

Wei D, Kanai M, Huang S et al (2006) Emerging role of KLF4 in human gastrointestinal cancer. Carcinogenesis 27:23-31.

Wei X, Xu H, Kufe D (2007) Human mucin 1 oncoprotein represses transcription of the p53 tumor suppressor gene. Cancer Res 67:1853-1858.

Whitney EM, Ghaleb AM, Chen X et al (2006) Transcriptional profiling of the cell cycle checkpoint gene Krüppel-like factor 4 reveals a global inhibitory function in macromolecular biosynthesis. Gene Expr 13:85-96.

Wood LD, Parsons DW, Jones S et al (2007) The genomic landscapes of human breast and colorectal cancers. Science 318:1108-1113.

Yamaguchi T, Iijima T, Mori T et al (2006) Accumulation profile of frameshift mutations during development and progression of colorectal cancer from patients with hereditary nonpolyposis colorectal cancer. Dis Colon Rectum 49:399-406.

Yang Y, Goldstein BG, Chao HH et al (2005) KLF4 and KLF5 regulate proliferation, apoptosis and invasion in esophageal cancer cells. Cancer Biol Ther 4:1216-1221.

Yang Y, Goldstein BG, Nakagawa H et al (2007) Krüppel-like factor 5 activates MEK/ERK signaling via EGFR in primary squamous epithelial cells. Faseb J 21:543-550.

Yang Y, Tetreault MP, Yermolina YA et al (2008) Krüppel-like factor 5 controls keratinocyte migration via the integrin-linked kinase. J Biol Chem.

Yoon HS, Chen X, Yang VW (2003) Krüppel-like factor 4 mediates p53-dependent G1/S cell cycle arrest in response to DNA damage. J Biol Chem 278:2101-2105.

Yoon HS, Ghaleb AM, Nandan MO et al (2005) Krüppel-like factor 4 prevents centrosome amplification following gamma-irradiation-induced DNA damage. Oncogene 24:4017-4025.

Yoon HS, Yang VW (2004) Requirement of Krüppel-like factor 4 in preventing entry into mitosis following DNA damage. J Biol Chem 279:5035-5041.

Zhang H, Bialkowska A, Rusovici R et al (2007a) Lysophosphatidic acid facilitates proliferation of colon cancer cells via induction of Krüppel-like factor 5. J Biol Chem 282:15541-15549.

Zhang L, Wali A, Ramana CV et al (2007b) Cell growth inhibition by okadaic acid involves gut-enriched Krüppel-like factor mediated enhanced expression of c-Myc. Cancer Res 67:10198-10206.

Zhang W, Chen X, Kato Y et al (2006) Novel cross talk of Krüppel-like factor 4 and beta-catenin regulates normal intestinal homeostasis and tumor repression. Mol Cell Biol 26:2055-2064.

Zhang W, Geiman DE, Shields JM et al (2000) The gut-enriched Krüppel-like factor (Krüppel-like factor 4) mediates the transactivating effect of p53 on the p21WAF1/Cip1 promoter. J Biol Chem 275:18391-18398.

Zhang Z, Teng CT (2003) Phosphorylation of Krüppel-like factor 5 (KLF5/IKLF) at the CBP interaction region enhances its transactivation function. Nucleic Acids Res 31:2196-2208.

Zhao W, Hisamuddin IM, Nandan MO et al (2004) Identification of Krüppel-like factor 4 as a potential tumor suppressor gene in colorectal cancer. Oncogene 23:395-402.

Ziemer LT, Pennica D, Levine AJ (2001) Identification of a mouse homolog of the human BTEB2 transcription factor as a beta-catenin-independent Wnt-1-responsive gene. Mol Cell Biol 21:562-574.

Part 5
Diagnostic and Therapeutic Applications of Krüppel-like Factors

Chapter 17
Krüppel-like Factors KLF6 and KLF6-SV1 in the Diagnosis and Treatment of Cancer

Analisa DiFeo, Goutham Narla, and John A. Martignetti

Abstract Beyond their initial identification in prostate cancer, the tumor suppressor KLF6 and its alternatively spliced oncogenic isoform KLF6-SV1 have now been associated with a number of human cancers. Expression patterns of each have been linked to tumor stage, disease recurrence, chemotherapy response and overall survival. Most recently, inhibition of KLF6-SV1 has been shown to increase survival in a pre-clinical model of ovarian cancer. This chapter reviews the basic biology of KLF6 and KLF6-SV1 as it relates to cancer, summarizes the published studies to date, and highlights the scientific rationale for therapeutically targeting KLF6 and KLF6-SV1 as a novel treatment paradigm.

Introduction

A role for Krüppel-like factor (KLF) family members in tumorigenesis was originally predicted based on knowledge of their critical functions in growth-related signal transduction pathways, cell proliferation, apoptosis, and angiogenesis (reviewed in Bieker 2001; Black et al 2001). Indeed, since the initial description of KLF6 as a tumor-suppressor gene in prostate cancer (Narla et al. 2001), reports have linked a number of the other KLFs with cancer development, and the list will likely continue to increase. To date, the broad range of implicated tumor types, possession of a germline single nucleotide polymorphism (SNP) associated with an increased lifetime prostate cancer risk, and alternative splicing of the gene into a functionally antagonistic, oncogenic isoform

A. DiFeo, G. Narla, and J.A. Martignetti
Department of Genetics and Genomic Sciences, Mount Sinai School of Medicine, New York, NY, USA

G. Narla
Department of Medicine, Mount Sinai School of Medicine, New York, NY, USA

J.A. Martignetti (✉)
Department of Oncological Sciences, Mount Sinai School of Medicine, 1425 Madison Avenue, Room 14-70, New York, NY 10029, USA
e-mail: john.martignetti@mssm.edu

R. Nagai et al. (eds.), *The Biology of Krüppel-like Factors*,
DOI 10.1007/978-4-431-87775-2_17, © Springer 2009

overexpressed in multiple cancer types distinguishes the tumor-suppressor KLF6 from other KLF family members. What had not been predicted and is potentially the most exciting aspect of KLF-related cancers is that KLF6 and KLF6-SV1 expression patterns are linked to tumor stage, disease recurrence, chemotherapeutic resistance, and overall survival in a number of cancers. Taken together, these studies provide the scientific rationale for therapeutically targeting KLF6 and KLF6-SV1 in a number of human cancers and provide a paradigm for cancer-related studies of the other KLF family members.

KLF6: Tumor-Suppressor Gene Inactivated in Multiple Human Cancers

The original evidence defining KLF6 as a tumor-suppressor gene, a gene sustaining loss-of-function mutations in the development of cancer, was provided by analysis of a cancer known to harbor a high degree of allelic imbalance at the gene locus. The KLF6 gene maps to human chromosome 10p15, a region previously reported to be deleted in approximately 55% of sporadic prostate adenocarcinomas (Ittmann 1996; Trybus et al. 1996). Accordingly, we examined a collection of well annotated primary prostate tumor samples for loss of heterozygosity (LOH) of the KLF6 gene locus (Narla et al. 2001). To finely map the minimal region of loss, we designed a series of microsatellite markers directly flanking the KLF6 locus. In total, 16 of 22 informative tumor samples (73%) displayed LOH using these novel markers. The smallest region of overlap effectively narrowed the tumor-suppressor gene to an approximately 60-kb locus (Fig. 1).

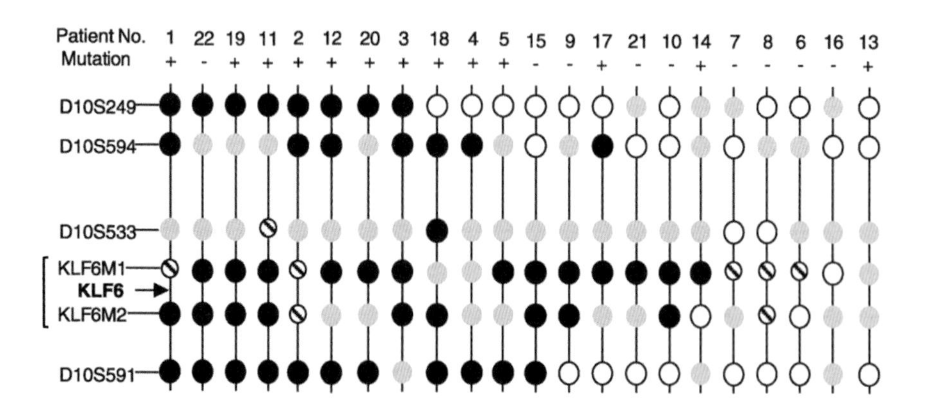

Fig. 1 Loss of heterozygosity (LOH) at the KLF6 locus in human prostate tumors. Summary of LOH patterns of 22 prostate tumors. *White circles* = retained microsatellite markers; *black circles* = markers demonstrating allelic loss; *gray circles* = noninformative markers; *hatched circle* = DNA that could not be amplified. Patient data were grouped according to the degree of LOH. Genetic map is not drawn to scale

All four KLF6 coding exons and intron/exon boundaries were then sequenced using genomic DNA extracted from microdissected and laser capture microdissected patient-derived tumors. Of 34 tumor samples, 19 (56%) were found to have tumor-specific KLF6 mutations. Unlike wild-type KLF6, these patient-derived missense mutations were unable to upregulate p21 expression or decrease cell proliferation. Interestingly, and highlighting the heterogeneity of prostate cancer and the possible existence of a subpopulation of KLF6-deficient cells, a number of patients had distinct mutations in different tumor foci whereas others had compound mutations. No mutations were present in the paired normal tissue genomic DNA, confirming that the prostate-derived mutations were somatic. These findings provided the first direct evidence for two inactivating events at the same genetic locus and were in accordance with Knudson's "two-hit hypothesis" for tumor-suppressor genes.

As shown in Table 1, subsequent studies from our laboratories and others have further highlighted the role of KLF6 in prostate cancer. These studies, albeit not all, defined inactivation of KLF6 through allelic loss and mutation, allelic loss and decreased expression, and (as discussed further below) changes in splicing patterns. Moreover, additional studies now support KLF6's general role as a tumor suppressor in a number of human cancers. As detailed in Table 1, the mechanisms of inactivation vary among cancers and include LOH and somatic mutation, LOH and/or decreased expression, and (as discussed below) increased alternative splicing. In turn, depending on cell type and context, KLF6's growth-suppressive properties have been associated with a number of highly relevant cancer pathways including p53-independent upregulation of p21 (Narla et al. 2001), disruption of cyclin D1 and CDK4 interaction (Benzeno et al. 2004), induction of apoptosis (Ito et al. 2004) and downregulation of E-cadherin (DiFeo et al. 2006a).

The broad range of frequencies of KLF6 inactivation among studies reporting on a number of cancer types appears curious at first and deserves mention for at least two reasons. First, these results highlight important differences between sample selection, the number of samples used in the analyses, tissue isolation techniques, and the specific analytic techniques used (e.g., DNA sequencing versus SSCP analysis; radioactive versus quantitative fluorescent LOH analysis; distance of microsatellite markers from the locus of interest). Unfortunately, approaches to validate the purity of tumor tissue and methods of analyzing mutations and deletions still have not been standardized. Second, another possible explanation for these differences is consistent with the concept of cancer stem cells/tumor-initiating stem cells (Visvader and Lindeman 2008). Specifically, the possibility, which has not yet been directly examined, is that only a subpopulation of tumor cells contain these KLF6-specific changes, and it is these cells that then drive the bulk of the associated tumor phenotype with regard to treatment, recurrence, and overall survival (see below). In accord with this hypothesis is the recent demonstration that a member of the KLF family of transcription factors, KLF4, is a critical reprogramming factor that can drive the induction of pluriopotency in mammalian cells. Therefore, an important question to address in the future is whether KLF6 plays a similar role in cancer cells.

Table 1 Cancer associations for KLF6 and KLF6-SV1

Prostate cancer

- Loss of heterozygosity (LOH) 77%; mutations 55% (A52T, S59R, W64R, T65I, R71Q, E78G, E100K, L104P, S110N, S112P, S113F, S116P, L119P, A123D, S137X, S142P, P166L, L169P, P172S, S180L, G187R, D194N, L217S, C265Y, D273G); mutants lose ability to transactivate p21 and KLF6 growth-suppressive properties (Narla et al. 2001)
- 24% LOH in tumors and cell lines; 15% somatic mutations (8% missense; T35I, L43F, D88N, P93S, S120F, K124M, G189S); decreased KLF6 expression in 20% of xenografts/cell lines (Chen et al. 2003)
- 27% LOH in tumors and cell lines (Hermans et al. 2004)
- ~1% Mutations (E227G) (Agell et al. 2008)
- No mutations (Mühlbauer et al. 2003)
- Decreased KLF6 expression is a predictor of poor clinical outcome/genome-wide microarray (Glinsky et al. 2004)
- Decreased KLF6 expression is associated with late-stage androgen-independent tumors/genome-wide microarray (Stanbrough et al. 2006)
- KLF6 expression correlates with Gleason score/genome-wide microarray (Singh et al. 2002)
- Overexpression of KLF6 in DU145 cells decreases cell proliferation, increases apoptosis, and shifts cell growth to the G_0/G_1 phase (Cheng et al. 2008)
- Overexpression of KLF6 induces apoptosis through upregulation of ATF3; KLF6 mutants are unable to induce apoptosis (Huang et al. 2008)
- *Germline KLF6 SNP (IVS1–27 G > A; rs3750861) is associated with increased lifetime PCa risk in both sporadic and hereditary prostate cancer*; identification of alternatively spliced, oncogenic isoform; KLF6-SV1 antagonizes tumor-suppressive function of KLF6 and is over-expressed in tumors (Narla et al. 2005a,b)
- *Inhibition of KLF6-SV1 decreases tumor cell proliferation, colony formation, migration, and invasion; intratumoral injection of siKLF6-SV1 decreases tumor cell growth in vivo.* (Narla et al. 2005a,b)
- *KLF6-SV1 overexpression is associated with markedly poorer survival and disease recurrence; KLF6-SV1-overexpressing PCa cells metastasize more rapidly in vivo; siKLF6-SV1 induces spontaneous apoptosis in cultured PCa cell lines and suppresses tumor growth in mice* (Narla et al. 2008)
- IVS1-27 SNP associated with decreased risk in Ashkenazi; germline mutations Q160X, P172L (Bar-Shira et al. 2006)
- No germline coding mutations in Finnish hereditary cancer patients; 11.6% of probands had silent sequence variants (Koivisto et al. 2004a,b)
- No association of KLF6 IVS 1–27 G > A germline single nucleotide polymorphism (SNP) with either prostate cancer or benign prostatic hypertrophy (BPH) risk in pCA risk in the Finnish population (Seppälä et al. 2007)

Liver cancer

- LOH 39%, mutation 15% (P149S, W162X, P166S, T179I, K182R); mutants cannot suppress growth of HepG2 cells (Kremer-Tal et al. 2004)
- LOH 36% (Wang et al. 2004b)
- LOH 6.8%; no mutations identified (Song et al. 2006)
- Mutation 8.7% (W162G) (Pan et al. 2006a)
- No mutations (Boyault et al. 2005)
- No differences in locus amplification between KLF6 in hepatocellular carcinoma (HCC) and surrounding tissue (Wang et al. 2004a)
- Decreased expression of KLF6 in ~ 35% of samples; W162G mutant has no effect on proliferation or p21 expression (Pan et al. 2006b)

(continued)

Table 1 (continued)

- Decreased expression of KLF6 in most of hepatitis B virus (HBV)- and hepatitis C virus (HCV)-associated HCCs; decreased KLF6 expression in dysplastic nodules compared to cirrhotic tissue; increased KLF6SV1/KLF6 ratio in 18% of HBV-related HCCs; ectopic KLF6 expression in HepG2 cells decreased proliferation and expression of related markers including cyclin D1 and β-catenin but increased cellular differentiation based on induction of albumin, E-cadherin, and decreased α-fetoprotein (Kremer-Tal et al. 2007)
- Decreased KLF6 expression in 38% of HCCs; KLF6 promoter fully methylated in hepatoma cell lines, HCCs, and corresponding non-HCC tissue with evidence of cirrhosis/hepatitis (Hirasawa et al. 2006)
- Correlation between decreased KLF6 expression and decreased p21 expression in HCCs (Narla et al. 2007)
- Decreased KLF6 RNA and protein expression in HCC tissue and hepatoma cell lines (Wang et al. 2007)
- Decreased expression of KLF6 in HCC precursor lesions (macronodules) detected by genome-wide array (Bureau et al. 2008)
- Increased expression of alternatively spliced KLF6 variant in HCC; splice-variant antagonizes KLF6 tumor-suppressor function in HepG2 cells (Pan et al. 2008)
- Inhibition of KLF6 in HCC; HepG2 and HuH7 cells strongly impaired cell proliferation-induced G_1-phase arrest, inhibited cyclin-dependent kinase 4 and cyclin D1 expression and subsequent retinoblastoma phosphorylation. Finally, KLF6 silencing caused p53 upregulation and inhibited Bcl-xL expression, to induce cell death by apoptosis (siRNA sequence silences both wt and sv1-JAM) (Sirach et al. 2007)
- The ras signaling in HCC samples correlates with increased KLF6 alternative splicing (Yea et al. 2008)

Ovarian cancer

- 60% LOH (histology-dependent) was significantly correlated with tumor stage and grade; KLF6-SV1 expression was increased approximately fivefold; KLF6 silencing increased cellular and tumor growth, angiogenesis, and vascular endothelial growth factor expression, intraperitoneal dissemination, and ascites production. Conversely, KLF6-SV1 downregulation decreased cell proliferation and invasion and completely suppressed in vivo tumor formation (DiFeo et al. 2006b)
- KLF6 and KLF6-SV1 antagonistic effects on E-cadherin expression, β-catenin subcellular localization, and c-myc expression in SKOV3 cell lines (DiFeo et al. 2006a)
- Decreased KLF6 expression is associated with increased cisplatin in the $PE01^{CDDP}$ ovarian cancer cell line (Macleod et al. 2005)

Colorectal cancer

- LOH 55%; mutation 42% of sporadic and inflammatory bowel disease (IBD)-related tumors; mutations result in loss of transactivation of p21 and growth-suppressive properties (T179I, S155G, P172L, P166S, P148L, P149S, S136G, E100K, D185N, G163D, G131S, A191T, K74R, Y97C, W162X) (Reeves et al. 2004)
- 43% LOH; mutations in 4% of samples (S155N, G163S, G163D, P183L, and G195S) (Cho et al. 2006a)
- LOH 48%; mutation 6% (Mukai et al. 2007)
- Differential and progressive increase in LOH: 4% in adenomas, 0% in intramucosal carcinomas, 35% in invasive carcinomas, and 33% in liver metastases; no mutations; identification of germline silent SNPs at codons 198 and 205 (Miyaki et al. 2006)
- No mutations (Lièvre et al. 2005)
- Decreased KLF6 expression in 37% of samples; direct association with tumor size (Cho et al. 2006b)

(continued)

Table 1 (continued)

Astrocytoma, glioblastoma, meningioma

- 77% LOH in samples with mutations; 12% mutation frequency in glioblastoma multiforme (GBM), 6% astrocytomas, 0% oligodendroglioma (Q60L, S77P, E78K, S92I, S112F, D114G, E132G, G167E, S180L, A191T, D194V, S215F, delta 715 frameshift) (Jeng and Hsu 2003)
- 62% LOH; translocation in one sample, but no other mutations identified; decreased expression in U373 MG cell line (Montanini et al. 2004)
- 88% LOH in GBM; no mutations; decreased KLF6/increased KLF6-SV1 expression; KLF6 and KLF6-SV1 affected cell proliferation in A235 and CRL2020 cell lines (Camacho-Vanegas et al. 2007)
- 8% Mutation frequency (D2G, M6V, E30G, S92R, V165M, P183I, G276A); decreased expression (GBM, astrocytoma, oligodendroma); 4.5× increased frequency of 5′ UTR SNPs (nt positions -4, -5, and -6) in patients than controls (Yin et al. 2007)
- No mutations; no evidence for decreased KLF6 expression (Köhler et al. 2004)
- No mutations detected in GBM, astrocytomas, or GBM cell lines (Koivisto et al. 2004b)
- Decreased expression in GBM cell lines and primary tumor samples; KLF6 represses tumorigenicity and inhibits oncogene-induced cellular transformation (Kimmelman et al. 2004)

Gastric cancer

- 43% LOH; four missense mutations (S155R, P172 T, S180L, R198K) (Cho et al. 2005)
- 53% LOH; four missense mutations (T179I, R198G, R71Q, S180L); decreased KLF6 expression was associated with loss of the KLF6 locus (48%); mutants fail to suppress proliferation (Sangodkar et al. 2008, in press)
- IVS 1–27 G/A polymorphism not associated with an increased risk for gastric cancer in Korean population (Cho et al. 2008)

Nasopharyngeal carcinoma (NPC, Chen et al. 2002)

- Mutations in 3 of 19 NPC tissues (16%; E75V, S136R, R243K)

Pancreatic cancer

- No LOH; no mutations; *increased splicing; KLF6 splicing correlated significantly with tumor stage and survival*

Barrett's esophageal cancer (Peng et al. 2008)

- Decreased expression of KLF6 (proteomic study)

Pituitary

- Three germline changes identified (5%): Val165Met, 523G→A, Ser77Ser) (Vax et al. 2003)

Head and neck squamous cell carcinoma

- 30% LOH, which *was strongly correlated with cancer progression, tumor recurrence, and decreased patient survival;* 10% mutations; altered subcellular protein localization in 64% of tumors

Lung cancer (Teixeira et al. 2007)

- High degree of LOH (> 40%) in KLF6 locus in both small-cell and non-small-cell cancer cell lines [Girard et al. 2000)
- 34% LOH; decreased expression of KLF6 in most samples (85%); KLF6 overexpression induced apoptosis in two cell lines (NCI-H1299, NCI-H2009) (Ito et al. 2004)
- Loss of KLF6 locus in two-thirds of patient-derived cell lines (Zhu et al. 2007)
- Germline missense mutations in 3% of samples (Yin et al. 2007]
- Decreased KLF6 expression (Wikman et al. 2002)
- Decreased KLF6 (Copeb) expression by array (Kettunen et al. 2004)
- Decreased expression of KLF6 in lung cancer cell lines (Lam et al. 2006)
- *Increased KLF6-SV1 expression is associated with decreased survival*; siKLF6-SV1 induces apoptosis both alone and in combination with cisplatin (DiFeo et al. 2008a)

(continued)

Table 1 (continued)

- *GWAS* (Genome Wide Association Study)-*detected decreased relative risk of lung cancer in patients with G/A or A/A allele (rs3750861) in an Italian population*; not confirmed in a Norwegian population (Spinola et al. 2007)

Breast

- KLF6 overexpression correlated with increased survival/tumor growth in drug-induced mammary adenocarcinomas in rat experimental model (Kovacheva et al. 2008)
- Increased methylation of TFPI-2 promoter KLF6-binding site in the invasive MDA-MB-435 cell line (Guo et al. 2007)
- *Intratumoral injection of siKLF6-SV1 decreases intramammary MDA-MB-435 tumor cell growth and systemic vascular endothelial growth factor (VEGF) expression in a mouse model* (DiFeo et al. 2008b)

Sections detailing associations between genotype or expression patterns with disease risk or prognostication are underlined.

Germline KLF6 SNP: Increases Prostate Cancer Risk?

Given our original findings of KLF6 inactivation in sporadic prostate cancer, we next investigated the possibility that germline KLF6 mutations or polymorphisms present in the general population are associated with increased prostate cancer risk. Underlying these studies was the theoretical assumption that if the KLF6 gene was an important contributor to prostate cancer development, and possibly progression, it could be expected that some men with increased risk for the development of prostate cancer would harbor germline mutations or polymorphisms. In addition, a possible genetic link between the KLF6 gene and familial prostate cancer had already been suggested (Xu et al. 2003). In the search for novel prostate cancer susceptibility genes, a genome-wide scan was undertaken of members of 188 prostate cancer families who had at least three first-degree relatives affected with prostate cancer. A marker on chromosome 10p15.3 (D10S249) was identified as one of the regions having the highest LOD score (1.39) when all men were analyzed regardless of age at diagnosis, number of affected family members, or race. When stratified by number of affected family members, this locus had the third highest LOD score for the entire analysis (LOD 3.0). The D10S249 marker lies approximately 4 Mb telomeric to the KLF6 gene.

To directly investigate the possibility that inherited polymorphisms in the KLF6 gene are associated with PCa, we first sought to identify and characterize biologically relevant SNPs (Narla et al. 2005a,b). Therefore, we directly sequenced genomic DNA isolated from blood samples from an "exploration set" of 142 men with a family history of prostate cancer. From among a number of silent, conserved, intronic gene changes, a relatively common intronic KLF6 gene polymorphism, IVS 1–27G > A, the IVSΔA allele (rs3750861), was identified as the most frequent. In total, 26 of 142 men (18.3%) possessed the IVSΔA variant (Fig. 2).

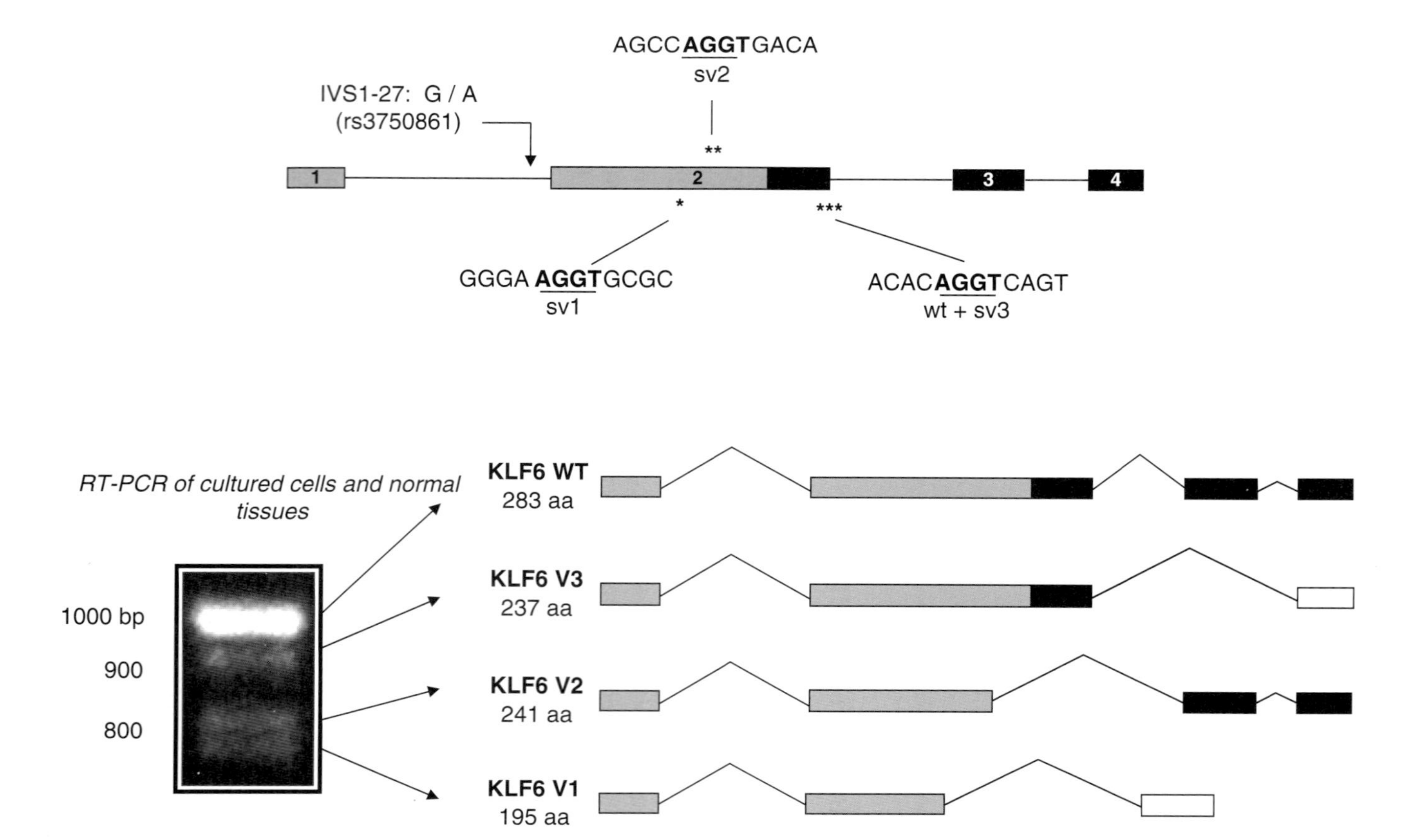

Fig. 2 Genomic organization of the KLF6 gene and its alternatively spliced products. The four coding exons, the location of the KLF6 IVS1 -27 single nucleotide polymorphism (SNP), the cryptic splice sites, and alternatively spliced isoforms are shown. The exons encoding the activation domain are shown in *gray* and the zinc finger elements in *black*. *RT-PCR* = reverse transcription-polymerase chain reaction

Using this SNP to interrogate our samples, we performed a sequential series of association studies by genotyping germline DNA from men from three large, independent centers. In total, genomic DNA from 3411 geographically diverse men was analyzed from the combined resources of the Johns Hopkins University, Mayo Clinic, and Fred Hutchinson Cancer Research Center Prostate Cancer registries. The samples represented men divided into three groups: 1253 men with sporadic prostate cancer, 882 men with familial prostate cancer from 294 unrelated families (3 men from each family), and 1276 controls.

Overall, the presence of this single germline SNP increased the risk of prostate cancer by approximately 50% in all men studied. The relative risk was 1.42 [$p = 0.01$; 95% confidence interval (CI) = 1.10–1.80] in men with sporadic PCa and 1.61 [$p = 0.01$; odds ratio (OR) = 1.61; 95% CI = 1.20–2.16] in men with familial prostate cancer. Among all men with a positive family history for prostate cancer, the carrier frequency was higher among men with an earlier age of diagnosis (< 65 years of age; $p \leq 0.03$). Although the increased risk offered by this single germline SNP is modest, it is in line with what would be expected for a relatively common polymorphism influencing a complex genetic disease wherein many genes are believed to play a role in cancer development and progression.

Notably, the KLF6 IVSΔA allele is the first highly prevalent, low-penetrance allele associated with an overall lifetime increased prostate cancer risk. Nonetheless, the use of this single SNP as a predictive biomarker for prostate cancer risk has not been evaluated in the general population. One factor that must be considered is the relatively broad range of heterozygosity frequencies between different populations (Table 1). The practical consequences of these findings are also highly translatable in the genomic era as large-scale association studies intensify to identify and characterize biologically relevant variations. In this instance, as discussed below, we have shown that a seemingly neutral polymorphism associated with prostate cancer risk is linked to a novel form of tumor-suppressor gene inactivation through alternative splicing.

KLF6 Gene: index, alternative splicing to Produce an Oncogenic Variant, KLF6-SV1

The molecular basis by which the IVSΔA noncoding germline sequence variant resulted in increased risk was soon thereafter defined (Narla et al. 2005a,b). The KLF6 gene is alternatively spliced into three isoforms, KLF6-SV1, KLF6-SV2, and KLF6-SV3, in both normal and cancerous tissues. These variants arise from the use of native cryptic splice sites in exon 2 and, when translated, lack either parts of the KLF6 activation domain and/or parts or all of the DNA-binding domain (Fig. 2).

KLF6-SV1, generated by use of an alternative 5′ splice site, contains a novel 21-amino-acid carboxy domain resulting from out-of-frame splicing of exon 3. The IVSΔA SNP generates a novel binding site for SRp40, a member of the SR protein family that is involved in selecting and regulating splice site usage. Binding of SRp40 results in overexpression of KLF6-SV1. Of particular significance with regard to cancer, and beyond our initial description of regulation by the IVSΔA SNP, KLF6-SV1 expression is also significantly upregulated in a number of cancers, including glioblastoma, prostate, ovarian, and lung cancer (Camacho-Vanegas et al. 2007; DiFeo et al. 2008a; Narla et al. 2005a, 2005b; Teixeira et al. 2007), regardless of SNP status. Although the relative importance to all cancers is not yet known, one mechanism of altered splicing has been defined in hepatocellular carcinoma (HCC). Yea et al. demonstrated that KLF6-SV1 expression is increased in an oncogenic Ras/PI3-K/Akt-dependent manner, thereby increasing the relative KLF6-SV1/KLF6 ratio (Yea et al. 2008). The effect of this increase was shown to enhance cellular proliferation.

Increasing evidence from multiple experimental systems and cancer types demonstrates that KLF6-SV1 is biologically active and can antagonize the growth-suppressive properties of KLF6 (DiFeo et al. 2006a; Narla et al. 2005a, 2005b). Most notably, KLF6-SV1 has been shown to affect tumor cell proliferation, invasion, colony formation, angiogenesis, and E-cadherin expression (DiFeo et al. 2006a; Narla et al. 2005a). The expression level changes in these two antagonistic isoforms, KLF6 and KLF6-SV1, would therefore be predicted to play important roles in cancer through shared and independent pathways.

KLF6 and KLF6-SV1: Diagnostic and Prognostic Cancer Biomarkers

A number of independent studies across four distinct cancer types have linked expression levels of KLF6 and/or KLF6-SV1 with prognostic significance. In particular, several studies have demonstrated KLF6 dysregulation with a more aggressive prostate cancer phenotype. Thus, evolving from its original description as a "candidate" tumor suppressor in prostate cancer, KLF6 has now evolved to be an independently verified biomarker of cancer progression and patient outcome with this disease. The first of these three independent studies identified decreased expression of KLF6 with poor prognoses (Glinsky et al. 2004). Based on the analysis of a genome-wide expression array, Glinsky et al. identified KLF6 as part of a five-gene cluster that could accurately predict relapse-free survival following radical prostatectomy. Following this report, Stanbrough et al. identified decreased KLF6 expression as part of a gene signature associated with metastatic androgen-independent, or hormone-refractory, prostate cancer. We recently examined the expression levels of KLF6-SV1 at the time of prostatectomy using quantitative reverse transcription-polymerase chain reaction (qRT-PCR) analysis and KLF6-SV1-specific primers. Notably, these samples all had associated clinicopathological correlates, including disease-free survival. KLF6-SV1 expression was strongly associated with poor survival, and the median survival difference was more than 4 years. Patients with high levels of KLF6-SV1 expression had a median survival of approximately 30 months compared to 80 months for patients with low KLF6-SV1 expression (Fig. 3).

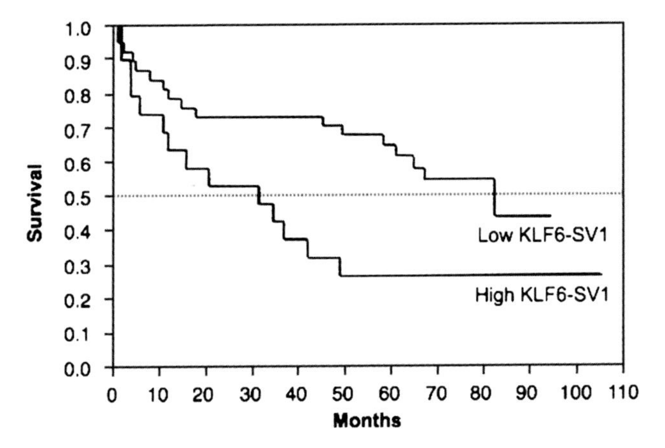

Fig. 3 Kaplan-Meier survival analysis of men with high and low levels of KLF6-SV1 expression. The median survival, as measured using biochemical recurrence and assessed by a reverse transcription-polymerase chain reaction (qRT-PCR), in men whose localized prostate tumors expressed high levels of KLF6-SV1 was 30 months compared with 80 months in men with low KLF6-SV1-expressing tumors ($p < 0.01$)

As dramatic as the survival difference is suggested by these results in prostate cancer, it is not unique. In a study of lung cancer patients, DiFeo et al. have now demonstrated that KLF6-SV1 expression levels were associated with a 6.5-year difference in median survival between patients in the lowest and highest tertiles. To date, two other studies have also demonstrated an association between KLF6/KLF6-SV1 levels and disease outcome. Intriguingly, mechanisms of loss in these two cancers were quite distinct. In head and neck squamous cell carcinoma (HNSCC), although mutation was infrequent, LOH of the KLF6 locus was highly correlative with median survival (Teixeira et al. 2007). Median survival of patients with LOH was less than half that of those without loss (19 vs. 41 months). Moreover, risk of death for patients with LOH was eight times greater independent of tumor size, nodal status, tobacco smoking, or treatment modality [hazard ratio (HR) = 7.89; 95% CI = 1.9–32.4). In contrast, although mutation and LOH were infrequent in pancreatic cancer, increased alternative splicing was a notable feature of these tumors (Hartel et al. 2008). In pancreatic cancer, patients with a high KLF6/KLF6-SV1 ratio had significantly longer survival (median 21 months, range 14–19 months) than patients with a lower ratio (median 9 months, range 6–10 months) ($p = 0.005$).

KLF6 and KLF6-SV1: therapeutics Targets in Human Cancer

Given the findings that decreased KLF6 and/or increased expression levels are correlated with worse prognosis and disease recurrence, and that accumulating evidence across many cancer types and at different stages of tumorigenesis and

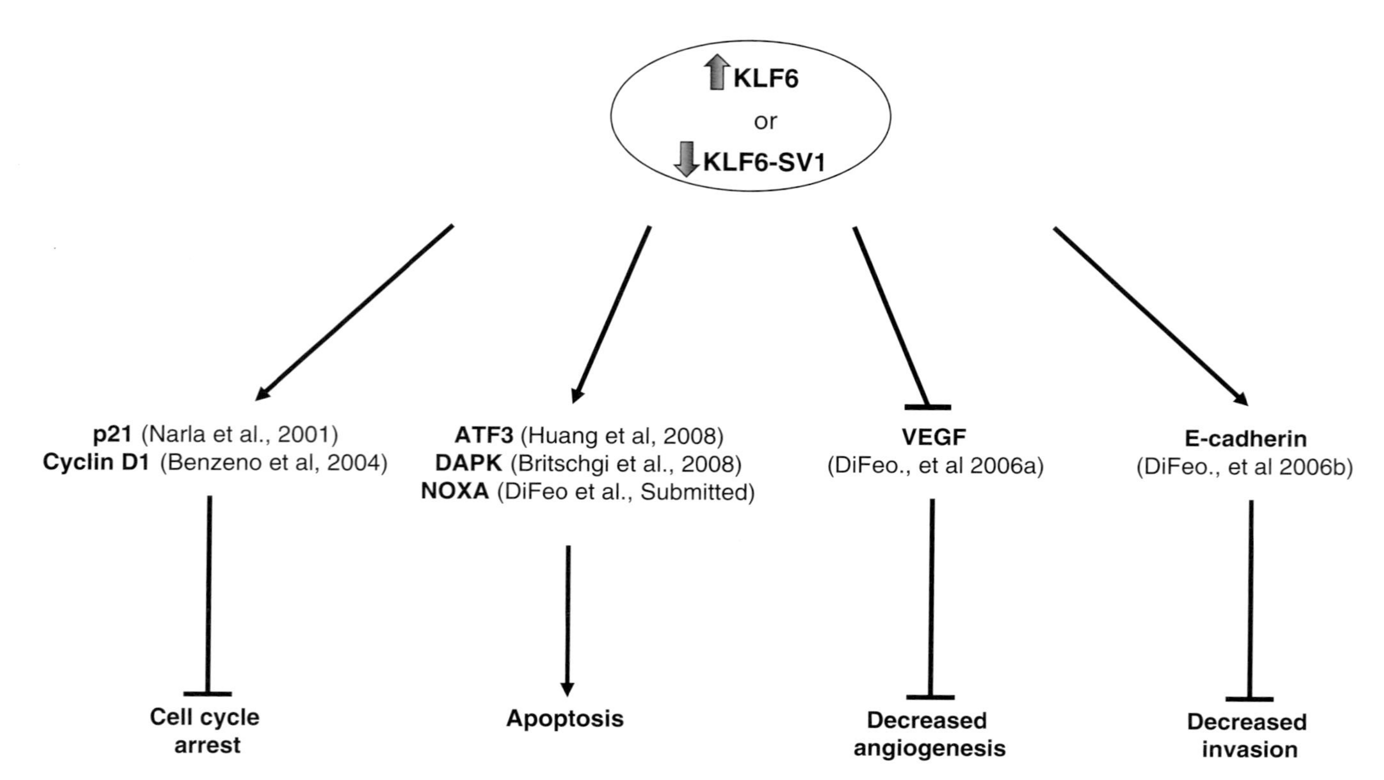

Fig. 4 Overview of molecular pathways regulated through KLF6 and/or KLF6-SV1

progression have now confirmed important functional roles and cancer-relevant pathways for KLF6 and KLF6-SV1 (reviewed in DiFeo et al. 2009) (Fig. 4), it is logical to consider targeting them as potential therapeutic agents. One approach would be to reconstitute KLF6 function in tumors shown to have KLF6 inactivation. In support of this, Kimmelman et al. originally demonstrated in a glioblastoma model that KLF6 expression leads to significant reversion of tumorigenic phenotypes (Kimmelman et al. 2004). In these experiments, KLF6 was introduced through retrovirus-mediated gene transfer into the DBTRG-05MG glioblastoma cell line, with the result that anchorage-independent growth and tumorigenicity were drastically reduced. This cell line was chosen for several reasons: KLF6 expression is virtually absent; chromosome 10 is monosomic; wild-type p53 alleles are present; and it is readily infected. Similarly, forced expression of KLF6 in lung cancer cell lines were shown to induce apoptosis (Ito et al. 2004). Ultimately, however, the limitation of this "gene therapy" approach will be achieving tumor-specific delivery.

A currently more feasible approach is to inhibit KLF6-SV1 in overexpressing tumors. Of particular note with regard to potential therapeutic targeting are the most recent studies demonstrating not only that overexpression of KLF6-SV1 is associated with decreased survival and metastatic spread of tumor but that siRNA-mediated inhibition of KLF6-SV1 has dramatic effects on tumor behaviour in vitro (DiFeo et al. 2006a; Narla et al. 2005a, 2008) and in vivo (DiFeo et al. 2006a; Narla et al. 2005b, 2008) independent of a tumor cell's p53 status. For example, in one of the first examples demonstrating the in vivo relevancy of KLF6-SV1 inhibition, the highly aggressive PC3M cell line was stably infected to express specific siRNAs to either luciferase, KLF6, or KLF6-SV1. These cells were then injected subcutaneously into nude mice, and after 8 weeks the mice were killed. Strikingly, reduction of KLF6 led to a more than twofold increase in tumor burden, whereas inhibition of KLF6-SV1 decreased tumor growth by approximately 50% (Fig. 5).

In preliminary studies, we have now extended these findings to an ovarian cancer dissemination model (DiFeo et al. submitted). The rationale for developing a preclinical ovarian cancer model is based on several unique aspects of the disease. First, we recently demonstrated that KLF6-SV1 is overexpressed in ovarian cancer (DiFeo et al. 2006b). All tumors except one (32/33, 97%), expressed KLF6-SV1; and, strikingly, KLF6-SV1 expression was increased, on average, nearly fivefold ($p < 0.001$) in the ovarian cancer samples relative to normal tissue. In addition, KLF6-SV1 upregulation was associated with poorly differentiated grade III tumors compared to well to moderately differentiated grade I or II tumors.

Second, ovarian cancer is one of the leading causes of death from gynecological malignancy, with more than 190,000 new cases each year worldwide. The high mortality rate is due to the fact that most patients present with advanced-stage disease wherein the tumor has disseminated/metastasized in the peritoneum, and the cornerstone of treatment is surgical debulking and platinum-based chemotherapy. Although initially responsive to chemotherapy, most of these women eventually succumb to recurrence and chemoresistance, including 50% of women who have no evidence of disease following primary therapy. Finally, as residual ovarian cancer recurrences are primarily confined within the peritoneum, from a treatment

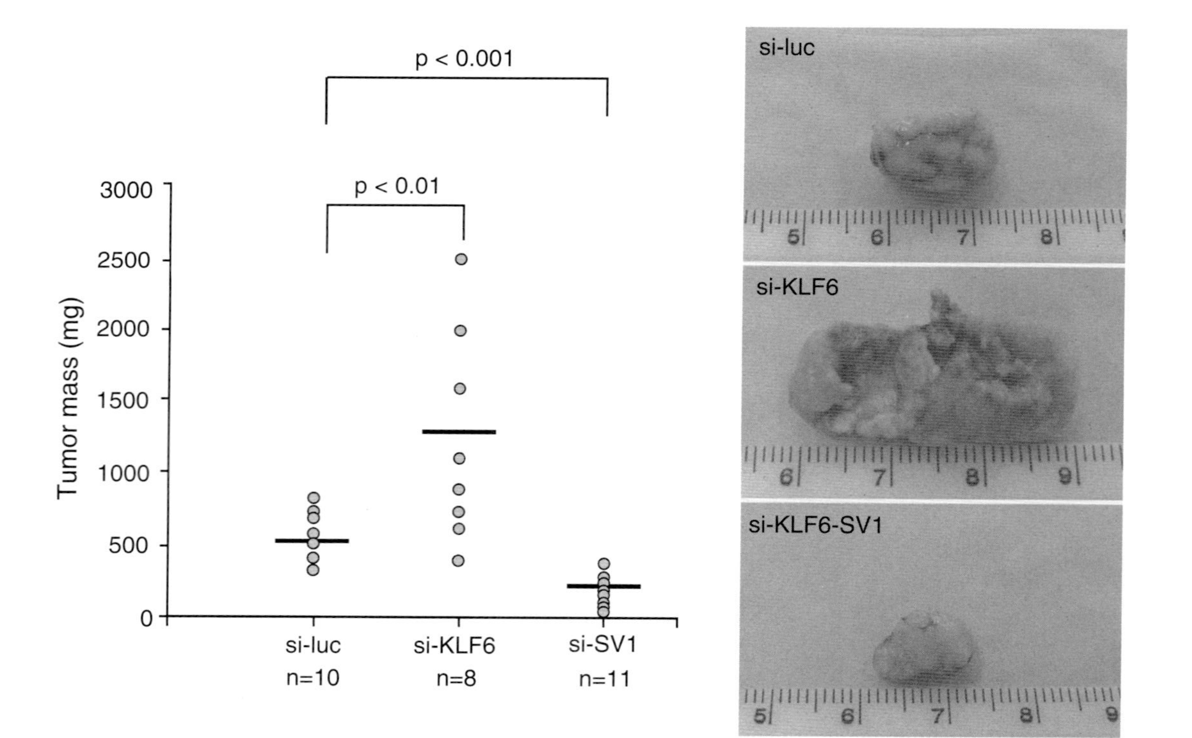

Fig. 5 Effect of siRNA treatment on prostate cancer xenograft growth in vivo. PC3M stable cell lines expressing siRNAs to luciferase (*si-luc*), KLF6 (*si-KLF6*), or KLF6-SV1 (*si-SV1*) were injected into nude mice, and the final tumor volume was measured. Inhibition of KLF6 resulted in > 90% increase in tumorigenicity, whereas PC3M cells expressing siKLF6 SV1 were ~50% less tumorigenic. Representative tumors are shown

perspective the disease can be considered a localized disease. In fact, intraperitoneal administration of chemotherapy was first proposed three decades ago, and currently used agents when given via the intraperitoneal route have been shown to have distinct pharmacokinetic advantages, including higher concentrations and longer half-life (Dedrick et al. 1978; Trimble et al. 2008). Most importantly, numerous studies have suggested that intraperitoneal dosing results in increased patient survival (Trimble et al. 2008). Thus, intraperitoneal delivery of siKLF6-SV1 presents a paradigm in ovarian cancer treatment. siKLF6-SV1, either as a single agent or combined with the gold standard platinum agent, could effectively be considered "localized" delivery achieved via a systemic route. The roadblocks to effective treatment would then be more focused on devising the correct chemical modifications to increase siRNA stability, tumor uptake, and prolonged silencing.

Therefore, to begin these studies, we generated a series of ovarian cancer (SKOV-3) stable cell lines using pSUPER-si-luc (si-luc), pSUPER-si-KLF6 (si-KLF6), and pSUPER-si-SV1 (si-SV1). In the si-KLF6-expressing cell lines, proliferation and invasive capacity were markedly diminished. Conversely, inhibition of KLF6 had the opposite effects, increasing proliferation and invasive capacity. The most striking effect of KLF6-SV1 inhibition was seen on tumorigenicity. The SKOV-3 stable cell lines expressing si-luc, si-KLF6, or si-SV1 were subcutaneously injected into nude mice, monitored weekly for tumor growth, and then sacrificed at the end of 6 weeks. Most conspicuously, all five mice transplanted with si-SV1 cells failed to form persistent tumors. The small tumors that initially developed regressed after 3 weeks. In contrast, reduction of the tumor suppressor KLF6 doubled the tumor growth rate and mass.

To investigate the potential in vivo antitumor effects of KLF6-SV1 inhibition on median and overall survival, we used a model of disseminated ovarian cancer to explore the effects of systemically delivered siRNA against KLF6-SV1. SKOV3 cells stably expressing luciferase (SKOV3-Luc) were implanted into the peritoneal cavity of nude mice and imaged after 48 hours to confirm the establishment of disseminated tumors (Fig. 6). We then administered by peritoneal injection three escalated doses of a chemically modified siRNA, specially formulated to resist degradation (Accell), targeting either a control sequence (siNTC) or KLF6-SV1 (siSV1) every third day for a total of six doses (DiFeo et al. 2009). Inhibition of KLF6-SV1 approximately tripled the median survival and more than doubled overall survival ($p = 0.0002$) in a dose-dependent manner. Thus, for the first time, KLF6-SV1 was shown to represent a therapeutically targetable control point in cancer.

Conclusions and Future Directions

As was originally predicted less than a decade ago (Bieker 2001; Black et al. 2001), at least one member of the KLF family has now been clearly implicated in the initiation, progression, and diagnosis of human cancer. Preclinical studies using siRNA, targeting KLF6-SV1, suggest that inhibition of this oncogenic KLF

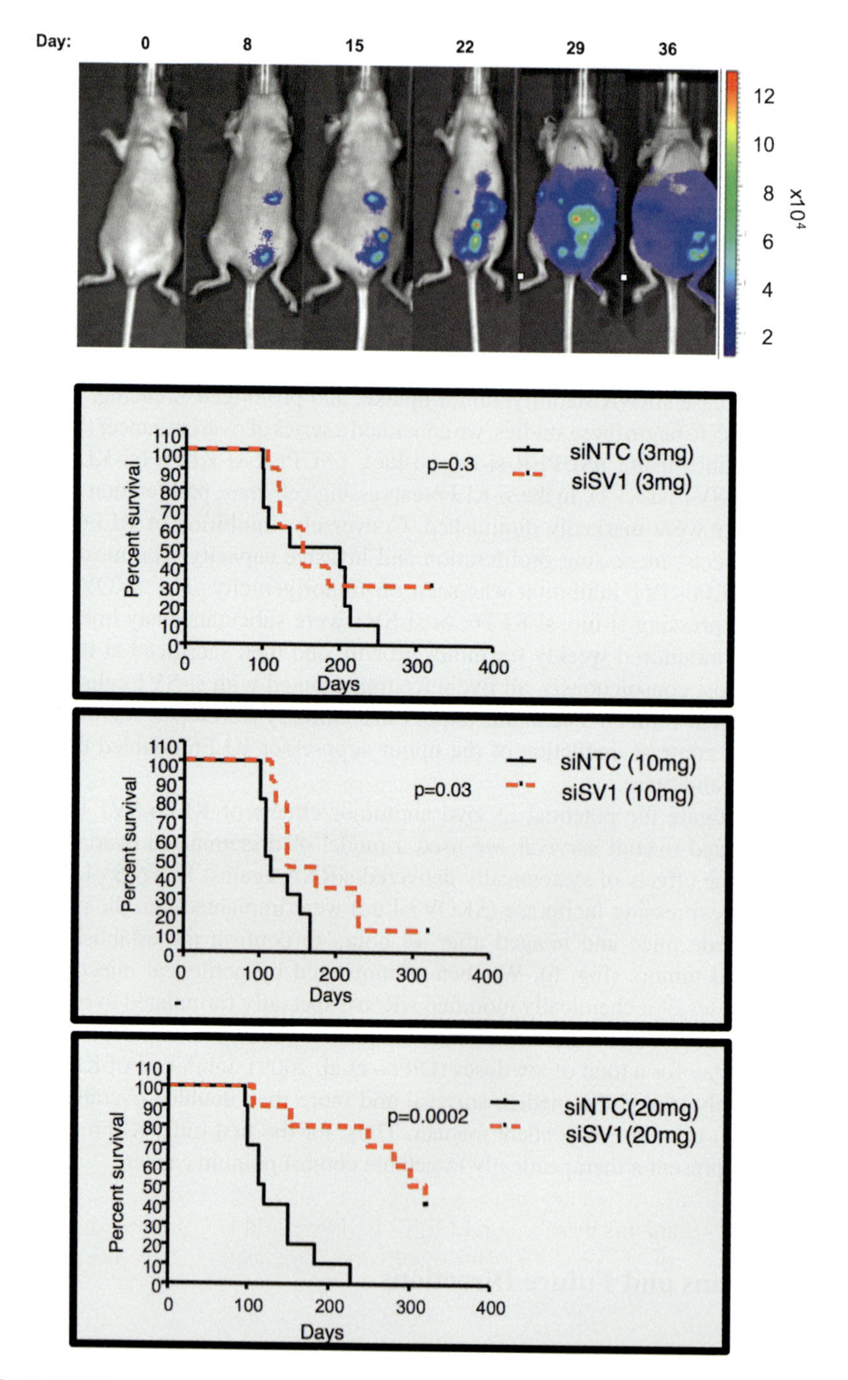

Fig. 6 Inhibition of KLF6-SV1 increases median survival in an ovarian cancer model. The total bioluminescent tumor signature of SKOV-3-luciferase-positive cells injected into the peritoneum is captured using whole-body luminescence imaging (Xenogen IVIS-200 system). Kaplan-Meier survival analyses were performed on mice treated with increasing doses of either siNTC or siKLF6-SV1

isoform may provide a powerful therapeutic target in ovarian cancer. Indeed, the horizon for the use of siRNA/RNAi seems to beckon with an unlimited potential for many human diseases (Gewirtz et al. 2007). Already a number of recent studies have begun to highlight the potential efficacy and specificity of siRNA/RNAi-based therapies and have moved well beyond its earliest demonstrations of gene silencing in mammalian cell culture (Elbashir et al. 2001). Beyond their use in animal models, several siRNA therapeutics have advanced to clinical trials for treating diseases such as age-related macular degeneration (AMD), respiratory syncytial virus (RSV), hepatitis B virus, and acute renal failure (reviewed in Whitehead et al. 2009). Although there are not many clinical trials using siRNA to treat cancer, the prospects for treatment of human cancers using siRNA has also been long recognized (Pai et al. 2006) but again hampered by the issue of delivery. Theoretically, targeting disorders wherein the siRNA can be delivered directly into the diseased tissue, as in our model of ovarian cancer, represent the most attractive option. The direct siRNA, delivery route ensures a high concentration of the siRNA and would be predicted to have the least likelihood for unanticipated effects systemically.

For other cancers wherein KLF6-SV1 overexpression has been linked to poor prognosis and disease recurrence, most notably lung (DiFeo et al. 2008a) and prostate (Narla et al. 2008) cancers, other delivery routes would need to be devised. For lung cancer, a particularly intriguing route could be via aerosol or intranasal delivery. For prostate cancer and many other solid cancers, new methods are needed for tissue-specific delivery. Although delivery remains a major hurdle in use of RNAi therapeutics clinically, ongoing studies focusing on siRNA-based formulations such as synthetic nanoparticles composed of polymers, lipids, lipidoids, or conjugates (Whitehead et al. 2009) may offer significant promise for the advancement of siRNA therapeutics.

Beyond an siRNA-mediated approach to inhibit KLF6-SV1 directly, a number of other strategies exist. For example, altering the splicing ratio of the gene to favor KLF6 expression and decrease KLF6-SV1 can be predicted to decrease tumor growth. A proof-of-concept clinical trial for Duchenne's muscular dystrophy uses this "exon-skipping" approach (van Deutekom et al. 2007). In that study, four children were treated with an intramuscular injection of a small antisense oligonucleotide favoring exon skipping over the disease-causing mutated exon. All patients had evidence of at least partial restoration of dystrophin expression in the treated muscles. Based on this model, the cryptic splice site within exon 2 of the KLF6 gene could be targeted as a "mutation."

Finally, a completely different approach to "inhibit" KLF6-SV1 function would be through the use of small molecules that lead to its degradation. Therefore, a number of mechanisms by which KLF6-SV1 RNA could be silenced, its expression decreased, or its protein activity subverted are available in the research laboratory. Based on an ever-expanding appreciation of the relevancy of KLF6 and KLF6-SV1 to human cancer, the immediate translation of these strategies first to animal models and then to the clinical arena is of high priority. We believe that, ultimately, developing the methods to therapeutically regulate KLF6 and KLF6-SV1 will play an important role in treating a number of localized and metastatic cancers.

References

Agell L, Hernández S, de Muga S, Lorente JA, Juanpere N, Esgueva R, Serrano S, Gelabert A, Lloreta J. KLF6 and TP53 mutations are a rare event in prostate cancer: distinguishing between Taq polymerase artifacts and true mutations. Mod Pathol. 2008 Dec;21(12):1470–8.

Bar-Shira A, Matarasso N, Rosner S, Bercovich D, Matzkin H, Orr-Urtreger A. Mutation screening and association study of the candidate prostate cancer susceptibility genes MSR1, PTEN, and KLF6. Prostate. 2006 Jul 1;66(10):1052–60.

Benzeno S, Narla G, Allina J, Cheng GZ, Reeves HL, Banck MS, Odin JA, Diehl JA, Germain D, Friedman SL. Cyclin-dependent kinase inhibition by the KLF6 tumor suppressor protein through interaction with cyclin D1. Cancer Res. 2004 Jun 1;64(11):3885–91.

Bieker JJ. Kruppel-like factors: three fingers in many pies. J Biol Chem. 276:34355–8; 2001.

Black AR, Black JD, Azizkhan-Clifford J. Sp1 and kruppel-like factor family of transcription factors in cell growth regulation and cancer. J Cell Physiol. 188:143–60; 2001.

Boyault S, Hérault A, Balabaud C, Zucman-Rossi J. Absence of KLF6 gene mutation in 71 hepatocellular carcinomas. Hepatology. 2005 Mar;41(3):681–2.

Bureau C, Péron JM, Bouisson M, Danjoux M, Selves J, Bioulac-Sage P, Balabaud C, Torrisani J, Cordelier P, Buscail L, Vinel JP. Expression of the transcription factor Klf6 in cirrhosis, macronodules, and hepatocellular carcinoma. J Gastroenterol Hepatol. 2008 Jan;23(1):78–86.

Camacho-Vanegas O, Narla G, Teixeira MS, DiFeo A, Misra A, et al. Functional inactivation of the KLF6 tumor suppressor gene by loss of heterozygosity and increased alternative splicing in glioblastoma. Int J Cancer. 2007; 121:1390–5.

Chen C, Hyytinen ER, Sun X, Helin HJ, Koivisto PA, Frierson HF Jr, Vessella RL, Dong JT. Deletion, mutation, and loss of expression of KLF6 in human prostate cancer. Am J Pathol. 2003 Apr;162(4):1349–54.

Chen HK, Liu XQ, Lin J, Chen TY, Feng QS, Zeng YX. Mutation analysis of KLF6 gene in human nasopharyngeal carcinomas. Ai Zheng. 2002 Oct;21(10):1047–50.

Cheng XF, Li D, Zhuang M, Chen ZY, Lu DX, Hattori T. Growth inhibitory effect of Krüppel-like factor 6 on human prostatic carcinoma and renal carcinoma cell lines. Tohoku J Exp Med. 2008 Sep;216(1):35–45.

Cho YG, Choi BJ, Kim CJ, Song JW, Kim SY, Nam SW, Lee SH, Yoo NJ, Lee JY, Park WS. Genetic alterations of the KLF6 gene in colorectal cancers. APMIS. 2006 Jun;114(6):458–64.

Cho YG, Choi BJ, Song JW, Kim SY, Nam SW, Lee SH, Yoo NJ, Lee JY, Park WS. Aberrant expression of krUppel-like factor 6 protein in colorectal cancers. World J Gastroenterol. 2006 Apr 14;12(14):2250–3.

Cho YG, Kim CJ, Park CH, Yang YM, Kim SY, Nam SW, Lee SH, Yoo NJ, Lee JY, Park WS. Genetic alterations of the KLF6 gene in gastric cancer. Oncogene. 2005 Jun 30;24(28): 4588–90.

Cho YG, Lee HS, Song JH, Kim CJ, Park YK, Nam SW, Yoo NJ, Lee JY, Park WS. KLF6 IVS1 -27G/A polymorphism with susceptibility to gastric cancers in Korean. Neoplasma. 2008;55(1):47–50.

Dedrick R, Myers C, Bungay PM et al. Pharmacokinetic rationale for peritoneal drug administration in the treatment of ovarian cancer. Cancer Treat Rep 1978;62:1–11.

DiFeo A, Narla G, Camacho-Vanegas O, Nishio H, Rose SL, Buller RE, Friedman SL, Walsh MJ, Martignetti JA. E-cadherin is a novel transcriptional target of the KLF6 tumor suppressor. Oncogene. 2006 Sep 28;25(44):6026–31.

DiFeo A, Narla G, Hirshfeld J, Camacho-Vanegas O, Narla J, Rose SL, Kalir T, Yao S, Levine A, Birrer MJ, Bonome T, Friedman SL, Buller RE, Martignetti JA. Roles of KLF6 and KLF6-SV1 in ovarian cancer progression and intraperitoneal dissemination. Clin Cancer Res. 2006 Jun 15;12(12):3730–9.

DiFeo A, Feld L, Rodriguez E, Wang C, Beer DG, Martignetti JA, Narla G. A functional role for KLF6-SV1 in lung adenocarcinoma prognosis and chemotherapy response. Cancer Res. 2008 Feb 15;68(4):965–70.

DiFeo A, Huang F, Leake D, Narla G, Martignetti JA. Inhibition of KLF6-SV1: Exploration of a novel breast cancer target. Abstract, CDMRP Era of Hope Meeting, Baltimore, MD 2008.

Difeo A, Huang F, Sangodkar J, Terzo EA, Leake D, Narla G, Martignetti JA. KLF6-SV1 is a novel antiapoptotic protein that targets the BH3-only protein NOXA for degradation and whose inhibition extends survival in an ovarian cancer model. Cancer Res. 2009 Jun 1;69(11):4733–41

Difeo A, Martignetti JA, Narla G. The role of KLF6 and its splice variants in cancer therapy. Drug Resist Updat. In Press.

Elbashir SM, Harborth J, Lendeckel W, Yalcin A, Weber K, Tuschl T. Duplexes of 21-nucleotide RNAs mediate RNA interference in cultured mammalian cells. Nature. 2001 May 24;411(6836): 494–8.

Gewirtz AM. On future's doorstep: RNA interference and the pharmacopeia of tomorrow. J Clin Invest. 2007 Dec;117(12):3612–4.

Girard L, Zöchbauer-Müller S, Virmani AK, Gazdar AF, Minna JD. Genome-wide allelotyping of lung cancer identifies new regions of allelic loss, differences between small cell lung cancer and non-small cell lung cancer, and loci clustering. Cancer Res. 2000 Sep 1;60(17): 4894–906.

Glinsky GV, Glinskii AB, Stephenson AJ, Hoffman RM, Gerald WL. Gene expression profiling predicts clinical outcome of prostate cancer. J Clin Invest. 2004; 113:913–23.

Guo H, Lin Y, Zhang H, Liu J, Zhang N, Li Y, Kong D, Tang Q, Ma D. Tissue factor pathway inhibitor-2 was repressed by CpG hypermethylation through inhibition of KLF6 binding in highly invasive breast cancer cells. BMC Mol Biol. 2007 Dec 3;8:110.

Hartel M, Narla G, Wente MN, Giese NA, Martignoni ME, Martignetti JA, Friess H, Friedman SL. Increased alternative splicing of the KLF6 tumour suppressor gene correlates with prognosis and tumour grade in patients with pancreatic cancer. Eur J Cancer. 2008 Sep;44(13): 1895–903.

Hermans KG, van Alewijk DC, Veltman JA, van Weerden W, van Kessel AG, Trapman J. Loss of a small region around the PTEN locus is a major chromosome 10 alteration in prostate cancer xenografts and cell lines. Genes Chromosomes Cancer. 2004 Mar;39(3):171–84.

Hirasawa Y, Arai M, Imazeki F, Tada M, Mikata R, Fukai K, Miyazaki M, Ochiai T, Saisho H, Yokosuka O. Methylation status of genes upregulated by demethylating agent 5-aza-2′-deoxycytidine in hepatocellular carcinoma. Oncology. 2006;71(1-2):77–85.

Huang X, Li X, Guo B. KLF6 induces apoptosis in prostate cancer cells through up-regulation of ATF3. J Biol Chem. 2008 Oct 31;283(44):29795–801.

Ito G, Uchiyama M, Kondo M, Mori S, Usami N. Maeda O, et al. Kruppel-like factor 6 is frequently down-regulated and induces apoptosis in non-small cell lung cancer cells. Cancer Res 2004;64:3838–43.

Ittmann M. Allelic loss on chromosome 10 in prostate adenocarcinoma. Cancer Res. 1996;56(9):2143–7.

Jeng YM, Hsu HC. KLF6, a putative tumor suppressor gene, is mutated in astrocytic gliomas. Int J Cancer. 2003 Jul 10;105(5):625–9.

Kettunen E, Anttila S, Seppänen JK, Karjalainen A, Edgren H, Lindström I, Salovaara R, Nissén AM, Salo J, Mattson K, Hollmén J, Knuutila S, Wikman H. Differentially expressed genes in nonsmall cell lung cancer: expression profiling of cancer-related genes in squamous cell lung cancer. Cancer Genet Cytogenet. 2004; 149:98–106.

Kimmelman AC, Qiao RF, Narla G, Banno A, Lau N, Bos PD, Nuñez Rodriguez N, Liang BC, Guha A, Martignetti JA, Friedman SL, Chan AM. Suppression of glioblastoma tumorigenicity by the Kruppel-like transcription factor KLF6. Oncogene. 2004 Jun 24;23(29):5077–83.

Köhler B, Wolter M, Blaschke B, Reifenberger G. Absence of mutations in the putative tumor suppressor gene KLF6 in glioblastomas and meningiomas. Int J Cancer. 2004 Sep 10;111(4): 644–5.

Koivisto PA, Hyytinen ER, Matikainen M, Tammela TL, Ikonen T, Schleutker J. Kruppel-like factor 6 germ-line mutations are infrequent in Finnish hereditary prostate cancer. J Urol. 2004 Aug;172(2):506–7.

Koivisto PA, Zhang X, Sallinen SL, Sallinen P, Helin HJ, Dong JT, Van Meir EG, Haapasalo H, Hyytinen ER. Absence of KLF6 gene mutations in human astrocytic tumors and cell lines. Int J Cancer. 2004 Sep 10;111(4):642–3

Kovacheva VP, Davison JM, Mellott TJ, Rogers AE, Yang S, O'Brien MJ, Blusztajn JK. Raising gestational choline intake alters gene expression in DMBA-evoked mammary tumors and prolongs survival. FASEB J. In Press.

Kremer-Tal S, Reeves HL, Narla G, Thung SN, Schwartz M, Difeo A, Katz A, Bruix J, Bioulac-Sage P, Martignetti JA, Friedman SL. Frequent inactivation of the tumor suppressor Kruppel-like factor 6 (KLF6) in hepatocellular carcinoma. Hepatology. 2004 Nov;40(5):1047–52.

Kremer-Tal S, Narla G, Chen Y, Hod E, DiFeo A, Yea S, Lee JS, Schwartz M, Thung SN, Fiel IM, Banck M, Zimran E, Thorgeirsson SS, Mazzaferro V, Bruix J, Martignetti JA, Llovet JM, Friedman SL. Downregulation of KLF6 is an early event in hepatocarcinogenesis, and stimulates proliferation while reducing differentiation. J Hepatol. 2007 Apr;46(4):645–54.

Lam DC, Girard L, Suen WS, Chung LP, Tin VP, Lam WK, Minna JD, Wong MP. Establishment and expression profiling of new lung cancer cell lines from Chinese smokers and lifetime never-smokers. J Thorac Oncol. 2006 Nov;1(9):932–42.

Lièvre A, Landi B, Côté JF, Veyrie N, Zucman-Rossi J, Berger A, Laurent-Puig P. Absence of mutation in the putative tumor-suppressor gene KLF6 in colorectal cancers. Oncogene. 2005 Nov 3;24(48):7253–6.

Macleod K, Mullen P, Sewell J, Rabiasz G, Lawrie S, Miller E, Smyth JF, Langdon SP. Altered ErbB receptor signaling and gene expression in cisplatin-resistant ovarian cancer. Cancer Res. 2005 Aug 1;65(15):6789–800.

Miyaki M, Yamaguchi T, Iijima T, Funata N, Mori T. Difference in the role of loss of heterozygosity at 10p15 (KLF6 locus) in colorectal carcinogenesis between sporadic and familial adenomatous polyposis and hereditary nonpolyposis colorectal cancer patients. Oncology. 2006;71(1-2):131–5.

Montanini L, Bissola L, Finocchiaro G. KLF6 is not the major target of chromosome 10p losses in glioblastomas. Int J Cancer. 2004 Sep 10;111(4):640–1.

Mühlbauer KR, Gröne HJ, Ernst T, Gröne E, Tschada R, Hergenhahn M, Hollstein M. Analysis of human prostate cancers and cell lines for mutations in the TP53 and KLF6 tumour suppressor genes. Br J Cancer. 2003 Aug 18;89(4):687–90.

Mukai S, Hiyama T, Tanaka S, Yoshihara M, Arihiro K, Chayama K. Involvement of Kruppel-like factor 6 (KLF6) mutation in the development of nonpolypoid colorectal carcinoma. World J Gastroenterol. 2007 Aug 7;13(29):3932–8.

Narla G, Heath KE, Reeves HL, Li D, Giono LE, Kimmelman AC, Glucksman MJ, Narla J, Eng FJ, Chan AM, Ferrari AC, Martignetti JA, Friedman SL. KLF6, a candidate tumor suppressor gene mutated in prostate cancer. Science. 2001 Dec 21;294(5551):2563–6.

Narla G, Difeo A, Reeves HL, Schaid DJ, Hirshfeld J, Hod E, Katz A, Isaacs WB, Hebbring S, Komiya A, McDonnell SK, Wiley KE, Jacobsen SJ, Isaacs SD, Walsh PC, Zheng SL, Chang BL, Friedrichsen DM, Stanford JL, Ostrander EA, Chinnaiyan AM, Rubin MA, Xu J, Thibodeau SN, Friedman SL, Martignetti JA. A germline DNA polymorphism enhances alternative splicing of the KLF6 tumor suppressor gene and is associated with increased prostate cancer risk. Cancer Res. 2005 Feb 15;65(4):1213–22.

Narla G, DiFeo A, Yao S, Banno A, Hod E, Reeves HL, Qiao RF, Camacho-Vanegas O, Levine A, Kirschenbaum A, Chan AM, Friedman SL, Martignetti JA. Targeted inhibition of the KLF6 splice variant, KLF6 SV1, suppresses prostate cancer cell growth and spread. Cancer Res. 2005 Jul 1;65(13):5761–8.

Narla G, Kremer-Tal S, Matsumoto N, Zhao X, Yao S, Kelley K, Tarocchi M, Friedman SL. In vivo regulation of p21 by the Kruppel-like factor 6 tumor-suppressor gene in mouse liver and human hepatocellular carcinoma. Oncogene. 2007 Jun 28;26(30):4428–34.

Narla G, DiFeo A, Fernandez Y, Dhanasekaran S, Huang F, Sangodkar J, Hod E, Leake D, Friedman SL, Hall SJ, Chinnaiyan AM, Gerald WL, Rubin MA, Martignetti JA. KLF6-SV1 overexpression accelerates human and mouse prostate cancer progression and metastasis. J Clin Invest. 2008 Aug;118(8):2711–21.

Pai SI, Lin YY, Macaes B, Meneshian A, Hung CF, Wu TC. Prospects of RNA interference therapy for cancer. Gene Ther. 2006 Mar;13(6):464–77.

Pan XC, Chen Z, Chen F, Chen XH, Zhou C, Yang ZG. Mutations of the tumor suppressor Kruppel-like factor 6 (KLF6) gene in hepatocellular carcinoma and its effect of growth suppression on human hepatocellular carcinoma cell line HepG2. Zhonghua Gan Zang Bing Za Zhi. 2006 Feb;14(2):109–13.

Pan XC, Chen Z, Chen F, Chen XH, Jin HY, Xu XY. Inactivation of the tumor suppressor Krüppel-like factor 6 (KLF6) by mutation or decreased expression in hepatocellular carcinomas. J Zhejiang Univ Sci B. 2006 Oct;7(10):830–6.

Pan XC, Chen Z, Ji F, Guo ZS, Chen M, Fu JJ. [Effect of KLF6 and its splice variant KLF6V on proliferation and differentiation of human hepatocellular carcinoma HepG2 cells]. Zhonghua Gan Zang Bing Za Zhi. 2008 Sep;16(9):683–7.

Peng D, Sheta EA, Powell SM, Moskaluk CA, Washington K, Goldknopf IL, El-Rifai W. Alterations in Barrett's-related adenocarcinomas: a proteomic approach. Int J Cancer. 2008 Mar 15;122(6):1303–10.

Reeves HL, Narla G, Ogunbiyi O, Haq AI, Katz A, Benzeno S, Hod E, Harpaz N, Goldberg S, Tal-Kremer S, Eng FJ, Arthur MJ, Martignetti JA, Friedman SL. Kruppel-like factor 6 (KLF6) is a tumor-suppressor gene frequently inactivated in colorectal cancer. Gastroenterology. 2004 Apr;126(4):1090–103.

Sangodkar J, Shi J, Difeo A, Schwartz R, Bromberg R, Choudhri A, McClinch K, Hatami R, Scheer E, Kremer-Tal S, Martignetti JA, Hui A, Leung WK, Friedman SL, Narla G. Functional role of the KLF6 tumour suppressor gene in gastric cancer. Eur J Cancer. In Press.

Seppälä EH, Autio V, Duggal P, Ikonen T, Stenman UH, Auvinen A, Bailey-Wilson JE, Tammela TL, Schleutker J. KLF6 IVS1 -27G>A variant and the risk of prostate cancer in Finland. Eur Urol. 2007 Oct;52(4):1076–81.

Singh D, Febbo PG, Ross K, Jackson DG, Manola J, Ladd C, Tamayo P, Renshaw AA, D'Amico AV, Richie JP, Lander ES, Loda M, Kantoff PW, Golub TR, Sellers WR. Gene expression correlates of clinical prostate cancer behavior. Cancer Cell. 2002 Mar;1(2):203–9.

Sirach E, Bureau C, Péron JM, Pradayrol L, Vinel JP, Buscail L, Cordelier P. KLF6 transcription factor protects hepatocellular carcinoma-derived cells from apoptosis. Cell Death Differ. 2007 Jun;14(6):1202–10.

Song J, Kim CJ, Cho YG, Kim SY, Nam SW, Lee SH, Yoo NJ, Lee JY, Park WS. Genetic and epigenetic alterations of the KLF6 gene in hepatocellular carcinoma. J Gastroenterol Hepatol. 2006 Aug;21(8):1286–9.

Spinola M, Leoni VP, Galvan A, Korsching E, Conti B, Pastorino U, Ravagnani F, Columbano A, Skaug V, Haugen A, Dragani TA. Genome-wide single nucleotide polymorphism analysis of lung cancer risk detects the KLF6 gene. Cancer Lett. 2007 Jun 28;251(2):311–6

Stanbrough M, Bubley GJ, Ross K, Golub TR, Rubin MA, Penning TM, Febbo PG, Balk SP. Increased expression of genes converting adrenal androgens to testosterone in androgen-independent prostate cancer. Cancer Res. 2006 Mar 1;66(5):2815–25.

Teixeira MS, Camacho-Vanegas O, Fernandez Y, Narla G, DiFeo A, Lee B, Kalir T, Friedman SL, Schlecht NF, Genden EM, Urken M, Brandwein-Gensler M, Martignetti JA. KLF6 allelic loss is associated with tumor recurrence and markedly decreased survival in head and neck squamous cell carcinoma. Int J Cancer. 2007 Nov 1;121(9):1976–83.

Trimble EL, Thompson S, Christian MC, Minasian L. Intraperitoneal chemotherapy for women with epithelial ovarian cancer. Oncologist. 2008 13:403–9.

Trybus TM, Burgess AC, Wojno KJ, Glover TW, Macoska JA. Distinct areas of allelic loss on chromosomal regions 10p and 10q in human PCa. Cancer Res. 1996;56(10):2263–7.

van Deutekom JC, Janson AA, Ginjaar IB, Frankhuizen WS, Aartsma-Rus A, Bremmer-Bout M, den Dunnen JT, Koop K, van der Kooi AJ, Goemans NM, de Kimpe SJ, Ekhart PF, Venneker EH, Platenburg GJ, Verschuuren JJ, van Ommen GJ. Local dystrophin restoration with antisense oligonucleotide PRO051. N Engl J Med. 2007 Dec 27;357(26):2677–86.

Vax VV, Gueorguiev M, Dedov II, Grossman AB, Korbonits M. The Krüppel-like transcription factor 6 gene in sporadic pituitary tumours. Endocr Relat Cancer. 2003 Sep;10(3):397–402.

Visvader JE, Lindeman GJ. Cancer stem cells in solid tumours: accumulating evidence and unresolved questions. Nat Rev Cancer. 2008 Oct;8(10):755–68.

Wang S, Chen X, Zhang W, Qiu F. KLF6mRNA expression in primary hepatocellular carcinoma. J Huazhong Univ Sci Technolog Med Sci. 2004;24(6):585–7.

Wang SP, Chen XP, Qiu FZ. A candidate tumor suppressor gene mutated in primary hepatocellular carcinoma: kruppel-like factor 6. Zhonghua Wai Ke Za Zhi. 2004 Oct 22;42(20):1258–61.

Wang SP, Zhou HJ, Chen XP, Ren GY, Ruan XX, Zhang Y, Zhang RL, Chen J. Loss of expression of Kruppel-like factor 6 in primary hepatocellular carcinoma and hepatoma cell lines. J Exp Clin Cancer Res. 2007 Mar;26(1):117–24.

Whitehead KA, Langer R, Anderson DG. Knocking down barriers: advances in siRNA delivery. Nat Rev Drug Discov. 2009 Feb;8(2):129–38.

Wikman H, Kettunen E, Seppänen JK, Karjalainen A, Hollmén J, Anttila S, Knuutila S. Identification of differentially expressed genes in pulmonary adenocarcinoma by using cDNA array. Oncogene. 2002 Aug 22;21(37):5804–13.

Xu J, Gillanders EM, Isaacs SD, Chang BL, Wiley KE, Zheng SL, Jones M, Gildea D, Riedesel E, Albertus J, Freas-Lutz D, Markey C, Meyers DA, Walsh PC, Trent JM, Isaacs WB. Genome-wide scan for prostate cancer susceptibility genes in the Johns Hopkins hereditary prostate cancer families. Prostate. 2003 Dec 1;57(4):320–5.

Yea S, Narla G, Zhao X, Garg R, Tal-Kremer S, Hod E, Villanueva A, Loke J, Tarocchi M, Akita K, Shirasawa S, Sasazuki T, Martignetti JA, Llovet JM, Friedman SL. Ras promotes growth by alternative splicing-mediated inactivation of the KLF6 tumor suppressor in hepatocellular carcinoma. Gastroenterology. 2008;134:1521–31.

Yin D, Komatsu N, Miller CW, Chumakov AM, Marschesky A, McKenna R, Black KL, Koeffler HP. KLF6: mutational analysis and effect on cancer cell proliferation. Int J Oncol. 2007 Jan;30(1):65–72.

Zhu H, Lam DC, Han KC, Tin VP, Suen WS, Wang E, Lam WK, Cai WW, Chung LP, Wong MP. High resolution analysis of genomic aberrations by metaphase and array comparative genomic hybridization identifies candidate tumour genes in lung cancer cell lines. Cancer Lett. 2007 Jan 8;245(1-2):303–14.

Chapter 18
Drug Development and Krüppel-like Factors

Ichiro Manabe and Ryozo Nagai

Abstract Recent advances in our understanding of the disease biology of KLFs have spurred considerable interest in their potential to serve as therapeutic targets. Results obtained with small molecules and nucleic acids (e.g., siRNA) targeting KLFs in vitro and in vivo strongly support the feasibility of therapeutic modulation of KLFs for the treatment of cancer as well as cardiovascular and metabolic diseases. Nonetheless, a better understanding of the precise mode of action of KLF in its transcription network, particularly its interaction with other transcription factors and cofactors and its posttranslational modification, would further facilitate development of KLF therapeutics. Moreover, development of improved drug delivery systems would increase the number of diseases that could be targeted by siRNA against KLFs. KLFs have also been used to induce pluripotency in induced pluripotent stem (iPS) cells, suggesting that pharmacological modulation of KLFs may be a useful approach to tailoring iPS cells and their derivatives.

Introduction

A growing body of research is now focused on clarifying the important roles played by Krüppel-like factors (KLFs) in various diseases, including cardiovascular and metabolic diseases and cancer. Moreover, the rapid expansion of our understanding of the disease biology of KLFs has spurred a strong interest in translating the basic science of KLFs into a clinically useful application. In this review, we first present an overview of the general strategies aimed at exploiting KLFs as therapeutic targets, focusing mainly on the use of small molecules and RNAi. We then take KLF5 as an example and discuss the development of novel therapeutics.

I. Manabe (✉) and R. Nagai
Department of Cardiovascular Medicine, The University of Tokyo Graduate School of Medicine, 7-3-1 Hongo, Bunkyo, Tokyo 113-8655, Japan

R. Nagai et al. (eds.), *The Biology of Krüppel-like Factors*,
DOI 10.1007/978-4-431-87775-2_18,

Development of Small Molecules Targeting KLFs

In response to environmental cues, multiple intracellular signals often converge on the transcriptional machinery to determine a cell's fate. Moreover, because a single transcription factor may control a whole group of functionally related genes, transcription factors have long been attractive targets for drug development. Unfortunately, transcription factors are not particularly "druggable" molecules. Indeed, with the exception of nuclear receptor ligands, no transcription factor is presently the primary binding target of any approved drug (Imming et al. 2006; Landry and Gies 2008). However, as our understanding of the molecular networks by which transcription factors control gene expression has increased, so has the number of pathways via which we may target transcription factors using small molecules. In addition, recent progress in nucleic acid therapeutics and gene delivery is enabling direct targeting of transcription factors without the use of small molecules. Such technologies, including siRNA, represent another approach by which to establish KLFs as clinical targets.

To date, there have been no reports identifying small molecules that directly bind KLFs to modulate their function, although progress in high-throughput screening technology may yet enable identification of small molecules do so (Emery et al. 2001). On the other hand, it is now well established that transcription factors, including KLFs, carry out their regulatory functions through close interplay with other molecules, including other transcription factors and various co-regulators (Rosenfeld et al. 2006). Such interplay enables transcription factors to respond dynamically to environmental and cellular cues, and the intricacy of these interactions broadens the array of potential target molecules useful for modulating transcription factor activity. Transcription factors are also controlled through post-translational modification mediated via intracellular signaling pathways that also can be targeted. This means that in addition to using small molecules to target KLF molecules themselves directly, the co-regulators and other transcription factors that interact with them, as well as the signaling cascades that affect their activities, including phosphorylation and proteolytic cascades, are all potential therapeutic targets.

There is considerable evidence that drugs, including clinically approved drugs, affect expression of KLF family members. For instance, statins [3-hydroxy-3methylglutaryl coenzyme A (HMG-CoA reductase) inhibitors] increase KLF2 expression in vascular endothelial cells via mouse embryo fibroblast-2 (MEF2)-dependent activation of the *KLF2* promoter (Parmar et al. 2005; Sen-Banerjee et al. 2005). Statins exert strong lipid-lowering effects, which is beneficial for the cardiovascular system, but there is also evidence that statins have beneficial effects on vascular endothelial cells that are independent of their effects on lipid levels. Indeed, recent studies have clearly demonstrated that KLF2 is essential for maintenance of healthy vascular endothelial cells (Atkins and Jain 2007). Thus, the beneficial effects of statins may reflect, at least in part, the upregulation KLF2 expression. In addition, synthetic PPARγ agonists derived from glycyrrhetinic acid and betulinic acid reportedly induce KLF4 expression (Chintharlapalli et al. 2007a, 2007b). Similarly, the natural PPARγ ligand 15d-PGJ_2 also upregulates KLF4 expression, although not via PPARγ; instead, 15d-PGJ_2 appears to act via the mitogen-activated protein kinase (MAPK) pathway (Chen and Tseng 2005). Thiazolidinediones, a class of antidiabetic PPARγ ligands, exhibit anticancer effects in vitro and in several animal models (Voutsadakis 2007), although it is not yet clear whether the drugs are clinically beneficial or deleterious in cancer (Koro et al. 2007). Given that KLF4 functions as a suppressor of several cancers (Ghaleb and Yang 2008), the anticancer activity of thiazolidinediones might reflect upregulation of KLF4 expression. Finally, selenium has been shown to suppress KLF4 expression in prostate cancer cells (Liu et al. 2008), suggesting that other anticancer drugs might also target KLF members.

A substantial number of transcription factors and cofactors that interact with KLFs have already been identified. For instance, histone deacetylases (HDACs) have been shown to interact with KLF1 (Chen and Bieker 2001), KLF4 (Yoshida et al. 2008), KLF5 (Matsumura et al. 2005), KLF6 (Li et al. 2005), and KLF13 (Kaczynski et al. 2001). Notably, several small-molecule HDAC inhibitors are presently in development, and some are already undergoing clinical trials as anticancer drugs (Dokmanovic et al. 2007), although the molecular basis for the antitumor effects of these HDAC inhibitors is still not well understood. If it is determined that some act by modulating KLF function, it may be possible to use those molecules as the basis for developing HDAC inhibitors that selectively affect the interactions between HDACs and KLFs.

The interactions between KLFs and nuclear receptors are also important from a therapeutic viewpoint, and several small-molecule drugs are currently in use or in development. As we will see with KLF5, some of these drugs may exert their biological effects through modulating KLF function. Another KLF known to interact with a nuclear receptor is KLF13, which interacts with the progesterone receptor (Zhang et al. 2003).

Strategies Using Nucleic Acids

The strategies developed so far to utilize genes and nucleotides as therapeutic agents involve the use of viral and nonviral gene delivery systems, ribozymes, antisense oligonucleotides, and small interfering RNA (siRNA). Because of its efficacy and specificity, RNA interference (RNAi)-mediated gene silencing using siRNA and short hairpin RNA (shRNA) is a particularly promising approach to suppressing expression of selected genes, including genes whose products are not considered to be practical drug targets (e.g., transcription factors). For instance, Martignetti's group reported that PC3M prostate cancer cells stably expressing an siRNA specific for a splice variant of KLF6 (KLF6 SV1) formed smaller tumor masses in mice than control cells (Narla et al. 2005), suggesting that it may be feasible to use siRNA as an anti-prostate cancer agent targeting KLF6 SV1.

There is certainly a strong potential for the rapid development of gene-oriented therapies targeting KLFs using siRNA. Achieving effective gene knockdown in vivo requires highly efficient delivery of siRNA to target tissues; and several methods for local siRNA delivery to the eye, liver, lung, brain, and tumors have been described. Systemic administration of siRNA also has been reported, although successful systemic administration has been largely limited to reticuloendothelial system organs, such as the liver and blood cells. Development of delivery systems that enable systemic administration of siRNA to a wider range of organs will greatly enhance the therapeutic potential of siRNA (Akhtar and Benter 2007).

KLF5 as a Therapeutic Target

We identified KLF5 as a positive regulator of *SMemb*, a marker gene for activated smooth muscle cells in vascular disease (Watanabe et al. 1999). As expected, KLF5 is involved in phenotypic modulation of vascular smooth muscle cells (SMCs); and haplo-insufficiency of *Klf5* reduces neointima formation in vascular injury models (Fujiu et al. 2005; Shindo et al. 2002). *Klf5*$^{+/-}$ mice also exhibit greatly reduced cardiac hypertrophy and fibrosis (Shindo et al. 2002), indicating that KLF5 is a key regulator of cardiovascular tissue remodeling in response to injurious stress. However, KLF5's activities are not restricted to the cardiovascular system. For instance, KLF5 plays a crucial role in various cancers, including gastrointestinal, bladder, and breast cancer (McConnell et al. 2007); and by contributing to the regulation of adipogenesis and energy metabolism, KLF5 also plays a role in metabolic disease (Oishi et al. 2005, 2008). Thus, KLF5 is an attractive target for the development of novel therapies for the treatment of cardiovascular and metabolic diseases and cancer. Furthermore, levels of KLF5 expression reportedly correlate with prognosis in some cancers (Tong et al. 2006), suggesting that they may be a useful marker for diagnosis.

We sought low-molecular-weight compounds with which to modulate KLF5 function and found a synthetic retinoid, Am80. Am80 (tamibarotene) is a specific agonist of retinoic acid receptors (RARs) α and β and is approved in Japan for the treatment of acute promyelocytic leukemia (Ohnishi 2007). Am80 selectively inhibits both the proliferation and migration of cultured SMCs without affecting endothelial cells. In vivo, Am80 acts at least in part by suppressing KLF5, which inhibits neointima formation in both a mouse femoral artery injury model and a rabbit femoral artery stenting model (Fig. 1) (Fujiu et al. 2005; Shindo et al. 2002). Given the selectivity of Am80's effects on SMCs and its high stability, Am80 seems to be attractive for delivering drugs from drug-eluting coronary stents.

Although Am80 clearly inhibits KLF5-dependent trans activation in SMCs, it does not directly bind the KLF5 molecule. Instead, Am80 binds to RARα, thereby disrupting the active transcriptional complex containing KLF5 and RARα. KLF5 forms a transcriptionally active complex with unliganded RARα and retinoid

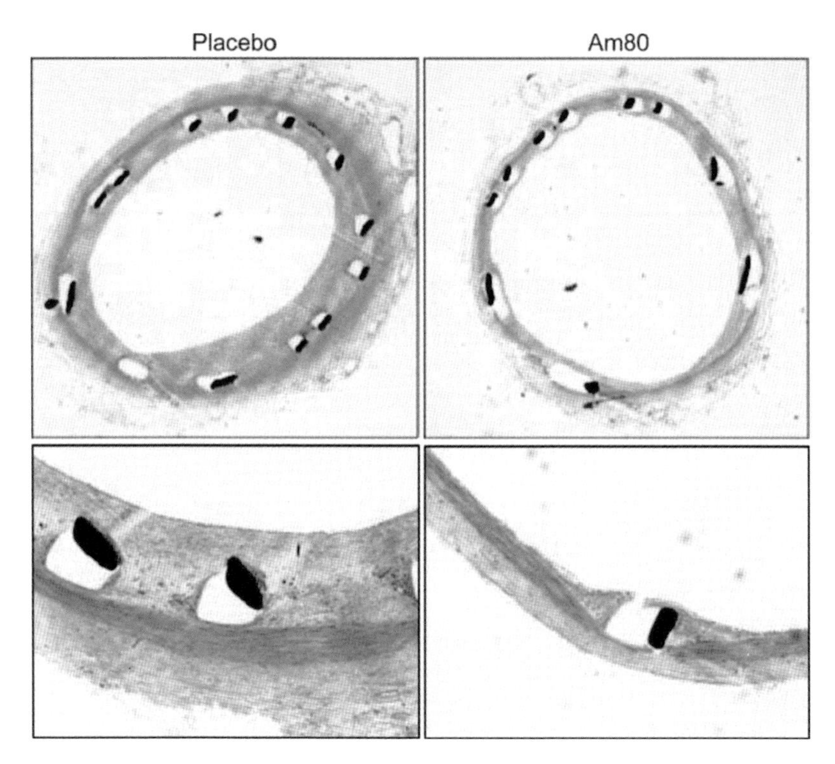

Fig. 1 Effects of Am80 on in-stent neointima formation. Metal coronary stents were placed in the iliac arteries of rabbits that were given Am80 at a dose of 1 mg/kg/day. Shown are low-power and high-power photomicrographs of representative sections from stent-implanted arteries from the placebo and Am80-treated groups harvested 28 days after stent placement. (From Fujiu et al. 2005.)

X receptor (RXR) α on the promoter of *PDGF-A*. Because the promoter does not contain a canonical retinoic acid response element, it is likely that RARα/RXRα associate with the promoter via their interaction with KLF5, which directly interacts with unliganded RARα. Am80 inhibits that interaction, thereby disrupting the trans-activating complex on the *PDGF-A* promoter and inhibiting KLF5-dependent trans-activation (Fig. 2).

KLF5 also interacts with other nuclear receptors. For instance, it interacts with PPARδ and is essential for PPARδ-agonist-dependent transcriptional regulation in skeletal muscle (Oishi et al. 2008). Under basal condition, KLF5 is SUMOylated and inhibits genes involved in energy expenditure and lipid catabolism by forming trans-repressive complexes with unliganded PPARδ and co-repressors. Upon agonist stimulation of PPARδ, KLF5 is deSUMOylated and forms trans-activating complexes with the liganded PPARδ and various co-activators (Fig. 3). There has been an extensive effort to develop synthetic ligands for nuclear receptors, which has already produced a wealth of potentially useful molecules. Although it may be difficult to develop low-molecular-weight drugs that directly bind to KLFs, the

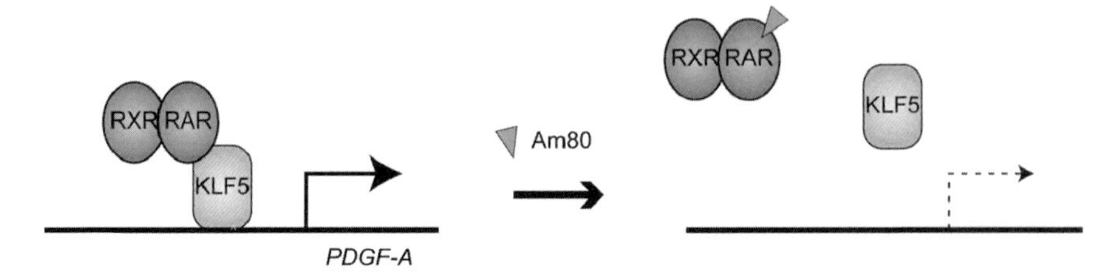

Fig. 2 Am80 disrupts the KLF5-RAR-RXR complex and inhibits KLF5-dependent transcription. KLF5 forms a transactivating complex with unliganded RAR and RXR on the *PDGF-A* promoter. When RAR is liganded with Am80, the transactivating complex is disrupted, and KLF5 is removed from the *PDGF-A* promoter, resulting in inhibition of transcription

Fig. 3 SUMOylation-dependent regulation of KLF5-PPARδ transcriptional complexes. In skeletal muscle under basal conditions, SUMOylated KLF5 and unliganded PPARδ interact with the co-repressor NCoR/SMRT to form transcriptionally repressive complexes. GW501516, a PPARδ agonist, initiates rapid local deSUMOylation that is followed by an exchange of co-regulators, chromatin remodeling, and activation of transcription. In transcriptionally active complexes, deSUMOylated KLF5 interacts with CBP and liganded PPARδ. (From Oishi et al. 2008)

fact that KLFs interact with nuclear receptors may facilitate development of drugs that modulate their functionality. It is known, for example, that nuclear receptor ligands differentially affect the interactions between nuclear receptors and their co-regulators (Kremoser et al. 2007). It therefore may be possible to develop nuclear receptor ligands that selectively affect the interactions with KLF members. Because nuclear receptors play divergent roles in multiple tissues, such target-specific ligands would improve efficacy and reduce the side effects of currently available nuclear receptor ligands.

Conclusion

Recent studies have established the feasibility of using KLFs as therapeutic targets in the treatment of various diseases. Rapidly expanding knowledge of the complex interplay between KLFs and other factors to control transcription, and how they are regulated posttranslationally in response to environmental and cellular cues, is enabling the identification of target points with the potential for modulation by small molecules. siRNA, when combined with efficient drug delivery systems for target tissues, could dramatically shorten the period required for developing KLF therapeutics (Yagi et al. 2009). Although we did not discuss it in this chapter, the generation of induced pluripotent stem cells (iPS) through overexpression of four transcription factors, including KLF4, highlights the importance of KLFs in stem cell therapy (Takahashi and Yamanaka 2006). Other KLF members have also been shown to be involved in the regulation of stemness, self-renewal, and early differentiation of stem cells (Parisi et al. 2008); and it may be that small molecules that modulate KLF functions will be useful for inducing stemness. KLF genes and small molecules that modulate KLF function may also be used to tailor the differentiation and function of iPS and iPS-derived cells to increase their therapeutic potential.

References

Akhtar S and Benter IF (2007) Nonviral delivery of synthetic siRNAs in vivo. J Clin Invest 117:3623–3632

Atkins GB and Jain MK (2007) Role of Kruppel-Like Transcription Factors in Endothelial Biology. Circ Res 100:1686–1695

Chen X and Bieker JJ (2001) Unanticipated Repression Function Linked to Erythroid Kruppel-Like Factor. Mol Cell Biol 21:3118–3125

Chen ZY and Tseng C-C (2005) 15-Deoxy-{Delta}12,14 Prostaglandin J2 Up-Regulates Kruppel-Like Factor 4 Expression Independently of Peroxisome Proliferator-Activated Receptor {gamma} by Activating the Mitogen-Activated Protein Kinase Kinase/Extracellular Signal-Regulated Kinase Signal Transduction Pathway in HT-29 Colon Cancer Cells. Mol Pharmacol 68:1203–1213

Chintharlapalli S, Papineni S, Jutooru I et al (2007a) Structure-dependent activity of glycyrrhetinic acid derivatives as peroxisome proliferator-activated receptor {gamma} agonists in colon cancer cells. Mol Cancer Ther 6:1588–1598

Chintharlapalli S, Papineni S, Liu S et al (2007b) 2-Cyano-lup-1-en-3-oxo-20-oic acid, a cyano derivative of betulinic acid, activates peroxisome proliferator-activated receptor {gamma} in colon and pancreatic cancer cells. Carcinogenesis 28:2337–2346

Dokmanovic M, Clarke C, Marks PA (2007) Histone Deacetylase Inhibitors: Overview and Perspectives. Mol Cancer Res 5:981–989

Emery JG, Ohlstein EH, Jaye M (2001) Therapeutic modulation of transcription factor activity. Trends Pharmacol Sci 22:233–240

Fujiu K, Manabe I, Ishihara A et al (2005) Synthetic retinoid Am80 suppresses smooth muscle phenotypic modulation and in-stent neointima formation by inhibiting KLF5. Circ Res 97:1132–1141

Ghaleb AM and Yang VW (2008) The Pathobiology of Kruppel-like Factors in Colorectal Cancer. Curr Colorectal Cancer Rep 4:59–64

Imming P, Sinning C, Meyer A (2006) Drugs, their targets and the nature and number of drug targets. Nat Rev Drug Discov 5:821–834

Kaczynski J, Zhang J-S, Ellenrieder V et al (2001) The Sp1-like Protein BTEB3 Inhibits Transcription via the Basic Transcription Element Box by Interacting with mSin3A and HDAC-1 Co-repressors and Competing with Sp1. J Biol Chem 276:36749–36756

Koro C, Barrett S, Nawab Qizilbash (2007) Cancer risks in thiazolidinedione users compared to other anti-diabetic agents. Pharmacoepidemiol Drug Saf 16:485–492

Kremoser C, Albers M, Burris TP et al (2007) Panning for SNuRMs: using cofactor profiling for the rational discovery of selective nuclear receptor modulators. Drug Discov Today 12:860–869

Landry Y and Gies J-P (2008) Drugs and their molecular targets: an updated overview. Fundam Clin Pharmacol 22:1-18

Li D, Yea S, Li S et al (2005) Kruppel-like Factor-6 Promotes Preadipocyte Differentiation through Histone Deacetylase 3-dependent Repression of DLK1. J Biol Chem 280:26941–26952

Liu S, Zhang H, Zhu L et al (2008) Kruppel-Like Factor 4 Is a Novel Mediator of Selenium in Growth Inhibition. Mol Cancer Res 6:306–313

Matsumura T, Suzuki T, Aizawa K et al (2005) The Deacetylase HDAC1 Negatively Regulates the Cardiovascular Transcription Factor Kruppel-like Factor 5 through Direct Interaction. J Biol Chem 280:12123–12129

McConnell BB, Ghaleb AM, Nandan MO, Yang VW (2007) The diverse functions of Kruppel-like factors 4 and 5 in epithelial biology and pathobiology. Bioessays 29:549–557

Narla G, DiFeo A, Yao S et al (2005) Targeted Inhibition of the KLF6 Splice Variant, KLF6 SV1, Suppresses Prostate Cancer Cell Growth and Spread. Cancer Res 65:5761–5768

Ohnishi K (2007) PML-RARα inhibitors (ATRA, tamibaroten, arsenic troxide) for acute promyelocytic leukemia. International Journal of Clinical Oncology 12:313–317

Oishi Y, Manabe I, Tobe K et al (2008) SUMOylation of Kruppel-like transcription factor 5 acts as a molecular switch in transcriptional programs of lipid metabolism involving PPAR-[delta]. Nat Med 14:656–666

Oishi Y, Manabe I, Tobe K et al (2005) Krüppel-like transcription factor KLF5 is a key regulator of adipocyte differentiation. Cell Metab 1:27–39

Parisi S, Passaro F, Aloia L et al (2008) Klf5 is involved in self-renewal of mouse embryonic stem cells. J Cell Sci 121:2629–2634

Parmar KM, Nambudiri V, Dai G et al (2005) Statins Exert Endothelial Atheroprotective Effects via the KLF2 Transcription Factor. J Biol Chem 280:26714–26719

Rosenfeld MG, Lunyak VV, Glass CK (2006) Sensors and signals: a coactivator/corepressor/epigenetic code for integrating signal-dependent programs of transcriptional response. Genes Dev 20:1405–1428

Sen-Banerjee S, Mir S, Lin Z et al (2005) Kruppel-Like Factor 2 as a Novel Mediator of Statin Effects in Endothelial Cells. Circulation 112:720–726

Shindo T, Manabe I, Fukushima Y et al (2002) Krüppel-like zinc-finger transcription factor KLF5/BTEB2 is a target for angiotensin II signaling and an essential regulator of cardiovascular remodeling. Nat Med 8:856–863

Takahashi K and Yamanaka S (2006) Induction of pluripotent stem cells from mouse embryonic and adult fibroblast cultures by defined factors. Cell 126:663–676

Tong D, Czerwenka K, Heinze G et al (2006) Expression of KLF5 is a Prognostic Factor for Disease-Free Survival and Overall Survival in Patients with Breast Cancer. Clin Cancer Res 12:2442–2448

Voutsadakis I (2007) Peroxisome proliferator-activated receptor γ (PPARγ) and colorectal carcinogenesis. J Cancer Res Clin Oncol 133:917–928

Watanabe N, Kurabayashi M, Shimomura Y et al (1999) BTEB2, a Kruppel-like transcription factor, regulates expression of the SMemb/Nonmuscle myosin heavy chain B (SMemb/NMHC-B) gene. Circ Res 85:182–191

Yagi N, Manabe I, Tottori T et al (2009) A Nanoparticle System Specifically Designed to Deliver siRNA Inhibits Tumor Growth in vivo. Cancer Res in press

Yoshida T, Gan Q, Owens GK (2008) Kruppel-like factor 4, Elk-1, and histone deacetylases cooperatively suppress smooth muscle cell differentiation markers in response to oxidized phospholipids. Am J Physiol Cell Physiol:00288.02008

Zhang X-L, Zhang D, Michel FJ et al (2003) Selective Interactions of Kruppel-like Factor 9/Basic Transcription Element-binding Protein with Progesterone Receptor Isoforms A and B Determine Transcriptional Activity of Progesterone-responsive Genes in Endometrial Epithelial Cells. J Biol Chem 278:21474–21482

Index